Science that Slept for

Sixteen Centuries

and

What it achieved When it Woke:

From Greek Science to Quantum Systems

Allan Karson

An Introduction to the Science of Physics

with

A Conceptual and Historical Approach as Contrasted to a Mathematical Approach

A Note for the Reader

I wrote this book for three classes of readers: First, the person who is a 'beginner in physics'—to introduce that person to the many fascinating topics in physics and its histories. If you are such a person, consider this to be your 'first date' with *The World of Physics*. Also, I wrote it for:

- The person who has taken one or two or more physics courses in high school or college—but may not know of the numerous "other" physics topics and histories that were not discussed in those courses, and
- High school or college students who have not yet taken a physics course—but who first want to learn what physics is all about.

The book is not written as a textbook; it is a 'standard' reading book. It does contain some equations, such as those that deal with fundamental, exciting issues, such as Newton's definition of Gravity and Einstein's incredible findings about light, as expressed in his Special Theory of Relativity. But such equations are always explained in straightforward English, and are accompanied by simple examples.

An important feature of this book is that the history of the Western World provides the roadmap for the presentation of physics' history—and it starts with the Ancient Greeks. I also believe that all the topics are presented clearly and you will find reading it to be an enjoyable, informative and stimulating experience.

More than sixty source books were used to help me write the book. These are listed at the end of the book and classified according to the book's appropriateness for the reader: Beginner, Intermediate or Advanced.

The book starts by explaining the history of what caused science to sleep for sixteen centuries and how that Big Sleep has affected us, now, in the 21st Century. This is shown in a list of specific items that might have been 'solved' by now, in the 21st Century, had we not had that 'Big Sleep.'

The book then concentrates on two classes of physics disciplines:

1. Those that are *fundamental* to our understanding our world and underlie many past and current issues in physics, such as Thermodynamics, the

four Basic Forces in the Universe—and the event that started it all, known as the Big Bang;

2. Those topics that are *current* in physics and will have a significant role in the 21st Century, such as Quantum Mechanics. As an aside, I predict that the Quantum Age will replace the current Electronic Age—in computers, communication, portable devices, plus many applications that are unpredictable.

The book also describes the histories of some of the women and men with whom you may or may not be familiar. To name only a few, in chronological order: Copernicus, Galileo Galilei, Isaac Newton, Ludwig Boltzmann, Max Planck, Albert Einstein, Karl Schwarzschild, Niels Bohr, Henrietta Leavitt, Erwin Schrödinger, Werner Heisenberg, Lise Meitner, George Gamow and Stephan Hawking.

Also, I've concentrated on providing an idea of 'What Physics is all about.' To do that, the book describes some of the more important topics in physics such as:

- The expanding Universe with its Planets and Stars

- General Relativity, String Theory, Dimensions (more than four), and

- Seventeen appendices that contain special topics and short histories, such as: What *actually* occurred between Copernicus, Galileo and the Church of Rome.

After you have read this book, you should be more at ease when you read a newspaper or magazine article about physics, watch a physics presentation on TV, or browse physics topics on the Web.

I hope you find this exploration of the world of physics to be informative, interesting and stimulating.

Sincerely,

Allan Karson

The Author

Allan Karson earned a bachelor's degree in electrical engineering from the City College of New York in 1951 and a master's degree from Columbia University in 1956, also in electrical engineering,

He entered the electronics industry in 1951 and within only a few years attained high-level technical and executive positions in electronic technology-based corporations. Some were small corporations; some were very large. These were the type of corporations that spurred the United States' and the world's growth in the last half of the Twentieth Century.

Allan often had the responsibility for the development of large computer-based systems. These were for a wide range of applications: U.S. military systems, worldwide communication systems and commercial industrial systems. His executive positions were located in the Northeast U.S. and Paris, France.

After leaving the industrial and technical industry in 2000, he concentrated on learning about the world of physics—to the point where his occupation is now "physics writer." He achieved this by studying books written by physicists that described their theories, plus scientific biographies of physicists and mathematicians. A list of some of those books is located at the end of this book under the title, Sources: Physics, Mathematics and Cosmology.

He continues his physics self-study program by continuing to read books of the genre described above, and attending physics colloquia at his *alma maters*.

Two Remembrances From My Undergraduate Years at the City College of New York

1. The author's belated wish:

My professor of English literature at the City College of New York dressed and looked very English, and he spoke in a plummy way (i.e., in a way typical of the English upper classes.) He would frequently remind us, "He who does not know Shakespeare and the Greek tragedies is not a fully educated person".

After taking my undergraduate courses in physics and mathematics, I would have wanted to say to him, "He who does not know physics and mathematics is not a fully educated person—and is missing out on so many wondrous concepts and ideas that exist in the world around us."

2. An anecdote of what occurred in my physics course on electromagnetism at the City College of New York.

In electromagnetism, there are two forces that are similar but subtly different. They are referred to as 'B', (representing magnetic flux), and 'H', (representing magnetic force). They both are fundamental to the study of electromagnetism.

The professor devoted a complete lesson to a description of these two. Toward the end of the lesson he realized that many of the students were still confused over their differences. He then interrupted his lesson and told us what happened to him when he first learned about 'B' and 'H' in his undergraduate class.

Our professor explained the differences—and explained them again, very clearly. But we just could not fully understand the differences. We felt like dummies. A day or two later, when we were all in study hall, our professor came in hurriedly and said that we must all immediately come into the class room, announcing, "Come in, at last I fully understand the difference between 'B' and 'H'."

PREFACE

A Greek Tragedy

A quote about Ancient Greece from the *Encyclopedia Britannica*, 15th Edition.

"Ancient Greece was the parent culture of Europe. No other people of antiquity exhibited such a wide range of genius or left such a vigorous legacy. The probing attitude of mind that enabled the Greeks to make major contributions to philosophy and the fine arts grew out of the emphasis—unique in the ancient world —on the participation of the individual citizen."

The Greeks established the sciences of physics, astronomy and mathematics, and philosophical thought. They were analytic and used numbers; they examined the heavens and asked what 'matter' is. They spoke of the atom—that is, the smallest form of matter. They asked questions about nature, government and the human body. Reason, coupled with criticism, dominated their thought processes.

The Greeks had the first modern civilization with scientific and philosophic leaders such as Pythagoras, Archimedes, Ptolemy, Euclid, Plato and Aristotle.

Now, a question, followed by its tragic answer: When we were in high school and learned about the wonders of the Egyptian/Babylonian/Persian and Greek civilizations, I wonder why we did not ask the question, "If those earlier Egyptian/ Babylonian and Greek civilizations were so flourishing and intelligent, why were they followed by the Dark Ages?"

The author's answer: Because in the First Century B.C., the conquering Roman Empire and its legions stifled the Greek civilization and its rational, search-for-the-truth thought processes. And later, with the establishment of the doctrinaire and controlling Church of Rome, Western civilization had to wait for sixteen centuries to pass by until science, with its broad base of knowledge about the world around us, was allowed to start again.

What a disaster! What a tragedy! What a horror! Ask yourself, "What would my life and my family's life be if that tragedy had never happened?" And how much more advanced would the world's population be in science—and in living in general, if that tragedy had never occurred?

This book describes some of what is now known about the world around us —essentially 'sixteen centuries late', and provides some answers to those questions.

Who led in this scientific reawakening? Johannes Kepler was chosen by the author to be the leader of that awakening. Galileo might have been chosen —as many other historians have done, but Galileo operated primarily under the aegis of the Church of Rome—and Kepler was independent of it.

Kepler was among the first to do what the ancient Greeks had done. He combined mathematics, physics and astronomy—along with human observation, to define precisely the laws of planetary motion within our solar system. Kepler, in 1608, helped put science back on the path from which the Greek flame had been extinguished.

How We Got to Where We Are Today

Before any discussion about the tragedy that occurred in Greece due to the Roman Empire, let us compare this early part of the 21st Century with all previous centuries.

We might ask, and our children might ask us, "Why is it that people who have lived during the last half of the 20th Century, and now, in the 21st Century, have this super abundance of high-technology products and systems—while previous populations had nothing to compare with those products and systems?"

A list of some of the 20th and early 21st Century products and systems:

- Computers, TVs, cell phones, cameras, telescopes and thousands of technology-based scientific and systems
- Systems for health care in health providers' facilities: Cardiograms, MRI, X-ray, electronic systems to test for DNA and body fluids

- Huge, fast commercial jet aircraft guided by electronic systems
- The worldwide Web, the Internet
- Worldwide electronic-based news dissemination and market systems
- TV, with access to hundreds of channels originating at worldwide locations
- Manned space systems and unmanned sensing system sent to 'observation posts' in our galaxy
- Nuclear systems (electronically controlled) for power generation, for research and for national defense

Let us go backward from where we are today, in The Electronic Era, as shown in time-chart 1.

The era before The Electronic Era was mainly built around the development of the ubiquitous steam engine. That era started in the early 1800s and is called the Industrial Era.

We might ask, "Why was the steam engine so important?" The answer is that in *all* eras before the steam engine, everything was built or moved manually or with the forced-assistance of animals, or depended on the weather. Wheel barrels, horse-drawn carriages, sailing-ships… The steam engine changed all that!

During the Industrial Era, the world changed significantly—more than it ever had in the past. Modern marvels were built on Earth—things never seen before, such as steam-engine driven factories, huge electric generating plants driven by steam engines, city electric-powered street cars, railroad trains driven by steam engines, continent-wide railroad networks, construction of buildings and *very* tall buildings—skyscrapers, international telecommunication networks, the Suez Canal and Panama Canal, huge dams to generate electricity, electrification networks, steamships, cars and highways, airplanes,—and on and on.

During the previous era, which was from the 1600 to 1800, there were essentially no significant changes to the physical world. The structures of buildings, roads, boats and wagons were essentially the same as they had been since the time of ancient Egypt and Greece. But there was a reawakening in science with people such as Copernicus, Galileo, Kepler,

Newton and Leibniz. And there was the ubiquitous printing press that was first introduced in *circa* 1450.

The previous era, from the beginning of the Christian era to about 1600 AD, was when Science Slept. That era's importance should also be remembered for what all of Mankind lost due to the sleep enforced on science in the Western World.

And before that Big Sleep, there was the flourishing ancient Greek Civilization, and before that, the Egyptian, Babylonian and Persian Civilizations.

Let us now enter into the physics world of today, to see where we are, how we got there—and the possible physics world of the future.

Allan Karson

Chart 1: The Six Eras of the Development of Scientific Knowledge

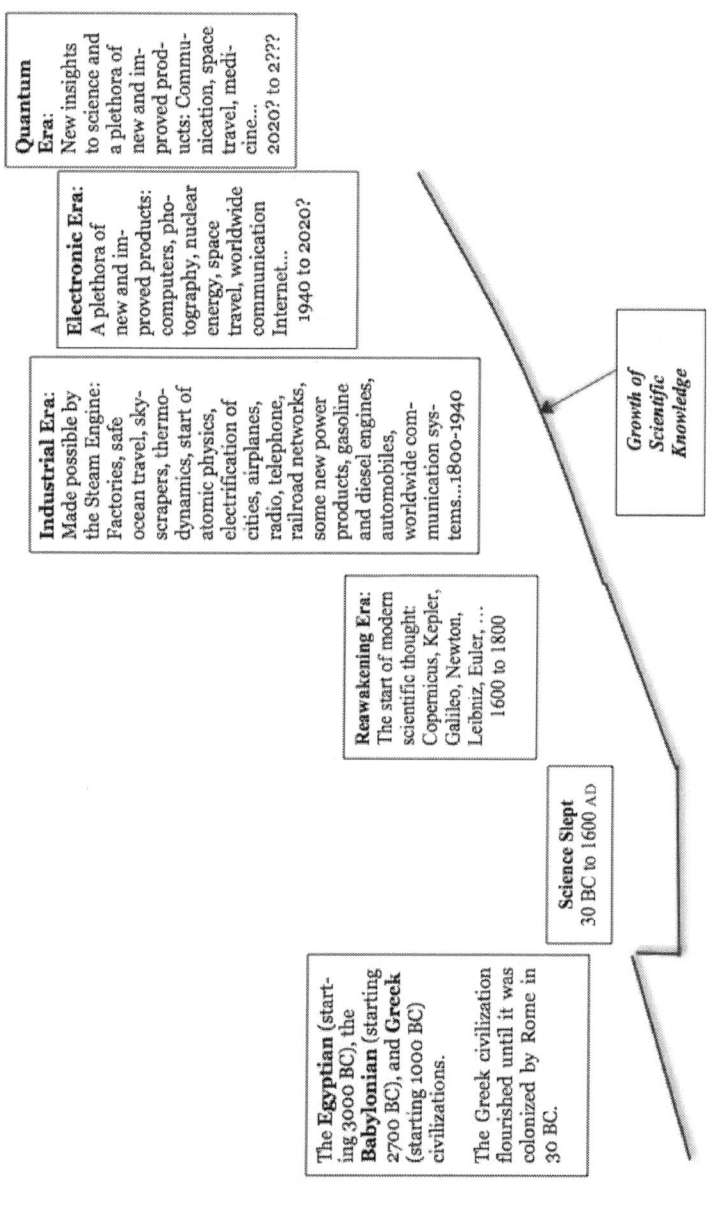

Quantum Era: New insights to science and a plethora of new and improved products: Communication, space travel, medicine... 2020? to 2???

Electronic Era: A plethora of new and improved products: computers, photography, nuclear energy, space travel, worldwide communication Internet... 1940 to 2020?

Industrial Era: Made possible by the Steam Engine: Factories, safe ocean travel, skyscrapers, thermodynamics, start of atomic physics, electrification of cities, airplanes, radio, telephone, railroad networks, some new power products, gasoline and diesel engines, automobiles, worldwide communication systems...1800-1940

Reawakening Era: The start of modern scientific thought: Copernicus, Kepler, Galileo, Newton, Leibniz, Euler, ... 1600 to 1800

Science Slept 30 BC to 1600 AD

The **Egyptian** (starting 3000 BC), the **Babylonian** (starting 2700 BC), and **Greek** (starting 1000 BC) civilizations.

The Greek civilization flourished until it was colonized by Rome in 30 BC.

Growth of Scientific Knowledge

2/17/12

5

Allan Karson

Table of Contents

7

8

- The Beauty or Ugliness of the Standard Model
- Unfulfilled Predictions of the Standard Model
- The Force of Gravity is Not a Part of the Standard Model

List of Time-Charts, Charts and Illustrations

** Denotes charts that are larger than one page (equivalently 2 pages x 3 pages, 3 x 2 and 5 x 3 pages respectively), are multicolored and information on chart is color-coded and are accessible at the book's web-site.

This book is dedicated to:

Heinz Pagels (1939-1988) Adjunct Professor of Physics at Rockefeller University. His book, *The Cosmic Code*, spurred me on to learn more about physics. It was Heinz Pagels who wrote, "If you know what physicists are doing in the research labs today, you will know what the future will look like in twenty years."

Abraham Pais (1918-2000) Professor at the Institute of Advanced Studies, and later, Professor of Physics at Rockefeller University. His insights, expressed in his many books about physicists and the first three-quarters of the 20th Century physics, provided significant assistance in writing this book.

Joseph Krieger, Professor Emeritus, Chairman of the Physics Department at Brooklyn College of the City University of New York, provided the author with a critical review of numerous physics topics presented in this book, and also helped me verify the accuracy of several sections.

My wife, Inge, and my brother, Marvin, Professor Emeritus of Statistics, University of New Hampshire—and my dear friend, Regina Sayres (1908–2012), a person with indomitable spirit.

And The City College of New York, known as CCNY, that provided me, a financially-needy student, with a level of education that enabled me to have a full, interesting and eventful life.

Introduction to the Structure of the Book

This introduction contains varied topics to help you understand subjects presented in this book:

- The order of the chapters.
- A brief description of some of the chapters.
- Numbers to get used to—Powers of ten.
- The system of measurements used in this book.
- A subjective topic: Physics Time.
- Ancillary topics such as how the author communicates directly with the reader.

The Order of the Chapters: Chapters 1 to 15 are presented in order of occurrence in world history.

A brief description of some of the chapters

Chapter 1 starts by describing the reason for the title of this book, "Science Slept for Sixteen Centuries" and introduces the idea that the world's scientific community went into a "Big Sleep." This is followed by an explanation of why that "Big Sleep" has had such a negative effect—even on our lives today. The chapter continues with a description of the work of one of the first modern scientists, Johannes Kepler, the first to apply mathematics to the study of the Solar System.

Chapter 2 describes how the invention of the steam engine changed just about everything on Earth. As said in the preface, the invention of the steam engine was *the* event that started the Industrial Revolution. And just as important, the steam engine brought about an enormous social change: These steam-driven transportation systems, on land and sea, enabled people who lived in distantly-located towns, cities and countries to meet with one another—a social event never dreamed of before, provided by the new, ubiquitous steam engine.

The steam engine is also the reason that the science of Thermodynamics started. Physicists, such as Clerk Maxwell (1831 – 1879) and Ludwig Boltzmann (1884 – 1906), were two of the important physicists who contributed to thermodynamics' success in developing the modern world of products, systems and infrastructure. (The list of those two, early physicists, is only complete with the additional name of Michael Faraday (1791 – 1867).)

Following the description of 'what first got physics going', (Chapters 1 and 2) the book then presents many of the presently important domains within physics, such as the Four Forces of the Universe, (Gravity, the Electromagnetic Force, the Strong Nuclear Force and the Weak Nuclear Force) the Universe itself, the Stars, Quantum Theory and String Theory.

Woven into those descriptions of physics are such descriptions of how Einstein developed his Special and General Theories of Relativity; how Planck, Einstein, Bohr, Heisenberg and Schrödinger developed Quantum Theory; and how the Large Hadron Collider (LHC) located in Switzerland, now provides physicists a deeper look into sub-atomic size matter.

Chapter 10 is somewhat different from the previous chapters. It is devoted to Niels Bohr, the leader in the development of Quantum Theory and his intense intellectual (*gedanken*, or thought) disputes with Albert Einstein. The book goes on, in Chapters 11 to 12, with biographical sketches of three contributors to Quantum Theory: Max Planck, Albert Einstein and Irwin Schrödinger, followed by a presentation of Quantum Theory's cultural history. Chapters 13 to 15 continues with discussions of ongoing studies within the physics community, such as the 'search for dimensions' and String Theory.

Physicists may call on mathematicians for assistance. You may have heard of the expression, "Mathematics is the language of Physics". This synergistic relationship between physics and mathematics is discussed in Chapter 16.

Chapter 17 presents the topic: 'How physicists measure or detect elements that are far away in the Universe, and how they can measure substances that are minutely small.' The chapter also presents the history of the evolution of the techniques and equipment that are used to make those measurements. The book ends with seventeen appendices that contain information that is related to information presented in the book.

Finally, at the end of the book is a list is of more than sixty books, referred to as "sources", that the author referred to in writing the book. The list of sources also provides the author's evaluation of each book according whether the source is for a 'beginner'—first learning about physics; a person with 'intermediate' knowledge—has taken two or more physics courses and possibly read a few books on physics, or, finally, is at an 'advanced' level.

Some of the Disciplines of Physics: Time-Chart 2 provides an overview of some of the disciplines in physics. Only nine are shown in the chart. The complete list of disciplines is greater, but for this introduction to physics, the names of nine should help you "get your bearing" in physics.

The names of the nine disciplines are on the top horizontal column of the chart. The vertical columns under each discipline show the years the events occurred. The start and advances in each discipline is shown according to the periods in which they are active.

As a comparative example, Classical Mechanics started with the ancient Greeks, and consisted of the study of weights, forces and the heavens, while String Theory started in the 20th Century. The chart also shows the names of some of the major contributors to the science of physics, and the approximate years of their contributions. There are, and were, many other famous and not-so-famous physicist and mathematicians who also contributed, but their inclusion in the chart would make it overly dense.

There are three such large time-charts associated with the book, and they are on the book's web site, available for printout. I suggest you print all the individual pages for each of the three charts, and then tape the pages of each chart together—to obtain three complete, time-subject charts.

Time Chart 2: Nine Disciplines of Physics and Their Major Contributors

Time Chart 2 is a 25" by 22" multi-color, chart. The Time Chart is located on the web at: http://bit.ly/akchart1 and is accessible for printing. After you access the chart it will available at the upper right corner of your computer screen, indicated by the words, *Show Downloads*.

This chart is equivalent to a chart consisting of three 8.5" x 11" pages horizontal and two pages vertical.

The pages in the chart should be printed on a color printer and taped together by clear tape to obtain one complete, overall chart of *Nine Disciplines of Physics*.

Nine disciplines of physics are presented on the top horizontal axis:

Classical Mechanics, Thermodynamics, Fluid Mechanics, Optics, Electromagnetism, Cosmology, Nuclear Physics, Quantum Theory and Quantum Mechanics and String Theory

Each discipline is represented by a different color. Entries of scientists are color-coded according to the discipline the scientist relates to.

The years from before 1500 AD to the present are presented on the left vertical axis. Physicists are presented in the appropriate location on the chart, by discipline and by time-period.

The names of the scientists are followed by the years of their life span and a brief indication of their contribution. The contribution can be a discovery, a new theorem, identification of a problem, etc.

Note: There are two additional multi-page charts for the book, which, like Chart 2 above, are located on the web at:

Time Chart: No. 3 Why Science Slept Sixteen Centuries, located at: http://bit.ly/akchart2:

Allan Karson

This chart consists of two horizontal pages, by three vertical pages. An explanation of the chart is located at page 27.

Time-Chart: No. 23 Quantum Fathers: *circa* 1895 to *circa* 1945+, located at: http://bit.ly/akchart3

This chart is five horizontal pages by three vertical pages. An explanation of the chart is located at page 146.

Numbers to get used to—Powers of Ten

We will be examining things that are tiny and also things that are very large. Similarly, we will be looking at very short times and very long times. To express both extremes, physicists frequently use the number ten, modified by an exponent. For example, the number, 10^8, represents 100,000,000, or '1' followed by eight zeros.

The positive exponent is used to express a large number; the negative exponent is used to express a small number. Here are some examples, first some positive exponents, then some negative ones:

Positive (Multiplied by)

$2 \times 10^3 = 2 \times 1000$ (Three zeros)

$2 \times 10^9 = 2 \times 1,000,000,000.$ (Nine zeros)

$3.25 \times 10^{12} = 3.25 \times 1,000,000,000,000$ (Twelve zeros)

Negative (Divided by)

$2 \times 10^{-3} = 2 /1000$ (Three zeros);

$2 / 10^{-9} = 2 /1,000,000,000$ (Nine zeros)

$3.25 \times 10^{-12} = 3.25 /1,000,000,000,000$ (Twelve zeros)

In the beginning of the book, both the full number form and the exponential form will be used in many places, waiting for you get used to the exponential form.

Note: In many physics, and other scientific texts, and some web sites, the number may be shown as such 3.25893(14). This signifies that the number is known for an additional 14 decimal places. That is, it may be: 3.22589325686321458215

Since the author does not consider it necessary to show all those lower decimal numbers, notations, such as (14), that notation will not be used. End of note.

The System of Measurements Used in this Book

In the world-at-large, here are basically two systems: the MKS system (meters, kilograms and seconds) and the U.S/Imperial system (feet, pounds and seconds.) We will use the MKS system. Note: The English system is included in the Imperial system, but there are some differences between the two systems. End of note

You may ask, "Why do we use the MKS system in this book?" The answer: Because scientists use the MKS system worldwide. The U.S. system is used within the U.S. by non-scientists and by only two other countries, Myanmar and Liberia.

Physics Time

What I call Physics Time is a personal, or subjective, view about physics. It concerns how quickly (or slowly) events occur, or when changes are made in the various physics disciplines. You will soon observe that 'Physics Time' moves at a much slower pace than is usually the pace of events in our daily life. It is also slower than the actions within most other domains, such as government, business and education. For example, there are ideas or theories presently in the world of physics that have been around for twenty, or even thirty years—and that are 'waiting' for a proof—or for a counter-theory that may disprove that idea or theory. Or, there are ideas that are not waiting, and are considered correct. Among the latter may be accepted theories that may be overturned in the far-off future. Hopefully, there are not many of those.

The 'slowness' of the comparative pace of new revelations in physics can be considered a blessing to us—and students of physics in general. I have read many physics books that were published ten to twenty years ago—and even thirty or more years ago, that are still near the forefront of physics thought and are still stimulating.

Digressions and Notes

I suggest that the chapters be read in the order that they are presented. The reason for this is that a later chapter may contain a brief discussion of a theory or a fundamental idea that had been previously been explained in detail.

The book contains no footnotes. When there is a 'side topic' that may interrupt the flow of the text, it is specifically called a 'Digression'. For example:

Digression: Professor Lotfti Zadeh, former Professor of Information Theory at Columbia University, frequently used 'digressions' in his lectures. He would first write 'digression' on the blackboard and then explain the topic of the digression; and at the end of the digression he would write: End of Digression.

When there is only a brief amount of additional information to the topic being discussed, this is introduced by the word "Note". "Note" is also used to give a 'heads up' about the topic. For example, it may be used to inform us that the topic will reappear in the text, or that we may see the topic frequently in other books on the subject.

The Author's Use of the first person, "I"

I consider that this book is closer to a classroom presentation than a formal book presentation. The pronoun, "I", is used frequently in such classroom lectures, and I will sometimes follow the classroom style.

The Use of the Internet/Web

Web pages are sometimes referred to within the text to augment the book's presentation. It is possible, however, that the pages that are referred to on the web may no longer be available on the web. If that occurs, I suggest you use your search engine to find similar documents.

I also suggest that you use the web whenever you are uncertain about a technical or nontechnical term in the text, or want to learn more about the topic or the theory—or about the person or persons who contributed to that topic or theory.

Capitalization and Italics. In order that the names of topics stand out, the names of theories, chapters, and charts are capitalized (Gravitation Theory, String Theory... Chapter 6...). Italics are also used to help recognize items that are important within the text.

One of the Reasons I Read About Physics.

I respect great heroines and heroes. I respect people who do things and accomplish things in their life. When I was a child, I would read about the life of Helen Keller, who was deaf and blind—but accomplished much; and Jane Addams, a social worker who founded Chicago's Hull House for children and adults who were having a hard time in life. And I would read about the 'world's heroes' such as Washington, Lafayette, Bolivar, Lincoln, Jefferson, Grant, Pershing and Teddy Roosevelt.

And I read the sea story, *Captain Blood,* by Raphael Sabatini, and the voyages and trials of Odysseus in Homer's *Odyssey* and those of Edmond Dante in *The Count of Monte Christo.* In films, I was inspired by actors such as Gary Cooper in his performances in *Meet John Doe, Sergeant York, For Whom the Bell Tolls* and *High Noon.*

During World War II period, I voraciously read the newspapers and books about the heroes of the war, starting early in 1941 with General Claire Chennault's American volunteer air force in China, called the *Flying Tigers.*

Later, I read of the soldiers, sailors, marines and merchant seamen who were at Pearl Harbor and Schofield Barracks on 7 December 1941; the PT boats in the Philippines, Bataan and Corregidor; the Canadian soldiers' tragedy at Dieppe, France, August, 1942; the merchant shipping vs. U-boat battles in the North Atlantic in 1942; and the military campaigns in the Pacific, North Africa, Europe. And later, I visited many of those battlefields in Europe and some in the Pacific.

Later, when I was working on the 'world scene', I read about its leaders— the good and the bad, Roosevelt, Churchill, de Gaulle, Stalin, Yeltsin, Gorbachev, Mao Zedong, Kennedy, Reagan, Johnson, and on and on. I continue to read about inspiring women and men—but now I read about them in the world of physics. The vast majority of these people have integrity, inquisitiveness, imagination, creativity and ingenuity, personal drive and conviction. They are my heroines and heroes of today.

I now invite you to learn about some of those people who decipher and explain the secrets of our Universe—and in so doing, make our individual lives more interesting and more aware of the impact of science on our everyday lives.

Finally, modest advice how you can handle the situation when you do not understand descriptions, explanations or the modest equations in the book: Physics involves mixing many disciplines, such as pure science (physics), mathematics/numbers—and sometimes, intuition. When you are stumped on an explanation or equation, follow the path that many college physics follow: Relax, read on, and come back later for a second or third review. It happens to all of us.

Part I

The Ideas and Theories of Physics, and Its People and Histories

Chapter 1. Science Slept for Sixteen Centuries

This chapter describes:

- Why Science Slept for Sixteen Centuries
- The Significance of the Big Sleep to the Individuals of the 21st Century.
- The end of the Big Sleep with Johannes Kepler: One of the First Persons to Apply Mathematics to the Study of the Universe

Two quotes, the first located at the bottom of page 1 and ending at top of page 2 in Chapter 1 of "Mathematical Statistical Mechanics" by Colin J. Thompson, 1972, recounting the development by the ancient Greeks of scientific concepts and relating them to the much later development of mathematical statistical mechanics:

"Apart from a poem by Lucretius (55 A.D.), expounding the ideas of Epicurus, nothing further was done essentially until the seventeenth century when Gassandi examined some of the physical consequences of the atomic view of Democritus."

Democritus was Greek philosopher (460-370 B.C.) who contributed to Greek thought concerning the atom.

The second quote is from the scientific biography of Albert Einstein; "Subtle is the Lord, The Science and Life of Albert Einstein" by Abraham Pais, author, former professor of physics and colleague of Einstein at Princeton University. (Page 79, the last sentence at end of the first (long) paragraph.), Pais writes:

"It is likely, however, that an imaginary dialogue between the Greek (of the early Golden Age of Greece) and the late eighteenth century philosophers might rapidly have led to a common understanding that in the two thousand years which separated them, very little had changed regarding the understanding of the basic structure of matter."

Why Science Slept for Sixteen Centuries

An important purpose of this chapter is to establish the following fact: There was a large time gap in the advancement of science from the time of the "end" of the Greek Civilization (*circa* 50 A.D.) to the years 1600 - 1610 A.D. A *BIG* Sleep.

After that is established, the answers to the question "Why did that occur?" will be provided. It will also be shown that the occurrence of that Big Sleep is important in considering the status of our present civilization.

In this book only the European and the Mediterranean World are considered, since that is where Western Civilization grew and fostered the heritage of much of the Western Hemisphere and parts of the Eastern Hemisphere. The advances and discoveries made in the Far East and in the Western Hemisphere, such as among the Aztecs and Incas, are not discussed.

Let us first consider the ancient Greek Civilization, which was the world leader in science and scientific thought from about 600 B.C. to about 50 B.C. Unlike most other societies, such as the Egyptians, the Persians and the Hebrews, which sought to explain the world around them by devotion to religion and Gods—the Greeks mainly used 'reason' (logos) to explain the world.

While many of the Greek Gods, as described in Homer's The Iliad and The Odyssey, are probably well known to most people, Homer (850 B.C.) was about two centuries earlier than the Golden Age of Greece.

The period under discussion is known as the Hellenistic period. It started to flourish in the time of Alexander the Great, about 350 B.C., and continued to the beginning of the domination of Greece by the Roman Legions, about 50 A.D. During the Hellenistic period the Greek language was the world's language. It was the *lingua franca* from Afghanistan to Spain—and of Rome itself, according to A. N. Wilson, a modern English author of critical biographies and cultural histories.

Greek mathematicians and scientists include such highly respected persons as Pythagoras, Aristotle, Plato, Euclid and Archimedes. The Greeks developed the sciences of astronomy, mathematics, physics—and reason. They also developed the arts such as sculpture, beautiful buildings, theatre, histories and epic poems, as well as being the founders of the Olympic games.

Examples of Greek science abound: There is Aristotle (384-322 B.C.), Greek philosopher, pupil of Plato and the tutor of Alexander the Great. He established the western scientific method that theory relied on empirical observation and logic.

There is also Euclid (~325-~265 B.C.), who invented geometry in the 3rd Century B.C. Geometry became the model for all deductive sciences, such as mathematics, physics and chemistry. Geometry became the fundamental mathematical tool for many mathematicians, and it continued to be a fundamental tool for many mathematicians even as late as the 18th and 19th centuries.

Geometry was also used to solve problems pictorially in arithmetic and algebra, and it was used in the seventeenth century as the basis for the development of calculus by its developers, Leibniz and Newton.

And there is Eratosthenes (276-195 B.C.) who determined the size of the radius of the Earth. He found it to be about 4,000 miles—and the now-known radius is 3,963 miles. And add to that feat, he calculated the angular tilt of the Earth as it makes its daily rotation around that axis. And add to that, the first person known to have proposed a heliocentric system (a system where of earth and planets revolve around the sun), was Aristarchus of the Greek island of Samos (*circa.* 270 BC).

While historians limit the title of 'Golden Age' to the Greek society of the 4th Century B.C., the complete period of the Greek Civilization could be called a Golden Age. Definitely, it was, when compared with any of the other civilizations of that time—and probably even with later civilizations that were functioning during, what is commonly referred to as, the Dark Ages.

An important note about the development of mathematics by the Greeks: Chart 3, *Why Science Slept Sixteen Centuries*, shows that the first subject within their field of scientific studies was mathematics. We may ask, "Why mathematics?" The stark answer: Because mathematics *does not depend on any other area of science*. The Greeks developed, what we call, *pure* mathematics.

It was only later, when civilizations applied their math to other studies, such as geometry, geography, and physics, that math became *applied* mathematics. Thus, there became two domains of mathematics—pure and applied.

The Greeks also developed physics. Its development was based on what they observed—at first, the stars and the sun; later, their environment— water, land... Hence physics is based on the general world around us; chemistry is based on physics, and on and on with politics, economics, medicine, geography—all based on something else. The only field of investigation *without* any base is *pure* mathematics.

Let us look at one of the many areas that their inquisitiveness led them into —numbers. Their 'play' with numbers became the basis for what is referred to today as 'Number Theory'. Here are two examples of their analysis and creativity:

$1*9 + 1 = 10$	$3*37 = 111$
$12*9 + 1 = 109$	$6*37 = 222$
$123*9 + 1 = 1108$	$9*37 = 333$
$1234*9 + 1 = 11107$	$12*37 = 444$
$12345*9 + 1 = 111106$	$15*37 = 555$

What fun! How cool. Who would have thought it?

In comparing the ancient Greek civilization with other civilizations such as the earlier Egyptian civilization, the following general assessment may be made: The Egyptians did produce studies into measurement of the Earth, astronomy, and construction of buildings such as their pyramids, but they never produced the momentous ideas and complete sciences that were developed by the ancient Greeks.

The Egyptian civilization lived on within the Greek civilization and did contribute to Greek civilization. An example is the famous, large library at Alexandria that became part of the Hellenic/Greek world. Also, many famous Greeks, such as Euclid, lived in Alexandria.

The library was known as a center of learning throughout the ancient world. It is said that it contained hundreds of thousands of historically and scientifically important papyrus scrolls. The exact number of scrolls is not known.

The library also sought texts from other civilizations and acted as a repository for their writings. An example of this is the storing of a document based on the Hebrew Scriptures. Ptolemy II Philadelphia (285-246 B.C.), king of Ptolemaic Egypt, had ordered a Greek translation of the Hebrew Scriptures and he commissioned seventy scribes to perform this task. The document became known to the world as "The Septuagint", or "The Seventy."

A review of the history of science shows there was a major hiatus from 50 B.C. to about 1600 A.D. That is, there is nothing in Western civilization during this long period comparable to the earlier tide/flow/surge of the analytic science of the Greeks.

Note: You may want to review the two quotations provide at the beginning of this chapter. Also, the Encyclopedia Britannica, 15th edition, Volume 16, page 367, contains a brief article describing this period, "Science in Rome and Medieval Science" at this "end" time period. End of note.

We now can ask, "What happened between 50 BC and 1600 AD when Kepler and Galileo and others started to contribute, once again, to science and math?" It is as if people mentally went to sleep from 50 B.C. to the 1600s, waiting to wakeup!

An explanation for this 'shutdown' of scientific thought for 1,600 years entails looking at what happened to the Greek Civilization and identifying the culprits. Time-Chart 3, Why Science Slept for Sixteen Centuries, provides insight into how the sleep was forced on civilization.

The chart shows the many contributions of the Greek civilization; the rise of a militaristic, dictatorship in Rome, Rome's fall, followed by the Church of

Rome; the reawakening of scientific thought, led by Copernicus, Galileo and Kepler; and what was going on in the rest of the European world and the parallel, emerging, Muslin world, The upper part of the left-most column of the chart shows where the Greek civilization stood among other scientific achievements in the European-Mediterranean world. It is clearly seen that that geographic area, the ancient Greek civilization, is THE birthplace of Western Civilization.

Look down the left column to 149 BC that shows the most significant (tragic, disastrous for the world-at-large) event—the conquest of Greece by the Romans. As contrasted to the Greeks, the Romans appeared to show no interest in science except for the construction of roads, aqueducts and buildings—and for conquering. (Note: In 50 AD the Roman Legions destroyed the library at Alexandria—which included "The Septuagint." End of note.

Conquer and plunder and stern governing, yes. Within the Roman Empire the military spirit abounds. Even in sports, there was a strong contrast: Gladiator sporting events (bloody, to the death) compared to the humane sports played on the ancient Greek plains of Olympia.

The Romans did build long-lasting roads, aqueducts, theaters, buildings, coliseums and sculptures. Admittedly, to build those marvelous, long-lasting edifices, they employed mathematics and the sciences of construction, building materials, the science of water/liquid flow and other sciences—but they had no interest in developing those sciences for science itself.

The politics and freedom of the thriving, ancient Greek city-states was terminated by that Roman invasion in 149 B.C, and more was to come to the Mediterranean Region. In 49 B.C. Julius Caesar, who started his career as a very successful military Roman leader, abolished the Roman Republic and established the Roman Empire. He was the first Roman Emperor and took the title *Augustus* Caesar. Long before becoming emperor, Caesar had been known as *Imperator*, which is a title given by the military to a leader who had shown extraordinary leadership and prowess in leading the military.

On Caesar's positive side, we should recognize that it was Caesar who installed the Julian calendar in 46 B.C. The Julian calendar was chosen by Caesar after consultation with the Greek astronomer and mathematician,

Sosigenes of Alexandria. The Julian calendar had been used worldwide until the 16th century when it began to be replaced by the more accurate Gregorian calendar.

Even long after he was assassinated in 44 B.C., Caesar's 'spirit' seemed to rule through the many emperors who followed him, such as Nero, Caligula and Titus—a heritage left by Caesar. If we look at http://www.scaruffi.com/politics/romans.html, a timeline of the Roman Empire, we will easily see what were its overall characteristics—such as 'conquer', 'destroy', 'subjugate' and 'killing'.

The Romans blocked the world of scientific investigation—(actually physically destroyed it) and did they continue the scientific investigations and thoughts of the Greeks! What a tragedy. Modern history books tell us that the hordes from the East later descended on Rome and helped cause the decline of the Roman Empire. A person might ask, "A decline from what?"

After the fall of the Roman Empire in about 476 A.D., a high level of culture (art, sculpture, and architecture) was attained in Europe from about 500 A.D. to 1600 A.D. Much of it, however, was to support, or tied to, the Church of Rome.

The Church of Rome continued what may be called the 'Spirit of the Empire", not necessarily opposing the study of mathematics, astronomy and the physical sciences, but not supporting such studies. Some science proceeded, such as learning about how the human body functions. But without a general scientific infrastructure, such work proceeded at a snail's pace and in limited localities.

There were, however, some leaders in the Church of Rome who supported science, such as Saint Albertus Magnus (before 1200AD–1280AD). Magnus was a strong advocate for the peaceful coexistence of science and religion. But it appears that his efforts and thoughts did not seem to have carried much weight in the Church-at-large.

During those eleven centuries, 500 A.D. to 1600 A.D., the Church of Rome did not appear to support the general advancement of science. The well being of the physical person, while on Earth, was not its prime concern. During this hiatus, the development of mathematics and science remained virtually stagnant. I suggest we look again at the left-most column. It speaks more than words. Just observe the "silence" in that column from the line "The Christian Era" to the line far, saying, "Science wakes up in the 16th Century."

We may learn more about what went on during this period by referring to Appendix 1, which describes the plight of Galileo (1564–1642) as he presented his ideas to the Church of Rome. Appendix 1 also describes how the leaders of the Church of Rome, including Pope Urban VII, liked Galileo and did not want to harm him. Even so, Galileo was called before the Roman Inquisition.

The reader is referred to Appendix 1, which recounts the events in Galileo's life in which he was in close contact with Church of Rome. Based on the results of those visits, the author's personal evaluation is that the Church of Rome, represented by Pope Urban VIII, 'gets a bad rap' for its 'supposed' bad treatment of Galileo. That is, Galileo's treatment by Urban VIII was not nearly as bad as commonly believed.

Now, what were other cultures doing at this time in scientific domains during these sixteen centuries? During this period, the earlier Hindu cultures in India and the emerging Moslem culture (starting in the 8th Century) did continue to make mathematical and scientific investigations. They contributed to algebra, geography, astronomy and optics. They also built institutions of learning for their people.

They, as contrasted to the European/Roman model, did not have religious doctrines that restricted scientific investigations, but it appears that they were not that interested in the structure of the universe and science, in general, as the Greek people had been.

Time-Chart 3: Why Science Slept for Sixteen Centuries

Time-Chart 3 is a 17" by 33" multi-color chart. The Time Chart has also been composed in MS.Excel.

The chart is located on the web at **http://bit.ly/akchart2**

(The book's two other time-charts are locate at

http://bit.ly/akchart1 and **http://bit.ly/akchart3**)

Allan Karson

The chart's size is equivalent to a chart consisting of 2 pages across and 3 pages down.

As with the previous time-chart, I suggest you print the chart on a color printer and then tape the pages together to obtain a 2-page by 3-page chart of 'Why Science Slept for Sixteen Centuries.'

Four sectors of World History are presented on the top horizontal axis:

- The Progress of Science in Europe, starting with the Greek Empire
- National and Church of Rome Events Affecting Science
- General World Events
- Progress of Science, Non-European

On the vertical axis: The years from 1000 BC to the 17th Century

Events are color-coded according to scientific discovery, action by a government policy or action, Church of Rome attitudes and rulings, other organizations' activities and general world activities such as exploration.

The Significance of the Big Sleep to People of the 21st Century

First, a note about how the hiatus in advancement of Science for 1,600 years affected me, the author. Later, you may also ask the same question for yourself. If you react the same way, you, also, must all make sure it does not happen again—in any small or large way.

My one personal example: My father died of cancer at the age of 38. He might have lived longer since cancer may have been cured by the 20th Century if science had not been forced to sleep. Have you ever thought about what might have been in *your* life if science had been allowed to continue along the Greek path of reason and inquisitiveness that it was going along in 50 B.C.?

The list below presents some of the successes and knowledge the world might have attained in the 21st Century—but has not attained, due to the Big Sleep. The list is mainly about humanistic or people-related events and knowledge. It does not attempt to predict any advances in physics.

31

- Cures for Disease: The Usual Suspects: ALS, Alzheimer's, Cancer, Downs Syndrome, Heart Disease, Malaria, Multiple Sclerosis, Muscular Dystrophy, Tuberculosis—and many others.
- Microprocessor controlled prosthetic devices to assist persons with muscular problems or loss of arms or legs to enable fully natural movement and control.
- Individual, almost-human robots, automatically directed to perform an individual's needs, to assist the elderly and incapacitated in performing daily functions such as dressing, bathing—and being there to be a friend to the lonely.
- Abundant fuel sources, either natural or man-made, reducing or eradicating carbon emissions into the atmosphere—probably enabling the avoidance of the present global warming.
- Efficient engines, in ranges greater than 60% efficiency. This would cover the gamut from large, stationary energy producing systems (power plants) to our public and private transportation systems (buses, trains, airplanes, ships—and automobiles).
- Worldwide, abundant pure drinking water and improved water-dispensing systems compared to our present self-indulgent water dispensing systems (wasteful showers, toilets...).
- Poverty would continue to be present in only small areas of the globe—or possibly eradicated.
- Abundant food production equitably distributed throughout the world—to all populations.
- The religious wars that the earth has suffered from over the centuries might not have occurred, since religion would not have gained the foothold it did during the sixteen centuries of the scientific hiatus.

I let you speculate about additional benefits and conditions that might have been available to us and to our children, today, if the Greek civilization had continued as it was before the Roman Hordes.

Before we leave this subject of 'could have been', I add one thought that is very different from the list above. It is about space exploration. The 'thought' is by the Astronomer-Scientist-Author Carl Sagan and is located on page 210 in his book, *Cosmos*, source 10. (In the following, the author introduces the measurement length, the light-year, which the distance that

32

light travels in one year, which is approximately 9.5 trillion kilometers or nearly 6 trillion miles.): According to Carl Sagan:

> If the Ionian (Greek) spirit had won, I think we—a different 'we', of course—might now be venturing to the stars. Our first survey ships to Alpha Centauri (4.4 light-years from earth) and Bernard Star (6 light-years from earth), Sirius (8.6 light-years from earth) and Tau Citi (12 light-years from earth) would have returned long ago.

The Reawakening Era (1600-1800)

The Reawakening Era, as defined in the preface, covers two centuries. For the purpose of this discussion, we zero in on the first of those two centuries, the 17th—and also the latter part of 16th Century.

The generalist historians might consider that what is called in this book, as being the Age of Enlightenment. But science historians might disagree with that assessment, since the historians might say the enlightenment had already begun in the time of ancient Greece—but was extinguished at the beginning of first millennium, A.D. The term, *Reawakening*, may be considered more applicable.

Chart 4 provides a list of some of the historic events, inventions, writers and thinkers who provided "new thoughts" (for good or for bad purposes) to the growing populations in European people during those selected years, 1550 to 1700,

From this list it is apparent that a reawakening of many types was on the horizon in different human disciplines. For example, in addition to the scientific reawakening, there were those that were on the political horizon, such as the 18th Century revolutions in the American colonies (1776…), in France (1789…) and also in smaller countries such as Corsica (1769, albeit, defeated) and Haiti (1791…).

Chart 4 Events Leading to the Reawakening Era

The following list shows that the Holy Roman Empire, which still existed in a much weaker form until the 17th Century, and the Church of Rome, are no longer the controlling forces in Europe.

- Invention of the first printing system in Europe, the Gutenberg Press, *circa* 1450—that instituted and promulgated the power of the uncontrolled, written word.
- Martin Luther breaks from the Church of Rome, first half of 16th Century.
- Henri VIII breaks England's ties with the Church of Rome, first half of 16th Century.
- The disruptive Thirty Years War within the Holy Roman Empire, which was said to be no longer holy, or Roman, or an empire, during the first half of 17th Century.
- Louis XIV becomes King of France in 1643 and establishes a long, powerful reign—independent of Rome.
- French Academy of Science established in Paris, 1666.
- Formation of new nations in Northern Europe, such as Prussia, 1667.
- Queen Elizabeth and the ascendancy of England as a European power, Second half of 16th Century.
- Birth of Mercantilism, which enabled small nations, such as Holland and Portugal, to compete with the larger nations, 17th Century.

The words and thoughts of many of the philosophers and leaders listed below were printed later than the dates shown, by more modern versions of the Guttenberg Press.

- Niccolò di Bernardo dei Machiavelli (1469–1527) Italian founder of modern political science
- Thomas Moore (1478–1535) English humanist
- Francis Bacon, (1561–1626) established inductive procedure for scientific investigations
- Hugo Grotius (1583–1645) Dutch jurist
- Cardinal-Duc de Richelieu (1585–1642) French promoter of the power of the nation-state
- René Descartes (1596–1650) French mathematician, scientist and philosopher
- Benedictus de Spinoza (1632–1677) Dutch philosopher
- Isaac Newton, (1643–1727) English physicist, mathematician, astronomer and natural philosopher.

- Gottfried Wilhelm Leibniz 1646–1716) German philosopher and mathematician
- John Locke (1632–1704) English political philosopher
- Jean Jacques Rousseau (1712–1778) French philosopher

Johannes Kepler (1571-1630) The First 'Modern' Scientist to Apply Mathematics to the Study of the Universe. (Mathematician and Data-Analyst *Extraordinaire* of Astronomical Data.)

Our journey into science's history starts with an examination of the contributions of Johannes Kepler. He is one of the first persons to discover the relationship between measurements of what is seen in the sky around us, the coming and going of the stars, the planets, the moon, the sun—and mathematics.

In addition to discovering a fundamental concept of physical nature, Johannes Kepler was one of the first individuals to kick-start the advance of science in the 17th Century—which has never stopped since then. Thus, I consider Kepler to be the first to *reawaken* Science after it dozed so many years, following the end of the Greek scientific era.

Kepler lived in the part of the world that was known as the Holy Roman Empire. The Empire consisted of territories now known as of Austria, Belgium, the Czech Republic, Denmark, Eastern France, Germany, Holland, parts of Northern and Central Italy, Slovakia and Switzerland.

The times were turbulent. The general population continually faced wars and famines and plagues. It is amazing that Kepler could concentrate and be left alone to do his work. He lived during, what is known as The Thirty Years War (1618–1648) that was being fought in the region where he lived. Even with modern warfare being considered the ultimate in destruction, the Thirty Years War is still considered a *very* destructive war. It forced Kepler to move a few times from country to country, due to both events and money-needs. We will now state what Kepler did, and this will be followed with 'how he did it':

Kepler was the first person to discover (or announce) that the Earth and planets move according to precise laws—as expressed by precise mathematical equations. He was also the first person to conclusively prove that the Earth and all the other planets in our Solar System orbit about the

Sun. That is, the Sun *is* the 'center' of our Solar System—and not the Earth, as most people believed at that time.

Kepler was a mathematician/scientist and worked for a very short time, about three years, for the wealthy and renowned Danish astronomer, Tycho Brahe (1546–1601), and it was Tycho Brahe's team of assistants who made the measurements. Kepler started working for Brahe in 1599—very near the end of Brahe's life. But what Kepler got, or took, from Brahe, is quite important, so let us first consider the accomplishment of Brahe, and then come back to Kepler.

Tycho Brahe started his life as the nephew of a wealthy and childless uncle. He was educated in the law, but spent his spare time observing the heavens —without any telescope. He shortly made observations with the naked eye that made him realize previous observations—or ideas made by others, were grossly inaccurate and perpetuated mistakes. He then pledged his life to correcting those mistakes he found and to expanding the astronomical data. He also observed a star not previously noted, which shook his belief in the Greek Aristotle's concept of a *non-changing* sky.

Brahe had good personal relations with the king of Denmark and was given the small Swedish island of Hven (now known as Ven) to build a large observatory, plus a considerable amount of monetary support. Brahe was then able to employ a staff of assistants who would watch the stars, the planets and the moons. Brahe and his staff kept a written record in which they logged every day the positions of these units in space. And this is amazing: They did not have a telescope to make their observations.

Notes: 1. The appearance of the night sky that most of us observe—black sky without any visible stars—made invisible by the outdoor electric lighting that most towns and cities use, is a far cry from the vibrant, vivid panorama that would have been seen in Brahe's time.

2. The first telescope is said to have be invented by Hans Lippershey in 1608 in the Netherlands, and one of the first well-known users was Galileo in 1609, almost a decade after Brahe's death. End of notes.

Brahe started his investigation of the sky using only a compass and sextant. While many web sites describe the dealings that Brahe had with the Danish king and the Czech king, some sites list the instruments that Brahe used were very few, giving the (false) impression that Brahe used only the compass and sextant. The lists also give the impression that a plethora of sightings could by made using such a minute instrumentation base!

This was not the case, however. A list of Brahe's full equipment is provided in the more-modern book, *The God Particle*, written by the Nobel laureate, Leon Lederman, with Dick Teresi, source 19. This source lists the full set of equipment that Brahe designed, built and calibrated and includes:

> "...Azimuthal semicircles, Ptolemaic rulers (an instrument for measuring the angle between the Moon and a point directly overhead, i.e. the Moon's zenith distance.), brass sextants, azimuthal quadrants and parallactic rulers. ...And he did this all with unprecedented precision."

Finding Brahe's equipment list in such a modern physics-related book is not surprising. Both Brahe and Lederman were looking for small-to-see objects: Brahe, the stars; Lederman, the very tiny, elusive Neutrino particle.

According to the *Encyclopedia Britannica*, "His (Brahe's) observations included a comprehensive study of the solar system and accurate positions of more than 777 fixed stars." Brahe's overall contributions to astronomy are listed below.

Some of Tycho Brahe's Contributions to Astronomy in the 17th Century

Prior to Tycho Brahe, most astronomical tables were either based on observations made by Ptolemy (a Greek) in the 2nd Century AD. Perhaps this was good enough for astrologers, but it was not sufficiently accurate for serious astronomers such as Brahe. Some of the 'updates' made by Tycho Brahe are as follows:

- Revolutionized astronomical instrumentation. He built huge, accurate instruments to perform his observations and he carefully calibrated them. Comparable instrumentation was only built fifteen years later.
- As opposed to observing planets at only certain times or certain parts of their orbit—which was standard practice at the time, Brahe and his assistants observed them *throughout* their orbit—changing significantly astronomical practices—and provided more accurate information to the astronomical tables of the time.

- Brahe's observations were accurate to what is called '2 arc minutes', whereas observations of his contemporaries were less accurate, to 15 arc minutes.

Overall, Brahe *significantly* increased the accuracy and the overall coverage of the astronomical tables in the 17th Century.

Brahe and his assistants did not attempt to analyze the data. They left that to others, such as Kepler. Brahe thought the planets went around the Earth in epicycles—and that the Sun also revolved around the Earth. Most people at that time—probably, including Kepler, believed that epicycles existed. It was the early Greeks who invented the idea of epicycles that were special paths, or patterns, for planets and the sun to follow in order to account for their belief that the Sun goes around the Earth.

And that had been the universal concept of how the Sun moved around the Earth—and how all the other planets moved around the Earth. This was the established line of thought, even among the ancient Greeks, who were the first recorded observers and astronomers in Western civilization.

In 1597 Brahe had a fallout with the King of Denmark. Thus, in 1599 Brahe settled in Prague where he became Imperial Astronomer, working for Emperor Rudolph II, who also hired Kepler to work for him. Brahe died shortly thereafter, in 1601, and Kepler obtained Brahe's data. In 1609, after carefully analyzing the data for more than ten years (some say sixteen years), Kepler determined that the orbit (its movement) of the planet Mars was an ellipse. (The reader can refer to Figure 5, below, for the various forms of an ellipse.)

An ellipse is a curved line, or a path, that can be described by specific mathematical equations. The curved line, or path of an ellipse, moves in a 'flattened circle'. Its flatness can go from almost being a circular circle, to almost being a line.

The equation of an ellipse indicates that an ellipse has two foci. Individually, they are called 'a focus'. Points, or bodies, or in this case, planets, move on the path of the ellipse and maintain a close mathematical relationship with these two foci.

And as important as Kepler's first determination—that the planets, including Earth, were following *elliptical* orbits, he concluded (conclusively) they were all going around the Sun—a discovery that went against current teachings of the Church of Rome.

Now look at Figure 6, which shows how Earth goes around the Sun— another finding by Johannes Kepler.

Do not think that Kepler's analysis work was an easy task. There were no logarithmic tables nor mechanical calculating aids or, of course, electronic computers. (Logarithmic tables were developed by Napier in 1614, too late to have been of use to Kepler.)

Kepler spent more about sixteen years going over the data and as Roland Omnès, Professor of Physics at the University of Paris, writes in his book, *Quantum Philosophy, Understanding and Interpreting Contemporary Science,* source 32, "The testing of each hypothesis then necessitated a tremendous amount of difficult calculations. Kepler may have been one of the first "modern scientists" to test and compare, laboriously, mathematics against observations—which is the essence of scientific investigation."

Figure 5   Various forms of an Ellipse

The mathematical equation of an ellipse:

The equation of an ellipse is $x^2/a^2 + y^2/b^2 = 1$

Where 'x' and 'y' are the horizontal and vertical coordinates of a point along the path, and '2a' is the length of the major diameter along the 'x' axis and '2b' is the length of the 'y' axis.

The enclosed area of the ellipse is Pi * a * b, where Pi = 3.14.....

Diameter = 2a

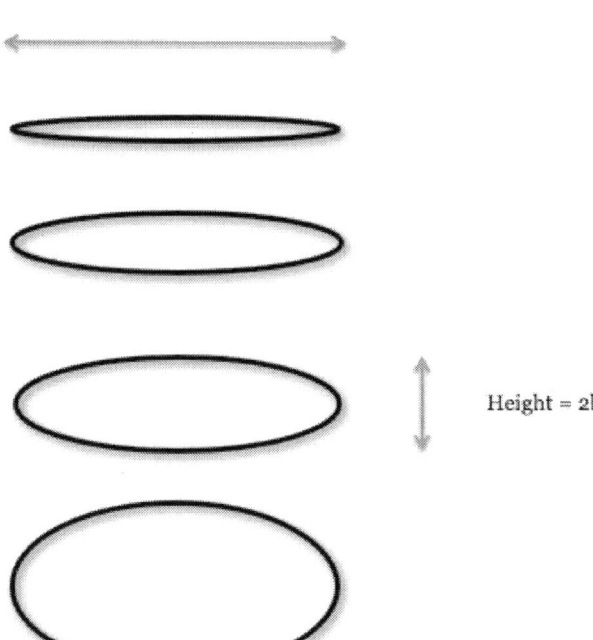

Height = 2b

40

Figure 6   THE LAWS OF PLANETARY MOTION
          (As Discovered by Johannes Kepler)

An ellipse has two foci; each one is toward either end. Individually, each is called a focus. In the ellipse drawn by Kepler, the Sun is located at one of the foci of the ellipse. (The second focus plays no role in Kepler's paths for Earth—or any other planets. The second foci, to the right in the ellipse, is not shown.)

The planet (Earth, in this case) that is following the path of an ellipse sweeps out equal areas in equal times as the planet travels around the Sun.

Path of Earth around the Sun

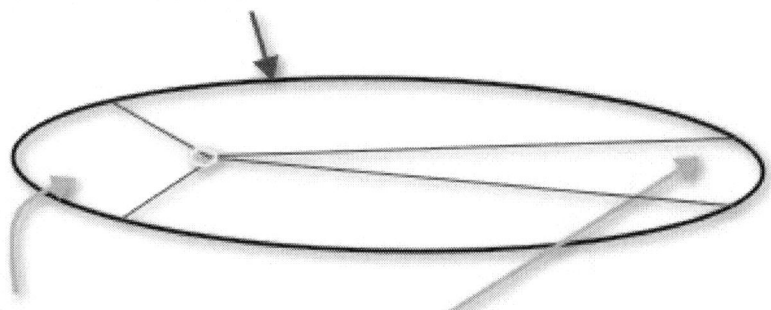

Equal Areas in Equal Times are swept out by
the planet, Earth, as it travels the path of an ellipse.

Note: The actual height of the ellipse that the Earth travels is closer to the length of the diameter of the ellipse than that shown above. The presentation above is meant to show the *possible differences* in the two equal areas swept out during the same time in the path of the Sun.

A Repeat: Kepler was the first person to discover that the Earth and planets move according to precise laws—as expressed by precise mathematical equations. He is also the first one to conclusively prove that the Earth and all the other planets in our Solar System orbit about the Sun. That is, the Sun is the 'center' of our Solar System.

Another 'modern' scientist of the time, the Florentine, Galileo Galilei (1564-1642) also spent considerable time in testing hypotheses in his many studies. His studies also included observations of the paths of the planets and his experiments of the effects of the force of gravity. He performed experiments on objects in free-fall, such as his Tower of Pisa experiments, and, as importantly, on objects rolling down inclines. (Gravity was not yet defined during Galileo's lifetime; Isaac Newton was the first to recognize it, and his recognition occurred during 1665-1666, during. what is known as, Newton's *Anni Mirabilis*, or *Year of Wonders*, or *Year of Miracles*.)

From then on, given these examples of activities by Kepler and Galileo, others set out to apply mathematics to other observed phenomena in the physical world and in the universe—people such as Michael Faraday, Clerk Maxwell, Ludwig Boltzmann, Max Planck, Niels Bohr, Albert Einstein, Edwin Hubble, Werner Heisenberg, Paul Dirac—and thousands of others.

An additional note about Kepler: In 1604 he wrote a book on Optics. Kepler's book was probably the first book on the science of optics, so Kepler could be called "The Father of Optics," although this title is usually accorded to Newton. In Kepler's book, he was the first to postulate 'the ray theory of light', to explain vision. We can see other contributors to the science of Optics on Time-Chart 2.

(The problem of identifying who is 'the first' or 'who should get this award' (or title) is difficult problem in physics, as it is in other fields of endeavor.)

If you would like to learn more about both Brahe and Kepler, both their *persona* and their science, I recommend pages 51 - 67 in source 10, *Cosmos*, by Carl Sagan.

We will now discuss the high-drama expressed in a relatively recent book written about Tycho Brahe, and Johannes Kepler. This book, may, if it is proven to be correct, detract from our evaluations of Kepler as a person, but should not effect our evaluation of him as a scientist. The book is *Heavenly Intrigue, Johannes Kepler, Tycho Brahe, and the Murder Behind One of the History's Greatest Scientific Discoveries*, and was published in 2004. Its authors are the married couple Anne-Lee and Joseph Gilder.

The book presents evidence that Kepler murdered his benefactor, Tycho Brahe. It also says that even while he was assisting Brahe, Kepler took Brahe's observational data. The book also describes, as mentioned previously, how Brahe was Imperial Mathematician to the Emperor Rudolph II of Austria. Before describing the events, you may wish to know more about Rudolph. He was Holy Roman Emperor, King of Bohemia (the name of Czechoslovakia at that time), King of Hungary and Croatia, Archduke of Austria, and member of the house of Habsburg. He also selected Prague as, what we would today call, the capital city of his kingdom. (Rudolph's reign, however, is considered to have been a disaster.)

Brahe had been working for Rudolph II, and Brahe introduced Kepler to the Emperor, citing the fact that Kepler was his assistant. This introduction gained additional funds for Brahe (and Kepler) for their joint research. Only a few days after Brahe died in 1601, the position of Imperial Mathematician was given to Kepler.

The question might be raised whether Kepler should get the full credit he typically does get in history for his findings. Perhaps the contributions of Tycho Brahe should be reexamined to evaluate whether there should possibly be a "joint award". The question then is, "What would Brahe have thought of such a sharing?"

A 'Truly' Final Note Concerning Brahe and Kepler:

We note that the subject of Brahe and Kepler was still active in the world of Brahe and Kepler as late as 2015—and in this case, we might say, "It was truly active." That is, earlier, in 2010, questions arose concerning the death of Tycho Brahe. Since Kepler truly benefitted from their joint work, and even more so, since his 'employer', Brahe, died, Kepler did not need to share any longer any monetary or personal-glory benefits. We now see that History never sleeps.

The following is an annotated copy of a BBC web report from November 15, 2010.

"Danish Astronomer Tycho Brahe Exhumed to Solve Mystery"

"Was the astronomer murdered? Tests may provide the answer. The body of a 16th Century Danish astronomer is being exhumed in Prague to confirm the cause of his death. Tycho Brahe was a Danish nobleman who served as royal mathematician to the Bohemian Emperor Rudolf II. He was thought to have died of a bladder infection, but a previous exhumation found traces of mercury in his hair. A team of Danish and Czech scientists hopes to solve the mystery by analyzing bone, hair and clothing samples."

Well, we can all rest and *NOT* incriminate anyone, including Kepler. In the November 16, 2012 issue of the magazine, *Livescience*, there was a posting that can be found and read on the web:

"Tycho Brahe Died from Pee, Not Poison"

"Two years after Tycho Brahe was exhumed from his grave in Prague, chemical analyses of his corpse show that mercury poisoning did not kill the prolific 16th-century astronomer. The results should put to bed rumors that Brahe was murdered when he most likely died of a burst bladder."

End of the discussion of Tycho Brahe, Johannes Kepler and the Emperor Rudolf II, King of Bohemia.

Chapter 2. The Steam Engine—the Prime Mover of the Industrial Era, and Thermodynamics, the First Modern Technology

This chapter describes:

- The Steam Engine, the machine that changed the world
- Thermodynamics
- Early Leaders in Thermodynamics: Daniel Bernoulli, Benjamin Thompson, Sadi Carnot, James Joule, Clerk Maxwell and Ludwig Boltzmann

Thermodynamics is the discipline in physics that studies the relationship between heat energy and mechanical energy, and the conversion of one into the other. That is, changing (or transferring) heat energy into mechanical energy, and changing mechanical energy into heat energy.

Thermodynamics involves many parameters that are studied and experimented with, such as pressure, volume, speed, force and temperature. Probably, the most ubiquitous parameter that is involved in both studies and experiments, is temperature—the way we measure heat. To make that measurement, we use a thermo-meter, or, thermometer.

To introduce you to Thermodynamics, we will first consider the ways heat is measured. To do this we will first mention its well-known measurements in the home.

We all are acquainted with the home temperature-measuring system, the 'ubiquitous' thermometer. Thermometers are used to measure the temperature of a child—or a person (or an animal) when he or she is sick, the temperature of a room, the home-freezer, the temperature of food as it is being cooked, and so on. Aside from its uses in the home, temperature-measuring systems are also used throughout the industrial world to measure the temperature of a car's engine, office spaces, computer rooms, factory equipment, room heaters and air conditioners—and so on. An almost 'infinite' list of applications is possible.

To examine a more complex system, let us see what NASA, the U.S. space agency, seeks to find in outer space in its one of its most important temperature-measuring operations. In later chapters we will see that "just measuring temperature in space" has a much broader application than 'just measuring temperature' in the systems used in the normal human-environment—and that it can provide significant information to our understanding of the Universe.

The following is a brief description of the goal of space satellite The Wilkinson Microwave Anisotropy (Space) Probe (WMAP). NASA launched WMAP in 2001, and NASA considered it was a 'top performer' in the annals of space exploration since thousands of technical papers were published following its return to Earth.

According to NASA (*temperature* italicized by author):

"The goal of WMAP was to map out minute *temperature differences* in the Cosmic Microwave Background (CMB) radiation in order to help test theories of the nature of the Universe." The CMB refers to a 'relic-in-space' that is the residue of the event known as the *Big Bang*—the moment of the beginning the Universe, some 13.75 billion years ago.

A further discussion of some of NASA's WMAP findings is delayed until Chapter 5, since we first must consider the *fundamental* aspects of thermodynamics before looking at the Cosmic Microwave Background. And it is in chapter 5 that we discuss the Big Bang—and, therefore, you will have, by then, a framework for understanding NASA's findings concerning temperature, the CMB—and the Big Bang.

Let us start talking about temperature with something that is more familiar to us—something that might, at first, be first considered a mundane system —the steam engine. We will learn and should appreciate that *the steam engine was a revolutionary system*—both in itself and in what it did for civilization.

46

The following describes the very beginning of the new science called *Thermodynamics*, and the product that fostered its development, the ubiquitous steam engine.

The Steam Engine, The Machine That Changed the World

I consider the steam engine to be the most important invention that man has invented in the past few hundred years. It was the first to enable people to control and change their individual destiny. It was the first tool or machine that enabled them to change the physical world.

Consider the following:

For the first time in Man's history, vehicles, called *trains*—that were *driven* (or powered by) a steam engine, enabled people to go on land faster than a horse. And ships, also powered by that (ubiquitous) steam engine, enabled people to move reliably on the oceans. And they could go just about anywhere, for as long as they wanted, and in just about any weather condition. Amazing! What a difference from the preceding, many centuries. Today, we might call that *Freedom*!

A steam engine heats 'something', such as wood or coal or oil. A *something* that burns and gives off heat. The heated or burning wood, coal or oil, heats a huge kettle of water (usually called a boiler) and turns the water into steam. The steam expands with high pressure and pushes 'pistons' that make 'something' move. Through the use of those basic ingredients, there are steam boats, steam railroads, steam shovels, steam pumps—and many more types of stream-driven power tools. All use the same principle: the system heats water, turns it to pressurized steam, the steam expands and pushes a piston, and the piston makes that something move.

Before the steam engine was invented, people were limited in their modes of travel. They could go to a neighboring town on foot, or by horse, or by horse-drawn carriage. If they wanted to go far, say along a river, or further, across oceans, they would go on a sailboat for what could be a long voyage. All these modes of travel are relatively slow compared to the means of travel that the steam engine provided. In the case of sailboats, the traveler had to rely on the wind being favorable to the ship's desired direction of travel.

Imagine going from where you live to a place a thousand miles away by horse-drawn carriage or by a large sailing boat. It would involve a long, tedious, expensive trip involving many stops at hotels and having many meals on the road, or at sea. If the trip involved a sailing boat, it could be dangerous. *The steam engine changed that dramatically.*

And all-important was the construction of the huge railroad networks that spanned continents—the United States, Canada and the rest of North America, South America, Europe, Africa, Asia and Australia. The steam engine also enabled people to meet together more frequently and, in so doing, brought about tremendous social, economic and political change. The following briefly describes the overall history of the development of that valuable steam engine-system—the major contributors to its system and to its base science, Thermodynamics.

The history of the steam engine may be considered to start with the French physicist, Denis Papin (1647 - 1712). Papin designed the first steam engine in 1687 and a second, much-improved version, in 1707.

In a parallel time period, the English inventor, Thomas Savery (1650-1715), patented the first steam engine in 1698. The purpose of Savery's engine was to pump water out of coalmines. It was a very practical invention and it was definitely important to England's important coal mining industry.

James Watt (1736-1819), also an Englishman, patented an improved version of the steam engine that enabled the steam engine to be used for many different applications, such as using it to move passenger wagons and freight wagons that are placed on rails (a railroad), or on a boat, by attaching a propeller to the to the shaft attached to steam engine.

Later, more advanced versions of the steam engine were used as drills to find the sought- after coal and metals (iron, copper, bauxite…), precious minerals (gold, silver, diamonds…), water and petroleum to produce fuels such as gasoline, diesel oil and natural gas. Thus, it had enormous financial effects and benefits worldwide. (And many 'anti-benefits' to the many coalmining persons and communities).

Appendix 3 lists some of the later developments of the steam engine, plus the entrepreneurs who commercially benefited from it. We will continue here with discussion of its impact on civilization-at-large.

The Watt steam engine enabled the start of the manufacturing industry, which we know as The Industrial Revolution. Before the steam engine, when people made cotton or wool textiles and transformed them into yarn, it was all done by hand or foot power (pedals) to make the looms turn. Manpower—or womanpower or, as likely, childpower. The steam engine let steam do the heavy work, the 'pushing', moving the looms quickly and easily—and changed how textiles, for example, were manufactured.

The steam engine enabled corporate enterprises to build the infrastructure for the railroad and steam ship lines, thereby starting, true, world trade. As an addition to that world trade, it enabled the beginning of the commercial tourist industry—both on land and sea.

Steam-driven shovels were used to dig huge holes in the Earth, and steam-driven hoists enabled tall buildings to be built, along with steam driven elevators to lift people within the buildings. Later, very tall buildings, now known as skyscrapers, became a reality. Huge, industrialized cities—with steam-driven local railroad systems, grew as never before.

Even today, the steam-driven turbine, that we might consider to be an advanced version of the piston-driven steam engine, produces about half the world's total electricity. The main difference, differentiating the two, is that the steam engine use *pistons* to drive the wheels or propellers; the steam turbine uses *fan-blades,* connected to a common axel, to drive their (equivalent) drive shaft.

Nuclear energy plants are used to generate electricity, and they continue to do this in the old-fashioned way—but at much higher temperatures than the traditional water-steam boiler. They heat water (or an equivalent fluid) to attain superheated temperatures that transitions to superheated steam, which is injected it into modern superheated-driven turbines.

The electric motor and the internal combustion engine, or gasoline engine such as the engine in your car, all came later—after the stream-driven system was established. When automobiles first appeared in the end of the 19th Century, the world already had in place a huge networks of steam railroads; big buildings built by the steam-driven construction industry, steam-driven factories and steamships going around the world, transporting people and products.

A prime example of the power and utility of the steam engine is the construction of the Panama Canal. The canal was built using steam engines to dig the canal and remove the Earth—its two major tasks. At the start of the project, the American engineers in charge of the building of the canal thought that the main problem would be the digging and cutting-out tons and tons of Earth to make way for the water to flow through the 'the big cut', the canal itself. As it turned out, the American team, found out that *moving the earth that was excavated away from the 'pathway' of the canal*, turned out to be a greater problem than the digging of the canal itself. The canal project was completed in 1913 and opened in 1914.

The chief engineer of the U.S. project was John Stevens (1853-1943), a former railroad man. It is an open question whether the project would have been successful if a person without a railroad background had been selected for that position. Stevens applied his railroad experience and recognized the problem was not as much a digging problem—as most people on the project originally believed, but rather a problem of transporting all that dirt away from the canal area!

This operation required that a huge railroad support network be built along the canal to haul the dirt away. Steam shovels did the digging and steam-driven railroad trains took the dirt away from the canal. The U.S. project was successful. The project started in 1904 under the overall direction of President Theodore Roosevelt, and it was completed in 1914. It is forty-eight miles long.

Digression: Prior to the American's building the Panama Canal, there was a French Panama Canal Project that was also located at the joining of the Northern and Southern Western Hemispheres, and the French Panama project was unsuccessful. The French team was under the direction of Ferdinand de Marie, Vicomte de Lessups, who had successfully built the Suez Canal that was completed in 1869. That project was managed as a *digging* project. As implied above, the French Panama project was unsuccessful. Also, many persons attributed its failure due to the presence of Malaria in the Panama jungle. End of digression.

Before we leave this discussion of the Panama Canal, look at figure 5, that appears on the next page—and feel the overall power that emanates from the artist's sketch of a steam shovel dumping dirt into railroad cars, building

50

the Panama Canal. The picture appeared in Joseph Pennell's *Pictures of the Panama Canal*, J.B. Lippincott, 1913. The description of the picture in that book is as follows:

XIV STEAM SHOVEL AT WORK IN THE CULEBRA CUT

This 'beast', as they say down there, "can pick up anything form an elephant to a red-bug"—the smallest thing on the Isthmus (of Panama). They also say that the shovel "would look just like Teddy (Roosevelt) if it only had glasses." It does the work of digging the Canal and filling the trains, and does it amazingly—under the direction of its amazing crews.

Figure 7 STEAM SHOVEL AT WORK IN THE CULEBRA CUT

Allan Karson

Thermodynamics—A New Science and Technology of the 19th Century

Let us first understand why Thermodynamics came into being as a science that eventually led, or provided information to, other scientific disciplines such as physics, chemistry, biology and cosmology.

When the steam engine became the workhorse of the 19th Century, scientists of varying disciplines, such as chemists, physicists, geologists and mathematicians, initiated the study of heat in all its manifestations. This was a major step in the study of the dynamics of heat and the varied scientists' studies and activities were integrated together and became known as *The Science of Thermodynamics*. Some of the activities that those scientists undertook were:

- Identification and study of the solids and liquids that are employed to heat the water to produce steam, such as wood, peat moss, coal, and later, petroleum products.
- Analysis of the solids and liquids resulted in numerical ratings representing their varied *heating abilities*. The ratings go by a number of measurements and terms, such as BTUs (British thermal units), joules (a unit of work done), calories (energy needed to raise the temperature of one gram of water by one degree Celsius) and ergs (also, a unit of work, but minute, compared to a joule. 1 erg = 10^{-7} joules).
- Analyses of the heat transfer process that goes on within the engine itself to determine the *efficiency* of the process.
- Development of the Kelvin temperature scale—needed to provide a fuller temperature range than the usual one associate with people and earth's modest temperature-range climate. (For example, the K scale starts at '0', or "infinite cold", or -273 Celsius, or, in Fahrenheit, -469.67 degrees.)
- Development of an understanding of the scientific concept of the always-present factor called *Entropy* —an indicator of 'the loss of energy.' (Entropy will be defined below.)
- Scientists frequently turn to the concepts the laws of Thermodynamics, as a 'starting point' for their investigation of non-thermodynamic problems or systems. A prime example of that, is that Albert Einstein started his investigation of his Theory of Relativity, that deals with Light, by examining concepts within Thermodynamics.

Textbooks on thermodynamics usually start by stating the three or four fundamental laws of Thermodynamics. For our purposes, and to avoid complicated explanations, we state here two of the most important of those laws:

1. Conservation of Energy: The Total Energy in a *closed* system remains constant. (The Universe, or a steam engine along with its environment, or a *closed* box, is a *closed* or *isolated* system). Also, Energy is a form of Heat, and visa-versa.

2. The *Entropy* of a closed system will increase in time, where entropy stands for *loss of energy*. In the present discussion of thermodynamics, energy loss implies heat loss.

Normally, this would be the place to discuss first, the law known as the Conservation of Energy. I would assume, however, that most readers know of it, so I only copy what Wikipedia succinctly says, which is, " ...total amount of energy in an *isolated* system remains constant over time. The total energy is said to be *conserved* over time." (Italics added for emphasis.) While that statement can imply *many* interesting meanings for certain applications, let us now go to the 'Law of Entropy'—which is a much more interesting and challenging topic—at least for this discussion.

The law of Entropy is applicable throughout the Universe. *I consider it to be the most interesting within the group of three or four laws of thermodynamics.* And, in addition to its universality, it has frequently been applied as an analysis tool to scientific areas that are *outside* of thermodynamics. For example, Professor Stephen Hawking, the well-known physicist and Cambridge University professor, employed the law of Entropy to his study of Black Holes in the Universe. (The results of his work will be described Chapter 6, in the part titled "Dying Stars and Black Holes.")

Note: Since this is the first time I mention Professor Stephen Hawking, I also wish to indicate that he held the Lucasian Chair of Mathematics at Cambridge University from 1979 to 2009, the same position held by Isaac Newton in the latter part of the 17th Century. End of note.

Another example of the importance of the concerns of Entropy concerns the German physicist, Max Planck (1858 – 1947), who was an early leader and 'appreciator' of the role of Entropy in scientific studies and experiments concerning Quantum Theory. Planck's efforts to disseminate the importance of Entropy to the world's scientific community are described in source book 1, *Planck's Scientific Autobiography*.

Now, Entropy: By definition, when there is a *loss* of energy, there is a *gain* in Entropy. Before a steam engine starts up, that is, before it gets heated, its entropy is 'zero'. But once it gets hot, even before it moves, there are *losses of heat* in the system—and *entropy* is said to increase. And we can assume, in general, entropy usually increases with time.

A note: The word 'usually' is used here, but it could well be replaced by *always*. 'Usually' is used for purists; 'always', for realists. Later in this chapter there will be more about the distinction between those two words, *usually* and *always*, and how they relate to entropy. End of note.

Entropy is with us in *all* actions where heat is exchanged—including outer space. Entropy is issue that physicists frequently consider in their evaluation of an experiment where a *transfer* of heat takes place.

To start our investigation of what entropy is, and to see how high-pressure steam is doing work, let us look again into the details of how a steam engine works. In a steam engine there is a wood or coal or oil-heated furnace, be it on a railroad steam engine, a ship's steam engine or the old-type steam shovel shown back in figure 7. The fire in the furnace heats up the water to transform the water into steam. A heat transfer.

The steam will eventually be pumped into the steam engine's pistons to push the piston to make it move. The heated steam has a considerable amount of pressure, and this pressure pushes the piston. (We might say that *steam is doing work*.) As the steam pushes and the piston moves, the steam's energy is *reduced*. This is seen by the measureable *reduction* of the steam's temperature and pressure. We might say, "The steam is worn out." Conversely, we can say, "Entropy has increased."

As the piston moves back to where it began its push, the much-cooler (weaker) steam is ejected from the cylinder. Refer to Figure 8 to see what a piston does. The figure shows the general role of a piston as it might be in a steam engine. (A note: The mechanical role of the piston in a steam engine is similar to its mechanical role in a gasoline-driven automobile engine. It is to turn a crankshaft—that has wheels, or fan blades, attached to the crankshaft. And similar heat transfers occur in electric-driven motors. End of note.)

There are many, many heat transfers in *all* engines. And in all of these heat transfers, there are losses of heat. That is, these additional heat losses do not contribute to the purpose of the steam engine, which is 'to make something (e.g., the piston) move'.

All heat transfers usually involve large losses of heat energy. Having any loss of energy in an engine means that it is not operating at 100% efficiency. Actually, most, if not all, engines can be considered to be inefficient—and in varying degrees of inefficiency. The following are some examples of various types of engines' efficiency:

- An ordinary steam engine's efficiency is less than 10%.
- A well-designed steam engine with special reheating features is in the 30% range.
- A modern steam turbine, such as are used in coal-generator power plants, is in the 50% range.
- Automobile engines are in the 20 to 30% efficiency range.
- Diesel engines are in the 40% range.

The meaning of those efficiency numbers: If the fuel being used (petroleum, diesel oil, coal, wood…) has the chemical capability to produce a certain amount of heat, which will be referred to as 'X' BTUs, the system will have as its maximum usable energy: 'X' BTUs multiplied by (or 'times') its efficiency. The efficiency numbers above, account for *only the fuel's efficiency*.

They *overall system efficiency* of the total system, taking into account losses (increase in entropy) in pistons, wheels, gears and other moving parts, is much lower. The data above is taken from a longer list of examples in Appendix 3, titled "Why Engines Are Inefficient." The appendix examines the engines in the list above and shows where other energy losses occur, to bring about very low, overall system efficiencies.

Figure 8 The Universal Role of the Steam, the Cylinder, the Piston, the Wheel, the Drive Wheel and Drive Shaft

Tank containing <u>Heated Steam</u> that is under High Pressure, and to be injected into Cylinder

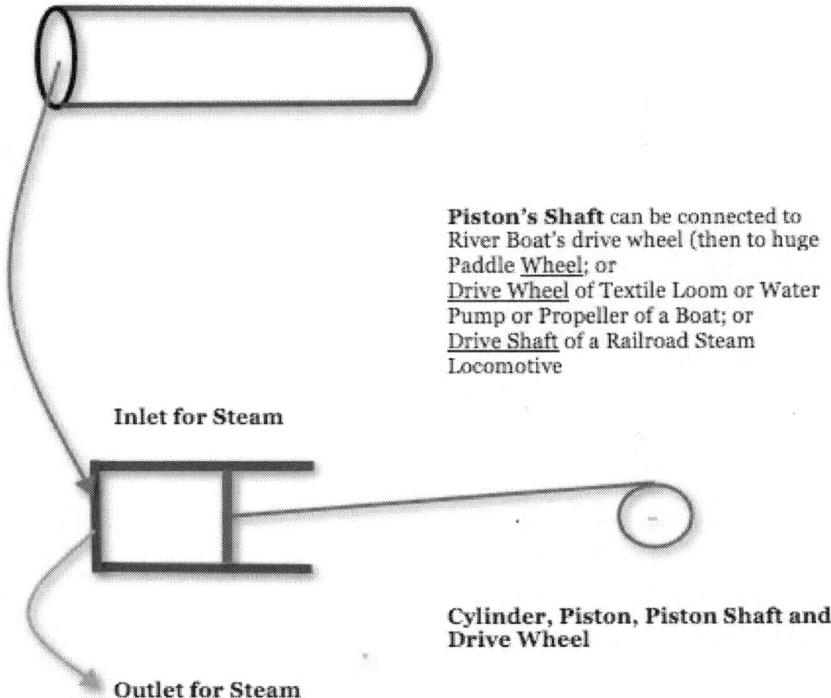

Piston's Shaft can be connected to River Boat's drive wheel (then to huge Paddle <u>Wheel</u>; or
<u>Drive Wheel</u> of Textile Loom or Water Pump or Propeller of a Boat; or
<u>Drive Shaft</u> of a Railroad Steam Locomotive

Inlet for Steam

Cylinder, Piston, Piston Shaft and Drive Wheel

Outlet for Steam

Depleted steam: Increase in Entropy

The concept of entropy can be extended to other domains within physics and associated science and engineering disciplines. Something that is orderly and then disordered is said to have *gained* entropy. For example, a deck of cards is usually 'orderly' when we open the pack. Any subsequent action to the deck, such as shuffling the cards, decreases order or increases disorder or entropy.

Similarly, if we take a book and remove the binding that holds the pages, and then throw the pages in the air, the order of the pages is now in disorder. Entropy has increased for this 'book system'.

In the 1940s, Dr. Claude Shannon (1916-2001), a senior scientist at the then-American Telephone and Telegraph Corporation's (AT&T) world-famous Bell Telephone Laboratories (BTL) and later a professor at MIT, extended the concept of entropy to information, such as the information sent on a telephone wire. Shannon established a mathematical relationship about information that stated the following: When information (voice or data) is transmitted in any way—on a wire or in space, there is always the possibility of an increase of uncertainty (disorder, or noise) in the message —which he referred to as *a gain in entropy*. In 1948 Shannon expressed this work in a BTL document, *Mathematical Theory of Communication* and thus founded the branch of science now known as "Information Theory."

While there are other so-called universal laws, this law—'entropy always increases in all transactions'—appears to be with us forever and to be everywhere. In all thermal processes there is a gain in entropy—a loss of power in an engine due to friction, the fuel (coal, gasoline, nuclear) combusting but not using all its built-in power. (Similarly, in the sending of information (be it voice, data, music or pictures) there is always an increase in entropy when there is any —even slight, mix-up or misunderstanding in a message.)

The concept of entropy has far reaching implications—especially in the study of the Universe. The cosmologist must observe the thermodynamic activity from a distance, as with the Wilkinson Microwave Anisotropy Probe. And by accounting for the direction of entropy—increasing or decreasing, the cosmologist is helped in her understanding of the overall action in the Universe. Entropy—in *all* its manifestations—energy, information, and, I will speculate, in some as yet undiscovered phenomena,.

Vast processes occur in the Universe all the time. Stars burn out, stars join together, and some stars become Black Holes. Each of these processes involves a transfer of energy, a *change* in energy or order. The standard thermodynamic model works the same all over the Universe—and no matter the event or the process, Entropy (usually) increases.

When cosmologists observe an event that does not conform to the standard rules of thermodynamics, but appear to follow a new set of rules, they seek to develop a theory that will explain the event. This is where the concept of entropy is so important.

For example, Cosmologists study the entire system of the Universe and develop laws that show the interrelationship of systems or discrete units in space. In the cosmologist's development of a new theory to describe this new event, cosmologists keep in the front of their thoughts that entropy must—or at least should increase in any closed system. If it does not increase, the theory is probably invalid. Thus, entropy can be a vital checkpoint to validate a theory. In summary, Entropy is considered in *all* theories, from the huge forces and distances encountered in cosmology, to the micro-distances encountered in studies of the atom.

A final note: Dying stars are discussed later in Chapter 5, where, it will be discussed how *considerations of entropy* led cosmologists to change basic notions concerning those often-discussed, very dense objects in space, Black Holes.

Digression: The Laws of Physics, The Direction of Time—and Entropy. Note: This digression may not interest many readers. I recommend it only for those readers who are *truly* interested in the 'exotic' subject of Entropy. Others should proceed to *Leaders In The Technology of Thermodynamics*, located about two pages later.

Entropy, again: Physicists frequently use mathematical equations to determine the result of a calculation. For example, they may want to calculate the speed of an airplane, a railroad steam engine, or a distance traveled. Such equations include 'time' in the equation. Time can be seconds, minutes…. The letter 't' usually represents time in such equations as: Distance traveled = velocity * t.

Those mathematical equations that represent the laws of physics—and which have within them the time variable 't', provide a valid, physical result whether 't' is positive or 't' is negative. (This condition is called time-symmetrical).

That is, if 't' is negative and it is used to calculate distance, we understand that the distance is 'backward' from where we started. Let us do that: Distance traveled=velocity * (-t), and the result is '-D ', a negative number. This means that the moving vehicle or moving person went in the opposite direction from the vehicle or the person's starting point. This is O.K. and understandable.

Usually, all calculations in physics 'accept' the negative result—*except one*, the calculation of Entropy. That is, if 't' is inserted in a physic's equation that expresses the laws of physics to calculate entropy, and the result comes out with *negative* entropy, we assume something is wrong. Negative entropy implies that something is going against the laws of nature that we all observe. That is, in all physical situations, total entropy should increase.

Leaves fall, ice melts to the ambient temperature, hot things cool down to ambient temperature, card decks get more unsorted, messages get garbled, and we grow older—as time goes on. It is highly doubtful that any process has been observed where any of those events went any other way than that time passes.

However, the equations that are used to calculate Entropy are not time-Asymmetric. Rather, they are Time-Symmetric. Time-Symmetric means:

- We cannot infer the direction of time from any of the other (present) laws of physics. (For example, the Laws of Gravity, the Law of Electromagnetism—and others.)
- BUT, we can infer the direction of time by observing Entropy.

To answer, "What is the cause of this difference between Entropy and all other laws of physics?" we can consider the following reason—and there may be more. (A *purist* physicist may not subscribe to the idea that law of

Entropy is a *true* law (a *fundamental* law) of physics. But, it still is a law of physics.)

Let us examine this issue. If Entropy is a fundamental law of physics, it should have some way to include the time variable and simultaneously account for negative time—as all other physics laws do. However, *the laws of physics may not be presently sufficiently developed to be able to incorporate correctly the time variable into equations of Entropy.*

I will repeat something that disputes what the purist would say:

> If Entropy increases, time is going in the right direction. If Entropy is decreasing, something in the experiment is wrong. In closed systems, entropy must increase—along with time. All of this fits in with our daily experience.

Try to think of an example where entropy decreases in a closed system. Remember, a closed system is where no outside device or person can enter the system. An example of such forbidden, outside devices is a cold source —in the form of a bag of ice cubes, to lower the heat in the system, a 'total' steam engine.

There is actually another method or rule to account for the direction of time when dealing with Entropy. That additional method is to follow a *probabilistic notion* of Entropy's time-direction. It was the Scottish mathematical physicist, James Clerk Maxwell (1831–1879) who, in 1859, introduced the idea of probability into the study of thermodynamics. It is also a fact that inclusion of probability within thermodynamics was the *first instance* of the application of probability within *any* discipline of physics.

In Maxwell's analysis of Entropy and the direction of time, he stated 'the Entropy law is *statistical* in nature' and that it is highly improbable that Entropy would ever decrease in time. (That is, Maxwell never said, "that Entropy would *never* decrease in time." A cautious, physicist.) The application of probability to thermodynamics shortly became known as *Statistical Mechanics*. Later, there were many more applications of probability into other domains of physics, outside of Thermodynamics.

Following Maxwell's contribution of probability to Thermodynamics, Thermodynamics became synonymous with the term, Statistical Mechanics.

We will learn, in later chapters, where probability's introduction in other parts of physics, notably in Quantum Mechanics, essentially caused a schism between certain physicists. End of note.

Leaders In The Technology of Thermodynamics

The late 18th century-early 19th Century was the era of the establishment and study of the new technology of Thermodynamics, or, Statistical Mechanics. It must have been a glorious time for scientists. A new technology, new ideas, all giving a strong sense that it was progress in its broadest sense—scientific progress in parallel with national and industrial development. And it was going to change civilization in a fundamental way, from the hundreds of previous years that were without *any* 'power' tools— and now, possessing *steam power*. What a transformation from *all* prior eras!

It was, however, a transformation that brought both good and bad times— especially in the lives of those 'hooked closely' to steam power. For such people, it was very bad. Just read the works of authors who wrote of working in the early factories and coalmines, and the associated poor quality of the home life in industrial and mining cities during We can assume similar living-problems were the norm throughout growing cities in the newly industrialized steam engine-driven world, such as in London, Glasgow, New York City and Chicago.

(For those reader not familiar with the sociological problems of industrialization—which was centered around and within large cities, it is suggested that you read about the problems, described on the web at: https://sites.google.com/site/5effectsofindustrialization/effects-in-the-1800s-1900s

The following describes some of the scientists who contributed to the field of thermodynamics during its infancy. They are presented in chronological order and their names are frequently referenced in modern studies and articles. If you are interested in any of them, you can learn more about their lives and contributions in some of the source books listed at the end of this book.

An addition note about the these people: They lived in the formative years of modern science when it was not always certain that the physicist/investigator could make a living studying and working in science. Thus, many of them started their careers in more predictable, non-scientific endeavors.

Daniel Bernoulli (1700–1782) a Swiss physicist and mathematician.

In 1738 Bernoulli published a text titled, *Hydrodynamic*. The text became the basis for describing the kinetic (active, or moving) theory of gas. He postulated that gas consists of molecules. Prescient!! When gas is heated, it is the impact of the molecules on a surface of the container that causes the gas pressure. Regretfully, Daniel Bernoulli's work was either forgotten or not was known by the world of physics until it was resuscitated in the latter part of the 19th Century.

The presence of molecules was contested strongly among physicists and scientists for a long time—actually, until the early 20th Century. It wasn't until 1908, that Batista Perrin, a French physicist (1870-1942), performed *the first experiment that was recognized worldwide to prove the presence of molecules*. Perrin received the Nobel Prize in 1926 for his experiment. (Two hundred years after Bernoulli's writings about molecules! Another demonstration of physics time.)

Digression concerning the family name, Bernoulli. You will find that there were many members of the Bernoulli family, many of whom were leading scientists of their time. And they made important contributions in many domains, such as mathematics, hydrostatic theory and physics. For example, there was Jacques Bernoulli, a Swiss mathematician (1654–1705) who was the first to use the term 'integral' in the study of calculus.

There are numerous web sites that provide highly interesting descriptions of the accomplishments of the members of the brilliant Bernoulli family. End of digression.

In spite of the late-recognition of Daniel Bernoulli's work and text on hydrodynamics, he became well known in his communities for his work on mathematical probability and differential equations. The following is an anecdote about Bernoulli that shows the reverence he (or, possibly, his family, due to his family name) received in general.

This anecdote is from a biography of Ludwig Boltzmann, whom we will discuss shortly, and is found in source 30. "On a journey, in his youth, he (Bernoulli) introduced himself to a foreigner… modestly saying, "I am Daniel Bernoulli", to which the foreigner replied, "And I am Isaac Newton."

Benjamin Thompson, an Anglo/American (1753-1814) was the first to recognize the equivalency of heat and energy. Thompson was born in Massachusetts, sided with the British during the American revolution, moved to England, then to Bavaria, where he was titled, Count Rumford. He had previously written documents on the force of gunpowder.

While he was boring out cannons for the Elector of Bavaria, he recognized that there was a direct relationship between the mechanical energy doing the *boring* and the *amount of heat* that was generated. He then proposed that heat is the irregular motion of minute particles of a body, which *collide* with one another. Thus, heat is a form of Energy.

Sadi Carnot, a French physicist (1796-1832). Sadi Carnot was an early contributor to Thermodynamics and is considered to be the founder of the first law of thermodynamics, *The Conservation of Energy.* (Given Sadi Carnot's numerous and important accomplishments in such a short lifetime —only thirty-six years, I consider that he should receive more recognition than is normally accorded him.)

Note: In the source 11, *Inward Bound*, there are quotes of Madame Curie, the Polish-born French physicist, discoverer of radiation from radioactive material (*circa* 1890s) and two-time winner of the Nobel Prize, as referring to the First Law of Thermodynamics as "Carnot's Law". End of note.

Carnot started his career as a military engineer in the French Army, inspecting fortifications. He later became interested in physics and developed the mathematics and graphs that showed the 'cycle of energy' that a heating agent (mainly steam) goes through as it went through its paces of pushing a piston. This cycle is known as the Carnot Cycle. Refer to figure 9 for a picture of the Carnot Cycle, which is located a few pages further in the book.

The Carnot Cycle is a four-step process. The first two steps are for doing the work, and the next two are for 'recovering' the moving element that did the work (or was pushed back). (I repeat: A piston is the moving element that does the work in both a steam engine and in an automobile. It is pushed strongly by steam (in the steam piston) or by high-pressure gasoline that had just been converted from the *exploding* liquid gasoline (in the automobile piston).

Carnot defined in detail the cycles that engines go through as they do their work. That is, they transform heat into energy—and in the case of most steam engines and automobile engines, make 'something' move, be it a huge boat, a railroad train, a car or an (piston-driven, non-jet) airplane. His documents also discussed *yet-unknown* types of engine. Only later was the 'true' Carnot Cycle steam engine built. This fact, alone, should qualify him for greatness and world renown.

Note: The type of engine that the Carnot Cycle describes is frequently referred to as a 'reciprocating engine'—its piston going back and forth. The earlier airplane engines all used piston engines that followed the Carnot Cycle. These engines were known as reciprocating engines, and they are the same type as used in most automobiles. In the 1950s, turbine and jet engines were introduced in aircraft (and in some automobiles, trucks and buses) and they follow a different cycle from Carnot's. End of note.

In the steam engine, the Carnot Cycle consists of:

- Step 1. Coal, or any other fuel, heats water in a water tank, turning it to steam. The *energy* that the steam has at any point in the cycle is called '*enthalpy*'. The steam is injected into the steam engine's cylinder that contains a piston. It is at this point that the steam has maximum energy, or enthalpy.
- Step 2. The steam pushes the piston. The piston moves all the way to the end of cylinder, at which point the steam has exhausted most of its energy (enthalpy). This push by the steam caused the piston to move— the system's goal.
- For simplicity, we combine Carnot's steps 3 and 4. The piston then comes back, pushing the 'exhausted' steam out of the cylinder.

The net amount of work that the engine performed is measured by the energy of the pushing steam (steps 1 and 2)—and subtracting from that the energy still available in the steam as it goes out of the cylinder (steps 3 and

4). We can see this net energy on the graph as being the area bounded by the four curves or four steps.

Similarly, if we were to apply the Carnot Cycle to the automotive engine:

- Step 1. Gasoline is injected into the car's engine cylinder that contains a piston. The gas is 'exploded'. (It is at this point that the gasoline vapors have maximum energy. (or maximum enthalpy).)
- Step 2. The heat from the exploded gas turns to pressure that pushes the piston, the piston moves all the way to the end of cylinder, at which point the gas has exhausted most of its energy (enthalpy). This push caused something to move—which is the system's goal.
- Again, for simplicity, we combine steps 3 and 4. The piston then comes back, pushing the 'exhausted' gas vapors out of the cylinder.

Carnot showed that the *higher* the initial temperature and the *lower* the 'exhausted' temperature, the more efficient the process. Also, he showed that no engine could be near 100% efficient. And, as we saw in the preceding section, it is *nowhere near 100%*.

Carnot's engine cycle is for a hypothetical engine; it represents the most efficient engine cycle. You will see variations on this in the literature and on the web. For example, you can enjoy observing the Carnot Cycle graphically and dynamically at various web sites that show this dynamic process.

Just enter the word "Carnot Cycle" in your search engine and you will find numerous sites showing and describing the cycle.

Allan Karson

Figure 9 The Efficient Carnot Cycle for Engines

V = Volume of Fuel Within Changing Cylinder Volume, due to Movement of Piston

P = Pressure of Fuel Within Changing Cylinder Volume

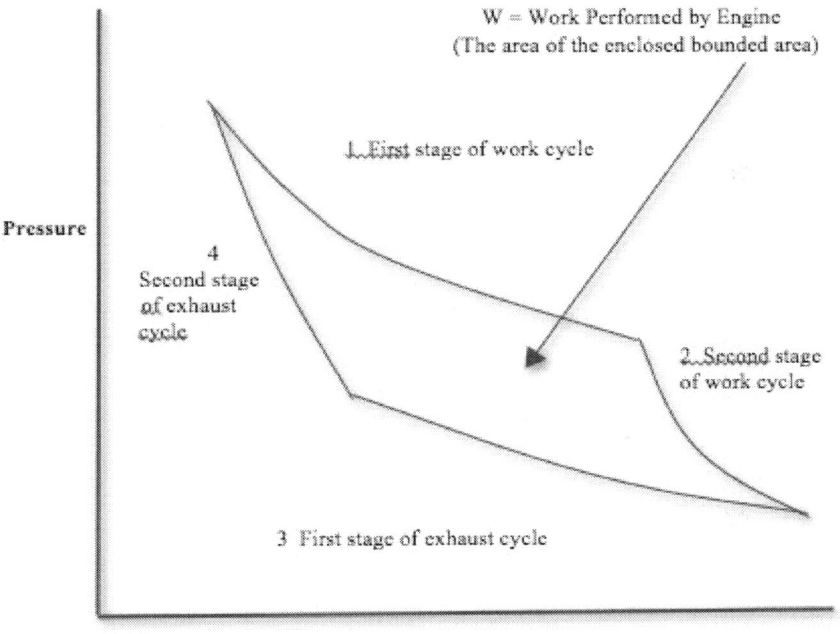

Note: Carnot Cycle above is for Steam Engines

67

James Joule (1818–1889) an English Physicist and Brewer, in 1858, named the relationship of the study of heat and power as *Thermodynamics*. A unit of energy is called the "joule". Joule's father was the owner of a brewery and James was the manager of the brewery. The study of heat and electricity was initially a hobby for Joule.

James Joule is considered to be the first person to provide the *basis* for what is known as the First Law of Thermodynamics, which is *The Conservation of Energy*. As cited above, this law states that for a closed system, the total energy remains constant, while *components* of the total system may change in form—fuel heat to friction heat, and so on.

You may ask, "What does a 'joule' represent." As written above, it is a unit of energy. For example, 'one joule' can be converted to the energy forms that are use to represent:

- The energy obtained from the food we eat, or .23 calories
- The electrical energy we use, or 1 watt-second
- The mechanical energy we use, or .1 meter-kilogram, or .737 pound-feet.

Prior to mid-18th Century, the heat of a gas was only examined by its overall measurements—volume, overall temperature and pressure. No thought or examination was considered concerning what actually was going on in the gas. This situation is related to the nature of physics at that time. Physics was primarily an experimental science and there was a minimum of theorizing.

This understanding of thermodynamics changed significantly in mid-century, in 1857, when the German physicist, Rudolph Clausius (1822-1888), published a paper titled *The Kind of Motion We Call Heat*. He hypothesized, as did Daniel Bernoulli did earlier, that heat was caused by the *kinetic energy* of the gas molecules. Thus was born the *Kinetic Theory of Gases*.

The next contributor is the Scottish mathematical physicist, **William Thomson (1824–1907), or 1st Baron Kelvin**, or, as he is more frequently referred to as, Lord Kelvin. He made many contributions to the field of Thermodynamics, but he is probably best known to the general public for developing the Kelvin scale to measure temperature. The Kelvin scale is primarily used in scientific measurements.

The Kelvin scale starts at zero, which is meant to represent the lowest temperature possible in the Universe. It is -273 degrees Centigrade. To convert from centigrade to Kelvin, subtract 273. As shown on Chart 10, at very high temperatures, Kelvin is equivalent to Centigrade.

Chart 10 Temperature Equivalence in Three Systems,

(Boiling and Freezing temperatures of water are shown in bold)

Fahrenheit	Celsius	Kelvin
1,799,508	999,727	1,000,000
1340	727	1000
212	**100**	**373**
32	**0**	**273**
461	-273	0

Conversion Equations:

Kelvin = Centigrade − 270

Fahrenheit = 9/5 * Celsius + 32

Note that Celsius degrees and the Kelvin degree have the same size and are both 9/5 times larger than the Fahrenheit degree. Also, when referring to temperature in Kelvin, the word 'degree' is not used, e.g., 44 kelvin.

Note: The web provides easy-to-use conversions between the Kelvin scale, the Centigrade scale and the Fahrenheit scale.

Clerk Maxwell (1831-1879), the world-famous Scottish mathematician-physicist, introduced the idea of 'organized' probability. Maxwell did this in 1859, shortly after Clausius's introduction of the idea of the kinetic energy of gases. As noted above, Maxwell's contribution was later given the name "Statistical Mechanics".

We shall meet Maxwell again when we discuss Ludwig Boltzmann in this chapter; and again, when we discuss Electromagnetism in Chapter 3—the topic for which he is best known. Given the many fundamental contributions that Maxwell made to physics—in many disciplines, Maxwell's contributions are comparable to those made by the two other giants of physics, Isaac Newton and Albert Einstein.

We now jump back to an earlier-named contributor, Rudolph Clausius. In 1866 Clausius did something in addition to his 1857 contribution, which—once again, can only be called, "gigantic". Clausius introduced the concept, idea and formulation of Entropy. (As previously indicated, Entropy is all around us. So here it is again.)

In deference to many persons who are not at ease with equations, I have resisted their inclusion in the descriptions of the previous contributors. But Clausius' calculus equation is so simple, so elegant, and so perceptive, that I could not resist showing it here:

$$dS = (d*Q)/T$$

where:

dS is the increase in Entropy for a system in an *infinitesimally short time interval*, (during which a time-reversal process occurs), and

d*Q is the heat supplied to the system during that brief time period, and

T is the absolute Kelvin temperature at which the system is operating.

Think of what the terms 'T' in the denominator means: The higher the temperature of the system, the less entropy (losses) increases. That is the reason why most heating system are designed to work at the highest possible temperatures.

There are more 'modern' versions of the Clausius equation, but his selection of the word 'entropy' is significant: It is from the Greek word for 'conversion, mutation, evolution, confusion and (also) shame. And it is also close to the English word, energy. Note that when Clausius *proposed* this equation about entropy, he had not determined a proof for it. But, almost immediately, in that same year, 1866, Ludwig Boltzmann developed the proof of that equation,

We are trying to stick with a chronological order, but Boltzmann's main presence on the scene begins later than 1866. That is, in 1866, Boltzmann was an unknown' on the thermodynamic scene.' Thus, even though Boltzmann provided a 'limited' proof of the Clausius equation, we must assume that many scientists did not learn about the proof, or even read it. Five year later, Clausius did publish a proof of his equation. When Boltzmann learned of it, he sent a note to Clausius, who returned an apology for not knowing of Boltzmann's proof of 1866. Boltzmann did not reply, but it appears that many scientists knew of the embarrassing situation.

We continue the discussion of thermodynamics by discussing the next major contribution, which was made by the **Austrian physicist and mathematician, Joseph Stefan (1835-1893).** Stefan's contribution, known as Stefan's Law, was developed in 1879.

Joseph Stefan taught physics at the University of Vienna. He was Director of the Physical Institute from 1866, Vice-President of the Vienna Academy of Sciences and member of several scientific institutions in Europe. He was a prolific writer of scientific articles, mainly for the bulletins of the Vienna Academy of Sciences.

Stephan had obtained experimental data from heat experiments that were performed by the **Irish physicist, John Tyndall (1820-1893).** Before doing these experiments, Tyndall had studied in Marburg, Germany, under the tutelage of **Robert Bunsen, the inventor of the Bunsen burner.** He also studied magnetism under the direction of **Heinrich Gustav Magnus**. Both Bunsen and Magnus were considered the best experimental instructors of the time.

Upon Tyndall's return to England, he set up a sizeable laboratory system to measure the heat arriving at the earth, provided primarily by the Sun. It was from this data that Stephan deduced the following law:

Radiated energy from a 'blackbody radiator' = (Temperature)^4 x 5.6 watts per square meter.

Definition: A blackbody radiator is an excellent *emitter* of heat radiation *and* an excellent *absorber* of heat radiation. Blackbody radiators will be described in greater detail later, in Chapter 9, when we will discuss Max

Planck and his analysis of data that led to recognition of, what I prefer to call, *Quantum Behavior.*

As stated in Wikipedia, (http://en.wikipedia.org/wiki/Joseph_Stefan) "With his law Stefan determined the temperature of the Sun's surface, which he calculated to be 5430°C. This was considered to be "the first sensible value for the temperature of the Sun." Stephan accomplished this in 1884.

Note that the *heat energy* is proportional to the *fourth* power of the absolute temperature!! What a power relationship! In the physical world, it is rare to find something that is proportional to the fourth power of something—or anything.

Now, imagine finding the number for radiation when we consider the temperature of the Sun. If we assume the temperature is 5000K, the effective radiated effect is due to 5000 x 5000 x 5000 x 5000 = 625,000,000,000,000 degrees, or, 625 x 10^{12} degrees!!!

Also, you will see that procedure used by Stefan—of using data derived by an experimentalist, is fairly typical in science: One or more experimental scientists (physicist, chemist…) perform a physical experiment and record date resulting from the experiment. The experimentalist then provides the data to one or more theoretical scientists, who analyze the data and/or develop equations.

We will see this procedure again when we discuss the work of Max Planck, who also reviewed the experimental results of blackbody radiation—but for a completely different reason—and about fifteen years later than Stefan's assessment. Planck's analysis of the experimental results led to the discovery of Quantum Theory.

The next major contributor to thermodynamics is the Austrian physicist and teacher, **Ludwig Boltzmann (1844 – 1906).** Note that Boltzmann was a student of Joseph Stefan.

Boltzmann made major contributions in numerous ways. Perhaps, his most important is that he expanded the use of statistics (and its fundamental building block, probability) in Thermodynamics (or, as we refer to it today, Statistical Mechanics). (As stated previously, Boltzmann's first contribution to Thermodynamics was his proof of Clausius's equation, but this proof did not involve Statistical Mechanics.)

Boltzmann's studies were first published in 1896 (at age fifty-two, quite late in his professional life) and were titled *Lectures on Gas Theory*. Much of Boltzmann's work, however, was done in years earlier to 1896. It appears he was not in a hurry to publish his results—which is not the typical attitude of most scientists. The following presents some of his concepts, contributions —and their highlights.

Boltzmann's work was strongly based on the fact that whenever he considered, or observed, a gas, he assumed that it consists of thousands, if not millions, of gas particles—or more correctly, atoms or molecules. These particles were assumed to be moving continuously in the gas container as the gas was being heated. It would be difficult, if not impossible, to know exactly where each one is and how fast that one is going without batching them together and describing their probabilistic condition by (1) statistical methods—and/or by (2) the physical conditions (pressure, temperature, etc.) of the gas under examination.

It was Boltzmann who refined the overall mathematics, or statistical systems, to study mechanics of those activities within the gas—the molecular activities. It was then that sector of physics became known as— and still is known as, *Statistical Mechanics.*

By significantly advancing the development of Statistical Mechanics within Thermodynamics, Boltzmann opened up a whole new application area for statistics—and for its adjunct, probability, within the field of science in general.

A note about the term, Statistical Mechanics. It appears that the deep involvement of statistics within the field of thermodynamics is the *first time that statistics was strongly introduced into any field of physics study*. We now know of the later introduction of statistics in many fields of study— fluid mechanics, earth sciences, climatology, meteorology population growth, economics and finance, voting patterns, etc.

(Imagine, if each time it was later introduced into a field of study, such as those listed above, the course books would be replete with subject names such as Statistical Fluids, Statistical Population and Statistical Finance.)

Rather than setting up numerous statistical groups for each subject, the study of Statistics matured rapidly and consolidated into one field of study —Statistics. It was then applied to other study areas, such as fluid mechanics and finance—and all the others. This was done without the necessity of having to develop and name a 'school of statistics' for each new application area. It appears, however, that its first application—that was in thermodynamics, retained its name, Statistical Mechanics.

I realize that I continue to advocate the use of the name, Statistical Mechanics, to stay close to historical references to it, but I will make a suggestion: When you read the words in this chapter, where it primarily deals with thermodynamics, I suggest you think of *Statistical Thermodynamics*—and not Statistical Mechanics. It should make reading and understanding easier.

It was the investigations of thermodynamics that occupied many great physicists from mid-19th Century to early 20th Century. Planck and Einstein were whizzes at applying Boltzmann's Statistical Mechanics (pardon this use) to the study of Thermodynamics. And it is from such studies of thermodynamics that Planck and Einstein opened up the Pandora's box that became known as Quantum Theory.

We will now be briefly discussing the famous Boltzmann formula, S = k Log W. But first, some background to the formula. You may remember that Boltzmann did his first work on Entropy in 1866 when he provided a proof to Clausius's fundamental equation. It is believed that Boltzmann formulated this more complete equation in the early 1870s—but it was only made known to other scientists when he published his work, *Lectures on Gas Theory*, in 1894.

The formula was engraved on Boltzmann's gravestone—and that has become a well-known picture in books (and on the web) on thermodynamics. (A sculpture of Boltzmann's head is located just above the vertical gravestone.) . Once again, the formula is:

$$S = k \text{ Log } W$$

In the above equation, 'S' is the (all-important) *entropy* of an ideal gas, 'W' is the number of ways the atoms or molecules can be arranged in the gas, and 'k' is called the Boltzmann constant (shown always as a small 'k') and its value is $1.380,648,52... \times 10^{-23}$ Joules/Temperature-Kelvin). Joules/Temperature represents, or is equivalent to Work/Temperature or Energy/Temperature.)

Boltzmann never used that value of the constant 'k'. He would assume it is '1'. For Boltzmann, it did not have any scientific value. He used it only to convert units. From source 30, "Apparently in the nineteenth century this was considered to be the last step in the calculation, not worth publication in a scientific paper." It was Max Planck who developed the present value of the constant 'k' that is shown above.

Given a large quantity of isolated gas particles, 'k' is the ratio of the average particles' energy, in joules, to its temperature, in Kelvin. (This can be written as J/T). Thus, it appears that there is a universal number to represent that *universal physical relationship* of: *Any* gas's average particles' energy/Temperature = k

Returning to the importance of Boltzmann's entropy equation, we now see that he added a new—but equivalent, interpretation to the meaning of Entropy. Whereby, before, it was limited to 'overall' parameters involving 'gross measurements' (pressure, temperature...), he showed its equivalence in the behavior of *atoms in the aggregate*, (i.e., a very complex systems)—a major finding for the new fields of both Thermodynamics and Atomic theory.

We now turn to another Boltzmann topic. Boltzmann is also identified with a set of equations that were originally developed by Clerk Maxwell (1831 – 1879) whom we will discuss in the next chapter, No. 3. These equations defined a function called 'The Velocity Distribution of a Gas in Equilibrium.'

Maxwell's equation, however, did not take into account possible, subsequent collisions. His definition of what could impact or affect the molecules was very simple, and consisted of 'standard' statistical analysis, 'standard' Gaussian to-be-expected graphs—all the 'off-the-shelf', expected happenings. We can assume that Boltzmann considered Maxwell's formulation to be incomplete.

Boltzmann's scientific life and career was mainly devoted to the study of thermodynamics. In contrast, Maxwell's life was more eclectic—but as intense, since he is mainly known for his brilliant, four *fundamental* equations describing the science of Electromagnetism. Thus, Boltzmann's concentration in Thermodynamics, therefore, provided Boltzmann with a deeper understanding and appreciation of *all the physical occurrences and conditions that could affect those distributions*, such as the change in pressure due to change in altitude.

Boltzmann's contribution was that he incorporated a plethora of external conditions and subsequent events into Maxwell's gas equation—and Maxwell's gas equation, conceived of *circa* 1859, became known, as it is today, as the Maxwell-Boltzmann Distribution. Boltzmann's contribution was made in 1872.

A historical note: According to source 30, Boltzmann had wanted to publish his additions to Maxwell's system in a short paper in the German journal *Annalen der Physik und Chemie*, to ensure his priority. His 'mentor superior', Joseph Stephan, however, advised him to present it first in a more formal paper, to the Academy of Vienna. Boltzmann abided by Stephan's advice and chose the latter route, the Academy route, which was also done in 1872. Presumably, if Boltzmann's report had first appeared in the German journal—which had a large circulation among scientists of many nations, it would have reached a much larger audience than it reached via the Academy route.) End of note.

You can observe the many graphics and curves that these equations generate by going to the web, looking for: 'Maxwell-Boltzmann Distribution.'

An additional contribution by Boltzmann: As indicated above, Boltzmann was a student of Joseph Stefan, and Boltzmann also developed modifications to *Stefan's Law*. Boltzmann's modification to the law was based on theoretical grounds, as opposed to Stefan's, which was based on experimental data. It is known today as the Stefan-Boltzmann Law. (A repeat: The original Stefan's Law states the *heat energy* is proportional to the *fourth* power of the absolute temperature.)

Now, a word about Boltzmann as a teacher: Boltzmann is known to have been a great, versatile and highly knowledgeable teacher and professor. The literature abounds about how he worked so closely with his students to help them understand science and help them develop their own philosophies of the science of thermodynamics. He was, we might say, in the field of physics, "A Man for All Seasons", as shown by the experience of one of his students, Lisa Meitner.

Lisa Meitner was a student of Boltzmann and went on to have a very successful professional career. Ms. Meitner developed the first experiment that proved the fission process, the process used in the first and early atomic bombs. The following is a quote from source book 21, a scientific biography of Lise Meitner, written by Dr. Ruth Lewin Simi. Page 12 contains a description of Meitner's second year at university:

"In her second university year, she (Meitner) began studying physics in earnest. Over the next six semesters, her *Meldungsbuch* (message book) lists analytical mechanics, electricity and magnetism, elasticity and hydrodynamics, acoustics, optics, thermodynamics, and kinetic theory of gases as well as mathematical physics each semester and a course in philosophy of science. A fairly typical curriculum, it was highly unusual in one respect: *All these courses were taught by one* person, Ludwig Boltzmann, the theoretical physicist."

Finally, we examine Boltzmann's personal problem with Ernst Mach, a fellow-physicist and fellow-Austrian of the time. Mach is known for his strong disbelief in the concept of molecules. The presence of molecules was fundamental to Boltzmann's work, and this strongly advocated belief by a colleague may have greatly disturbed Boltzmann.

Reports say Boltzmann had a bi-polar personality disorder and was in a continuous state of depression. These reports also hint that his difficulty with Ernst Mach may have contributed to the depression. Boltzmann committed suicide at the age of fifty-eight.

Note: We must realize that in addition to Mach being a non-believer of molecules, there were many others scientists and German philosophers who did not believe in the concept of molecules. It was not so much that they did not believe in molecules, but rather that they— and Mach, believed in 'only the things that could actually be observed.' This line of thought, called positivism, and was that positivists accept only the *immediate sense impressions* as being real. That is, if you cannot see it, it is not real.

As we learned earlier in this chapter, molecules were first detected in 1908 —but still were never accepted by many positivists, such as Ernst Mach. (The role of Mach and his fellow non-believer positivists is one of the main physicist-on-the-scene of that epoch and is described in source book 37, *Boltzmann's Atom* by David Lindley.)

This is the same Mach whose name and studies of sound waves is remembered when we refer to high speed aircraft and missiles traveling at speeds of Mach 1, Mach 2, and so on. The Mach number represents the speed of aircraft or spacecraft as a multiple of the speed of sound, which is 760 miles per hour. End of note.

We will be meeting Boltzmann and Mach again when we discuss the cultural background of Quantum Theory, since both were professors at one of the centers for Quantum Theory's development, the University of Vienna. We will also learn, further on in the text, how Mach provided guidance to physicists in general, as they did observe minute particles, and guidance to Einstein in Einstein's development of the General Theory of Relativity.

Chapter 3. The Four Fundamental Forces in The Universe

The preceding two chapters describe items you can see, touch, observe or hold: Planets, a hot steam engine, coal, oil, and so on. This chapter is different; it describes things you cannot touch or observe directly. It describes the *four basic forces* in the Universe.

But you experience them every day as you walk on Earth (Gravity), call a person on your cell phone (Electromagnetic), or look at the blazing Sun (Strong Nuclear Force). The fourth force (Weak Nuclear Force, or Weak Force) is much more subtle. It is all around us, but it is not that easy to perceive directly.

We might think of the Weak Force as being a "particle to particle" force. We might also say it is the 'mystery force.' It is the force associated with radioactivity. It plays a role, however, in many aspects of our lives. As an example, it plays a key, important role in the ubiquitous semiconductors that are used in cell phones and computers.

These are the four forces we observe in the world around us. They are basic, or fundamental forces. At present, we only know of there being 'four'. We will look at each of these forces and see what it is, what it does, and how we 'use' it.

We first discuss the force of **Gravity**. Gravity is the force that provides the attraction between two or more bodies, or what physicists refer to as "between two or more masses". We will be discussing three examples where Gravity comes into play. Since it is a subject that is frequently presented as a science topic to high school students, you may already be acquainted with these examples.

We observe and use the effects of Gravity every day, all around us. It appears to be a 'simple' force. I consider it is not simple, but rather subtle. It is different from the other three forces in numerous ways. It operates over very short *and* very long (astronomical) distances; it can be very strong and very weak, given the local circumstances. Take notice of its subtleties. Albert Einstein did that in the early 20th Century and developed the General Theory of Relativity that is based on Gravity—and he changed the way we perceive the Universe.

First, to demonstrate the variety that is possible in Gravity-in-Action, we look at three simple examples. These will be followed by a discussion of the history of the discovery—or, more precisely, the identification of what of Gravity is. We will then describe some of the issues and factors involved in its mathematical calculations to determine its force, or strength, and the universal units that are used to describe Gravity-in-Action.

First, the simple examples of Gravity acting between:

1. Celestial bodies (e.g., Earth and Mercury), or between a planet and a star, or between two galaxies, or between any two bodies having their own, independent paths in space. Since their geometry, or their vector-distance, is continually changing, this consists of a *continuously changing* situation for *All* participants in what may be called the Gravity-action model,

2. A body (a person, a rocket, a spaceship…) falling toward Earth and accelerating at a rate established by Gravity. This, too, involves a continuous, big change,

3. A person, such as the present reader, while standing sitting in a chair: The Earth is spinning, and you should be flying off as you would on a carrousel. But Gravity keeps you "tied down" to Earth. A similar case is when you determine your individual weight using the bathroom scale. We can consider this to be a 'stable' situation, even though there is a very slight, change of force, due to the person 'moving through space' on Planet Earth.

4. The oceans' tides that occur within the Earth's oceans are caused by the attraction between the Mass of the particular ocean being considered and the Mass of the Moon.

And there are many other manifestations of Gravity, such as identified by Albert Einstein (1879 – 1955). In his General Theory of Relativity, he shows how Gravity can cause the 'space coordinate lines" in space ('x, y, z'), *to bend*. (We will be discussing Einstein and Relativity in Chapter 11, Part 2.)

I would like to describe two effects of Gravity, while not being met in every-day situations, demonstrate the relationship between Gravity and the shape, or *Geometry,* of *distant* objects in spcace. The effects are known as the *Weyl effect* and the *Ricci effect*. I will present an example of each and you may think of other, possible instances where their effect comes into play.

But first, I must introduce a key part of the measurement of Gravity, which many readers may already know. The force between two bodies, Gravity, increases or decreases according to the reciprocal of the square of the distance between the two bodies, that is, $1/r^2$. The greater is the distance, 'r', from the source of gravity, the weaker is the force. And we assume that *all gravity force-lines emanate from Earth's geographical center,* located deep down in the Earth.

Now, imagine a large spherical air-filled gasbag, suspended over the earth, with flexible, soft, outer skin,. We first consider, what is called, the *Weyl Effect of Gravity*. The *Weyl Effect* will cause a *distortion* in the normally *spherical* shape of the flexible body so that it approaches the shape of an *elongated* dirigible (or blimp) standing on its head.

To understand 'why', consider the unseen 'lines of gravity' that run through the point-center of the spherical gasbag, that are perpendicular to the earth. The bottom of the gasbag is *closer* to the Earth than the top of the gasbag, so their respective $1/r^2$'s *are different*. Sometimes a large difference, sometime small one—but *always* a difference. The bottom is being *pulled* or distorted *closer* to Earth than the top, which is *farther* from Earth. The spherical gasbag now resembles a blimp standing on its head. This distortion effect is known as the *Weyl effect*.

There is another distortion on the *sides* of such bodies, due to the different *angles* between the individual Gravity lines' strength in 'approaching' a body. It is known as the *Ricci Curvature Effect*.

In the *Ricci* case we consider the *angles* between Gravity's unseen (but present) force lines. As in the Weyl effect, there will be a difference between the strength of each Gravity line—not so much, necessarily, as that cause by a difference in length, *but a difference* exists that is due to the small *angle-differences* in the Gravity's (unseen) lines. This results in a slight, but *differences* in strength, acting on the sides of the gasbag.

Now, in this case, we consider a six-sided empty cube in space ,with sides measuring a few feet. Once again, it is these *differences* of the force of gravity at the various point in the cube that *causes* distortion in the cube's body. In this case, the cube's vertical sides will implode *greater at the bottom of the cube* than at the top of the cube. That is, its sides slant in more on the bottom than on the top, making it narrower on the bottom than on the top.

End of introduction to Weyl and Ricci Effects.

The mass of a body is what determines the attracting force between the two bodies. Further on in this chapter, we will soon see how Mass is a *fixed* characteristic of matter, while weight is a force—and the weight-force changes according to *where* the mass is located *with respect to other bodies*.

The equation for the Gravitational force of attraction between two bodies also contains an important constant, called the Universal Gravitation Constant. In equations, it is represented as 'G'. G's value is 6.67384×10^{-10} meters³/(kg x seconds²). (One purpose of 'G' is to present Gravity in the appropriate gravitation-force units.)

The Force of Gravity = $(G*M_1*M_2)/(distance)^2$

The units of this equation are:

Newtons = (meters³/(kg x seconds²) x kilograms x kilograms)/meters²

When Isaac Newton originally identified the relationships that make up the Force of Gravity, he referred to it as the *Universal Gravitational Constant*. It is now frequently referred to as only 'G'. His measuring equipment of the time, however, was not sufficient to provide a completely accurate value of the value of 'G'. Newton died in 1727 and it was until 1798 that the British scientist, Henry Cavendish (1731-1810), also at Cambridge, derived the first *exact* numeric value of that constant, 'G'.

Note: Cavendish had set up a laboratory at Cambridge to perform this measurement. And it was Clerk Maxwell, in 1831, who continued to build on that laboratory, and Maxwell who named it the Cavendish Laboratory—the name it still has. End of note.

Case 2 The Force (Weight) of a Body on a Planet: When Gravity is measured to determine the weight (force) of an object, the body's force is also expressed in Newtons. Newtons are functionally equivalent to what is what are 'pounds' in the U.S. system of measurement. One pound is equivalent to 4.45 Newtons.

Scientists have assisted us in determining the weight of a person by simplifying the Gravity equation. That is, they have determined the specific values that are associated with Earth. Weight in both MKS and the U.S. system are shown below:

Weight = Mass (called 'slugs' and applicable to a person or an object) x 'a' (gravitationally-caused acceleration toward earth) which is 32 ft./sec^2 or 9.8 meters/sec^2. This equation-in-words is commonly seen as:

$$F_{weight} = M_{mass} * a \text{ or, } F = Ma$$

Thus, for example, if a person's weight, or F_{weight}, is 165 pounds, or 734 Newtons, his mass is 5.128 slugs in *any* system of measurement. On the Earth's Moon, which has 1/81 the mass of the Earth, he would weigh 27.5 pounds, but his mass would still be 5.128 slugs. And, if he were relocated to a different planet that had twice the mass of the Earth and the same radius as Earth, he would weigh 330 pounds; and his 'faithful' mass would still be 5.128 slugs. I jokingly say: Such universal constancy!

Case 3 is where a body (a person, a rocket, a spaceship…) is falling and accelerating toward Earth. The rate of fall is established by gravity. In this case, no bodies other than Earth are considered, nor is air friction. This case is similar to the experiments that Galileo performed when he dropped various items from the Tower of Pisa. And other experiment is when he rolled balls down incline planes.

Little 'g' is different. It represents the constant that accounts for gravity's attraction for a body *falling to Earth*. It applies *only* to Earth, or, in the unlikely case, to a celestial body with the same combined mass as Earth. The value of 'g' in the U.S. system of measurement is 32.1 feet/(seconds)^2, or 9.81 meters/(seconds)^2 in the MKS system. (If you were around when the

U.S. landed on the moon, which is a smaller body than earth, you might have seen members of the team 'jump'—to greater heights than they could on Earth. On the moon, 'g' is 5.3 ft/s^2 or 1.6 m/s^2)

The following is one of ways 'g' is used to determine the distance of the fall:

If I—or any one of us, were to fall from the top of the Tower of Pisa—or any tower, we would fall with an acceleration of 32.1 $feet/(second)^2$. 'D', the distance during the fall is a function of the square of time 't'; that is, t^2. The formula for the *distance* of fall is: $D = \frac{1}{2} gt^2$.

About Mass: For most practical uses, we can consider mass 'represents us'. Mass may also be considered to be what is called 'dead weight', or just something that hangs around—not pushing or doing on anything. As state previously, Mass has a *constant* value, according to the person or object—something that weight does not have. And as stated previously, we are always the same mass—no matter *where* we are in the Universe.

Thus, assuming we knew the mass of the person or object, to calculate the force that a person exerts (weighs) on Earth, we solve the classic calculation:

$$F = m * a, \text{ or } F = m * 32.1 \ feet/(second)^2$$

Note 1: Neglecting, for the moment, the distance that that Gravity can act over, (which is *astronomical*) it is a very weak force compared to the other three forces. This can be seen in chart 13, which compares the relative strengths of the four forces—and the distances over which they can exert their forces.

Note 2: Additional information sources about Gravity: The easiest and probably the most accessible source of information about Gravity can be found on the web. I have also used the *Encyclopedia Britannica* and sources 19. These two books contain interesting, but brief, discussions about mass and force—plus a wealth of other topics in physics. It was the Nobel Prize holder, Leon Lederman, who coauthored the article in the *Encyclopedia Britannica,* as well as and the book, *The God Particle*, sources 19. End of notes.

A digression about Isaac Newton, who is honored in the 'Science of Gravity' for stating that Gravity is a *Universal Force*: It is quite difficult to establish the specific date of Newton's realization that such a force as Gravity was Universal. Newton's *Anni Mirabiles* was in 1665, but it is known that he did not immediately write down many of the investigations he performed and their results. It was in 1684 that he wrote his *magna opus*, the *Principia*, that combined *all* his theories and thoughts into one publication, consisting of three books.

Richard Westfall, the author of an authoritative biography of Newton, writes that it was probably in 1666 that Newton developed a flash of insight concerning gravity. Westfall also makes the interesting observation about how iconic tales repeat across centuries: We all know the folklore of the apple falling and making Newton aware of the 'presence' of Gravity. Richard Westfall writes: "Small wonder that such an anecdote, redolent of the Judeo-Christian association of the apple with knowledge, continues to be repeated."

The Electromagnetic force, the second force: This force can be 'observed' as a wave in space and in wires, waveguides and fiber optic cables. The frequency range of these waves is shown on the next page, in Figure 10, The Electromagnetic Spectrum. It shows 'where' in the spectrum, some user of that spectrum locates his product or system within the spectrum. For example, our home-radio networks use the longer wave portion (or low frequency) part of the spectrum, while X-ray systems use the shorter wave portion (or high frequency) part of the spectrum.

Following Figure 10 is Chart 11's which provides the reader with some addition measurement-features that are usually associated with the Electromagnetic Spectrum. Note: Since Chart 11 is relatively short, the text of this chapter continues within Chart 11

Figure 10 The Electromagnetic Spectrum

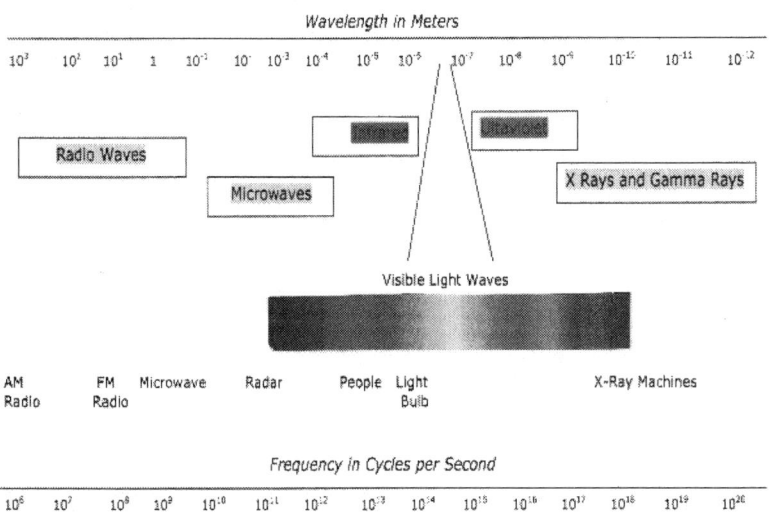

Note: Electric Motors and Electric Generator
Operate at 50 cycles (Europe) and 60 cycles (U.S.)

Chart 11 The Electromagnetic Spectrum

The following is an explanation of the terms associated with the Electromagnetic Spectrum, as can be found at numerous locations on the Web.

The original version/picture of the **Electromagnetic** Spectrum, displayed above, can be found on the web at:

http://www.lcse.umn.edu/specs/labs/glossary_items/em_spectrum.html

The following are aids to those readers who seek that chart on the web.

Legend:
mm = millimeters = 10^{-3}
μm = micrometer = 10^{-6}
nm = nanometers = 10^{-9}
cps = cycles per second

Numbers below are approximate:

Infrared	Visual light range	UV range
1mm - 700nm	700nm – 400nm	400nm – 10nm
300×10^{12} – 430×10^{12} cps	430×10^{12} – 800×10^{12} cps	800×10^{12} – 30×10^{15} cps

The electromagnetic force comes to us in many manifestations:

- The light energy we see—and do not see, such as radio waves, TV waves, radar, X-rays, infrared waves, ultraviolet rays, etc.
- The force that lights our houses (referred to as electricity).
- The force that drives our electric motors, that provides us with cell phones, computers, telephone and kitchen microwaves—and lots more.

End of Chart 11 and Resumption of Text.

The development of our total understanding of the electromagnetic force was done over many years, starting with the early Greeks. Chart 12, below, lists some of these early discoveries in this process. Those early discoveries were followed by the works/discoveries/theories of two highly eminent physicists: The English *experimentalist extraordinaire*, Michael Faraday (1791 - 1867), and the Scottish *theoretician extraordinaire*, Clerk Maxwell (1831–1879). Their work is now described.

The "total" Electromagnetic force was first investigated and described by Michael Faraday, an English physicist. Faraday is *The discoverer* of this combined force. The use of italics is to stress that Faraday is *The* discoverer of this combined force. This is done to differentiate him from, Clerk Maxwell who is also (frequently) equally associated with the Electromagnetic force, and, sometimes, more often than with Faraday.

Faraday was self-taught in physics and mathematics. But he was an expert in what he did. In spite of Faraday's lack of theoretical knowledge, he ferreted out—and discovered (or uncovered) the interrelationships of the magnetic force and the electric force—and how they combine to work together—to describe the electromagnetic force.

Faraday did this by performing experiments with fixed wires, moving wires, magnets and other rudimentary devices. (With modesty, I studied electrical engineering at the college level and can appreciate the *complexity* of the overall 'electromagnetic domain'. I developed a strong respect for Michael Faraday early-on in my studies of electromagnetism.)

I would like to cite Faraday's later work that, in a sense, can be identified with Albert Einstein's later work, when Einstein resided in the United States and was attempting to *unify* the four forces (electromagnetism and the other three forces). From the quote below, it appears that Faraday had similar 'grand' thoughts.

The quote is from what Faraday had written in his lab notes, and quoted from source 8, a scientific biography of Einstein, *Subtle is the Lord... The Science and Life of Albert Einstein.*

"After he (Faraday) experimentally unified electricity and magnetism, had not Michael Faraday tried to observe whether gravity could induce electric currents by letting pieces of metal drop from the tope of the lecture room.... When his experiments were not successful, he wrote, "They do not shake my strong feeling of the existence of a relation between gravity and electricity, though they give no proof that such a relation exists."

Chart 12 Earliest Steps in the Development of the Theory of Electromagnetism.

A Greek Shepherd *Circa* 800 B.C

Magnetic metal was mined in the Greek province of Magnesia in Thessaly. A legend attributes the first detection of magnetism to an elderly shepherd who felt an unseen force pulling the nails in his shoes. The metal that produced that magnetism was known, in English, as *Lodestone*.

Benjamin Franklin, American statesman, publisher, inventor and scientist (1706-1790)

- Discovered that lightning is electricity.
- Labeled 'charge' as being either positive or negative.
- Discovered the principle of conservation of charge. He was also the first to use the word 'battery' to describe an array of electrically charged glass plates.

Charles-Augustin de Coulomb, French physicist. (1736-1806)

- Defined the electrostatic force of attraction or repulsion between electric charges, called 'Coulomb's law'.
- The electrostatic force, F, is equal to $(K * q_1 * q_2)/(\text{distance apart})^2$ where K is Coulomb's constant $(8.987 * 10^9)$ and q_1 and q_2 are the respective charges of the two charges, expressed in Coulombs.
- The SI unit of charge is called the Coulomb.

Alessandro Volta, Italian physicist (1745-1827)

Invented the voltaic pile—the first electric battery, in 1800 and discovered the first practical way to generate electricity.

Hans Christian Oersted, Danish physicist and chemist (1777-1851)

Discovered that an electric current produced a magnetic field. Note: This is the first time a 'relationship' between two different electromagnetic phenomena was introduced in the overall subject of electromagnetism.

Finally, again, **Michael Faraday** (1791-1867). Faraday's discovery 'that a wire loop moving in a magnetic field induces a current' is the present basis for producing the electric energy that powers the Earth.

Now, to **Clerk Maxwell** (1831–1879). Maxwell was much more technically knowledgeable than Faraday—especially in mathematics. It was Maxwell who performed the 'translation' of Faraday's observations into mathematical form.

Maxwell's "translation", if we can call it that, was *momentous*. He used the most succinct and elegant form of mathematics known at that time—the Calculus, that was invented by Isaac Newton in the mid-1600s. *But the set of equations that Maxwell derived—specifically for electromagnetism, is probably the most sophisticated, precise and succinct that has ever expressed such a momentous series of simultaneous physical occurrences.*

There are only four equations that make up the set known as Maxwell's Equations. They are fundamental to our understanding electricity and magnetism. All four are represented in the most succinct mathematical form as a differential equation, based on an integral equation. (I knew them all four by heart, but that was a long time ago.) Readers who wish to gaze at these equations, should go to the web where there are a few sites showing these unbelievably short equations.

These four equations were actually based on the analysis and findings by Faraday and two other leaders in the physics of electricity and magnetism, Gauss (1777-1855) and Ampere (1735-1836). The role of each equation is listed below:

- Gauss's law that related electric charge to an electric field;
- Gauss's law for magnetism: there are no magnetic charges (i.e., equivalent to an electric charge);
- Faraday's law: A changing magnetic field can induce an electric field; and
- Ampere's Law, as modified by Clerk Maxwell: A magnetic field can be generated by electrical current and by changing electrical fields.

The modification to Ampère's law by Maxwell was the addition of a very small term, consisting of two factors. This was a momentous addition since it led to the result that the equations predicted the *creation of electromagnetic waves in space*. These are the waves that are associated with radio, television, and microwaves. And these waves enabled Maxwell —and mankind, to recognize *that light is an electromagnetic wave*!

We will now consider another, different impact of that one additional term in Ampere's Law, (the last on the list). Ampère's Law is used to analyze magnetic fields created by currents. But as Ampere meant it, it is valid only for non-time varying electric fields (We call such currents 'direct currents', or DC.) The genius of Maxwell is that he also modified the equation to include *time-varying* electric fields, such as the alternating currents (AC) we use in our homes and offices.

Maxwell's additional term is called the displacement current. The meaning of displacement comes from 'early science', the science of the 19th Century. The equations are valid for *any* time-dependent current, not merely ordinary alternating current, or AC.

Maxwell's complete formula for the fourth law also included another step of genius: Maxwell included two additional characteristics of the vacuum, the *permittivity* of free space and *permeability* of free space, both known at the time—and now, as universal constants.

Permittivity is the ability of space to store electrical energy that is generated by an electric field—in this case, an electromagnetic wave, and Permeability is how a substance (space, in this case) 'handles' magnetic flux in the region where there is a magnetic field.

The inclusion of those two universal constants by Maxwell provided him with the *ability to calculate the speed of all electromagnetic waves!* Maxwell then realized the light was also a form of an electromagnetic wave —as he then calculated the speed of electromagnetic waves *and of light*. The speed of light, as well as other electromagnetic waves (radio waves, ultraviolet waves, radar, loran…), all travel at the same speed. That speed, now known as a 'universal constant', is referred to as c. The numeric value of c is 299,798.458… kilometers per second, approximately 300,000 kilometers per second, or 186,000 miles per second.

Please pause now, to recognize the significance of Maxwell's finding of the speed of light in space. The speed of light is that ONE speed. This may well have been the starting point for Einstein's development of his Theory of Special Relativity!

Maxwell's equations involve calculus, which was probably too advanced for most physicists, even of the 19th Century. It was Isaac Newton who created calculus in order to solve his own, special, new equations. (In the previous century the great German mathematician, Gottfried Leibnitz (1646-1716), had also created Calculus, to solve *his* equations.) But it is doubtful that many physicists even of the 19th Century knew of the novel expressions, or mathematical operators, used by Maxwell—who was an *exceptional* mathematician.

Digression: Even if you have taken one or two courses in calculus, you will see on these web sites two calculus operators that you many never have come across before. They are known as the *Divergence* and the *Curl*. The divergence represents a partial derivative, as in 'simple' calculus. The *curl* is not as simple an operation. The *curl* is used to represent *a force acting at 90 degrees to another force*. In Maxwell's equations, the force can be an electrical current that generates a magnetic field that is at 90 degrees to the current, or, a magnetic field generates an electrical current that is at 90 degrees to the magnetic field. End of digression

Maxwell's equations apply to the electromagnetic force *wherever* the system is located. In outer space, in the Earth itself, and within any other substance or element. They describe the action of radio waves, television waves—and light itself.

Maxwell showed that a few relatively simple (but fundamentally complex) mathematical equations could express the behavior of electric and magnetic fields and their interrelated nature. Maxwell's four equations are one of the great achievements of 19th-Century physics.

Maxwell's equations represent how electromagnetic waves move and interrelate in space. The waves represent a force. The force acts at a distance. The waves do not need direct physical contact to exert their force. Hence, Maxwell's equations represent a second break from the, then, mechanical-only picture of classical physics, which required direct contact between two bodies. (The first break was Newton's postulation of the force of Gravity.)

Note: According to the author-physicist, Abraham Pais, again in source 8, Maxwell may have also considered that Gravity may also have its waves, by which it generates the gravitational force. Today's physicists are searching for the presence of that as-yet, undetected, massless gravity particles that are referred to as *Gravitons*. End of note.

Let us now examine what these waves can do. When you shine the light from a flashlight onto a wall, you are sending, or radiating, energy to that wall. In the case of the flashlight, it is probably an infinitesimal amount of energy, but energy, in any case.

A more important example of a radiating body is The Sun. It is a huge nuclear furnace. The furnace generates heat and radiates that heat to Earth and to other planets in the form of electromagnetic waves. The total spectrum of Sun's waves contains visible light (as seen by human beings) plus other parts of the light spectrum that are not seen, such as the ultraviolet band and the infrared band. The total amount of electromagnet radiation from the Sun is what heats our planet, Earth.

The study of the spectrum of bodies is called *Spectrum Analysis* and the description of various discoveries that were made by spectrum analysis will be considered as we move on. An important device used in spectrum analysis is called a Spectral Grating.

The most successful and important of these spectral gratings is the Rowland Spectral Grating. Henry Rowland (1848-1901) was the first Professor of Physics at Johns Hopkins University, in Baltimore, Maryland, and it was he who designed and built that grating in the late-1800s. Rowland's first system produced a map of the spectrum of the Sun that was ten times more accurate than any other previous spectral grating. And Roland's design was so good that astronomers continue to use Rowland-style gratings.

Digression: A description of the Rowland Gratings and the science of spectrum analysis are provided in Chapter 17. It is mentioned here since it is spectrum analysis that was an important method used by scientists starting in the late 1800s, and their spectrum-related activities are discussed before you reach Chapter 17. End of digression.

During the mid-19th Century physicists were studying the nature of the *transformation* from the radiation of the electromagnet wave to *Heat*

Energy. I use in italics and capital because you will now see the importance these studies had on 20th Century physics.) It took about fifty years (a typical physics-time delay) for their work to lead directly to the discovery of Quantum Theory. (As a 'premature' introduction to Quantum Theory, Quantum Theory is one of the *fundamental* physical concepts discovered in the 20th Century, and is introduced and discussed in Chapter 7.)

Back to Electromagnetism and varied items about that subject: 1. The Electromagnetic force is more powerful than gravity by a factor of 10^{38}. (Ten with 38 zeros following it.) It, too, can exert its still-modest force over astronomical distances.

2. Maxwell's formulation of the four equations enabled the start of the electronic age and the nuclear age. Physicists, such as Madame and Monsieur Curie, J. J. Thomson, Röntgen, Rutherford and Geiger developed electrical instrumentation that would have been impossible without Faraday's contributions and Maxwell's complete formulation of the electromagnetic system. (We will discuss these physicists in Chapter 6.)

3. A quote by Albert Einstein about Clerk Maxwell that appeared in the book, *James Clerk Maxwell: A Commemorative Volume*, 1931, Cambridge: University Press):

> Since Maxwell's time, Physical Reality has been thought of as represented by continuous (electromagnetic) fields (in space)...not capable of any mechanical interpretation. *This change in the conception of Reality is the most profound and the most fruitful that physics has experienced since the time of Newton.* (Italics added)

4. The experimental German physicist, Heinrich Rudolf Hertz (1857–1894), for whom the name/term *Hertz* is frequently used to represent 'cycles per second', was an important experimentalist in the early days of the study of electromagnetism. *He was the first to produce and receive electromagnetic waves (radio) in space*—the waves predicted by the earlier, Maxwell. Hertz also showed that light waves act similarly to the way electromagnetic waves act, thereby equating the two. And it was Hertz who first identified the photoelectric effect, which was later analyzed by Einstein and led to Einstein's Nobel Prize.

Hertz achieved this in only thirty-seven years of life. The Nobel laureate, Leon Lederman, considers that since Hertz confirmed all the predictions of Maxwell's theory, Hertz is a true hero in the physics of the early Maxwellian Age. (And just as a reminder, the electromagnetic force is the force that enables Radio waves, Sight/Seeing waves, Cellphone waves, TV waves, Radar waves, other specialized information-carrying waves—and waves that have not yet been conceived of by scientists.

The Strong Nuclear Force, the third fundamental force: The Strong Nuclear Force refers to the interactions between electrons, neutrons, protons and many other sub-atomic particles that make up the known-atoms in the Universe. These sub-atomic particles are ubiquitous; they exist in many different forms throughout the Universe. They go from being the constituents of: the dirt on the Earth's ground, to the constituents of our bodies, to the constituents of the stars in the sky.

Particles that are in the domain of the Strong Nuclear Force can be positively charged, negatively charged, or neutral, i.e., no charge. They can repel one another (both have similar charge); they can attract one another (both have dissimilar charge); or they can have no mutual attraction. The action is dependent on the charge that the particle may have on it, positive, negative, or none.

Two uses (out of many) of the Strong Nuclear Force are:

- Heats our Sun (a star) and all the stars (suns) in the sky (with their internal nuclear furnaces) and radiates this energy (via actual, low frequency heat waves throughout the stars' surroundings.
- Is harnessed by Man on Earth to provide:
  - Nuclear power plants on land, to generate electricity; at sea, to propel large ships (currently only naval ships); and, in space (currently, only planned)
  - The destructive power of atomic or hydrogen bombs.

The strong nuclear force is the strongest of the four forces. It is more powerful than gravity by a factor of 10^{41}. But, it is limited in distance; it can exert its force only over subatomic-size distances, on the infinitesimally small order of 10^{-17} centimeters.

95

The Weak Nuclear Force, the fourth fundamental force:

When scientists say that a radioactive element or chemical is decaying, they are referring to an action by the Weak Nuclear Force, or just the Weak Force. The Weak Force was first studied and identified by Madame and Monsieur Curie in the 1890s and early 1900s, for which they shared the Nobel Prize.

There is radioactive decay in many rock-minerals in the Earth. This decay is a manifestation of the Weak Force, and this decay contributes to making the core of the Earth very hot, in the range of 4000 to 5600 degrees.

Scientists have worked out the rate that radioactive material decays, and this is expressed as the "half-life" of the material. It is this half-life feature in certain elements that provides a *timing mechanism* within the elements. Measurement of this decay within a particular sample material enables scientists to calculate the approximate time the material came into being. Most bodies, such as building materials, chemicals, human remains, animal fossils, and items in space contain elements for which the half-life is known. This enables scientist to determine the 'birth-year' of just about any radioactive element by measuring its present Weak Force.

Geologists and cosmologists, therefore, no longer need to be "bedeviled by the theologians of the 19th Century in estimating the age of extinct animal, plants, rocks and the Universe." (The quote is from source book 31, *Einstein's Miraculous Year: Five Papers That Changed the Face of Physics*.)

The Weak Nuclear Force is more powerful than gravity by a factor of 10^{25}. But it is similar to the Strong Nuclear Force, since it can exert it force only over sub-atomic size distances, on the order of 10^{-20} centimeters.

Comparison of the Four Forces: Chart 13, located a three pages further on, shows a comparison of the strengths and ranges of the four basic forces. The strengths and ranges of these four forces are very different. To assist you in understanding the significant differences between the forces, the relative strengths of the four forces are shown in two ways. One way that is shown has gravity with a force of '1', and the other has the strong nuclear force with a force of '1'. The two presentations are essentially equivalent.

As will be discussed later, the disparity in the *character, or* features, of Gravity, is one of the causes for Gravity's 'not fitting in' with the other forces in creating what is called The Standard Model—which is an attempt to *put the four forces in one model*. (The Standard Model is described in Chapter 13.)

I consider that it is the Electromagnetic force that powers our world. I use the word 'power' in the broadest sense, such as actual power (electricity) and in a broader sense, *total* communications.

The Strong Nuclear Force is the strongest of all—*but with a very short, limited range*. (It is best known as being the basis of the atomic and thermonuclear bombs.) The Weak Nuclear Force is weak both in strength and in distance. It might be considered a *moderating* force. Further on in this book, when the Standard Model is discussed in Chapter 13, a specific example of the Weak Force is discussed.

So that you do not have to keep guess what that example is about, it is about the *Higgs particle*. That particle, named at the time-of-prediction in the 1960s, by one of its 'predictors', the British theoretical physicist, Peter Higgs, had been long sought. The Higgs particle had never been seen in any physicists system, and it was identified as being a real particle based *only on theoretical considerations*. In July 2012, however, it was finally clearly identified as a *real* (not-hypothetical) particle. This search for the Higgs was performed at CERN's Large Hadron Collider (LHC) in Switzerland—a momentous occasion, applauded worldwide.

Now, a completely different subject: The following addresses the problem "Why it is difficult to understand what (truly) happened at the initial moment of the Big Bang?"

As noted in the preceding discussion, in the Universe today—and for a long time in the past, there have been four basic, separate forces. At the initial time of the Big Bang, however, at Time = '0', the four forces are believed to have been *united into one force*. (At least three of them may have been united. It is still a question whether gravity was united with the other three forces.)

When they were united, it may have been that the strength *was all the same for each* of the three (or four) forces, and the amount of distance each force

97

could exert itself over was, also, the same for all. Cosmologists and physicist believe the unified forces split up during that initial period, referred to as T=0—and each force then 'went its own way'. As we now know, each force has a different strength and a different distance over which it can exert its force.

Modern physicists have been trying to unite these forces—as they presently exist, under one set of physical rules or laws or equations. In the 1970s, physicists 'unified' the Electromagnetic Force with the Weak Nuclear Force. This, of course, does not mean they have been united as they were at the time of the Big Bang. It means they can be analyzed together, and need only consider their present strengths and effective distance-ranges.

The Strong Nuclear Force and Gravity still stand alone, operating under their own set of 'individual' physical laws. Physicists have not been successful in understanding how they might be bound and work together. The Strong Nuclear Force, however, has been joined in certain ways with a distinguished 'outsider' to the four forces, Quantum Mechanics (Chapter 9).

(A note to the reader: There may be no reason for you to remember the *absolute* sizes of the forces. It may be advantageous, however, to remember the relative strength and effective distances of the forces.)

Finally to the subject of the four forces, a question for the Reader that has nothing to do with what we have been discussing, so you can relax about the question that will be asked:

The electromagnetic force is all around us—TV, telephone, cell phones, lights, computers and so on. But now think of the motor—which also operates on electromagnetic principles. Have you ever asked yourself how many electric motors are in your home? An answer to that may show you how the electromagnetic force has *fully* entered your life.

Take a few moments out, make a summation of all of electric motors you have at home and then refer to Chart 14, located below, for a list of possible motors that you may have overlooked. Then compare the two lists, yours and the books. The result can be surprising.

Allan Karson

Chart 13: Comparison of the Fundamental Forces

Name	Relative Strength	Range (Meters)	Relative Strength (Reverse)
Strong Nuclear Force	10^{38}	10^{-15}	1
Electromagnetic Force	10^{36}	Infinity	10^{-2}
Weak Nuclear Force	10^{25}	10^{-20}	10^{-13}
Gravitational Force	1	Infinity	10^{-38}

Strong Nuclear

Electromagnetic ━━━━━━━━━━━━━━━━━━━━━━━━━━ To Infinity

Weak Nuclear ▬

Gravitational ┄┄┄┄┄┄┄┄┄┄┄┄┄┄┄┄┄┄┄┄┄┄┄┄ To Infinity

Lines are shown to illustrate the very large differences in strengths and distances.

99

Chart 14 Lists of Electric Motors That May Be Found In A Person's Home

1. Electric clocks [not electronic]
2. Electric toothbrush [1 motor in each—even if battery-operated]
3. Electric razor
4. Electric drill and/or other electrical tools
5. Electric fan, Microwave fan
6. Refrigerator (2), a compressor to make food cold, a fan to disperse hot air.)
7. Window air conditioner (2)
8. Air conditioner to heat the entire house (2).
9. Computer hard drive
10. DVD (2) one to drive disc, one to open and close disc slot
11. CD player (2) one to drive disc, one to open and close slot
12. Cell Phones
13. Phonograph
14. Hand mixer or similar food processors
15. Tape recorder, large (2) or hand-held
16. Electric trains
17. Children's toys (dolls, cars, animals,) with electric motors
18. Battery-driven small cars,…. There is a battery driven motor in each engine or car.
19. Vacuum cleaner, electric lawn mowers
20. Clothes washer (2) one to spin washing unit, one to drain water)
21. Clothes dryer
22. Dishwasher (2), one to spin washing unit, one to drain water
23. Garage door opener, Electric chimes
24. …?
25. …?

Allan Karson

Chapter 4. The Universe—As We Understand It

This chapter discusses:

- Cosmology and the size of the Universe
- The Composition of the Universe

Cosmology and the Size of the Universe

Cosmology is the study and observation of the overall Universe—how and when the Universe was first formed, what constitutes the Universe—and how it is changing/evolving.

People who specialize in studying the overall Universe are called cosmologists. Many other types of professionals also study the Universe, such as astronomers, physicists and mathematicians. Astronomers usually specialize in the study of planets, suns and stars and may be less concerned with the overall system of the Universe.

The following describes the beginning and expansion of the Universe, starting with the beginning of the Universe—the epic occurrence that is referred to as the Big Bang. This description also examines other aspects of what is occurring in space, such as how the planets, the supernovas, the suns and the stars evolve.

The Concept of the Universe at the Beginning of the 20th Century

To get an idea of where scientists were in their thinking in the beginning of the 20th Century, you must realize that they believed:

- The planetary system they investigated through their optical telescopes appeared to be stable: It was neither expanding, nor contracting.
- The *entire* Universe consisted of the Milky Way galaxy.

We might say that the concept at that time had not advanced much since the time of Galileo, Kepler and Newton. But, as we shall now see, the American astronomer, Edwin Hubble (1889–1953) *overturned the idea of a stable and modest-sized Universe.* For this finding Hubble should be considered the most important astronomer of the 20th Century.

Up to 1923, most of the world's astronomers believed that the Universe has a set, non-changing, distribution of mass—as it had been for centuries. In that year, Hubble made his first Grand Announcement on the world's scientific stage: The Universe is *not* uniform in it distribution of matter and that 'our' galaxy, the Milky Way, is *not* the only galaxy in the sky—and that the Universe *is continually expanding.*

And in 1929 Hubble made his second momentous Grand Announcement. He announced that the Universe is expanding, and that the parts within it are expanding at a *rate proportional to the relative distance to the closed neighbor.* The following describes how he came to those conclusions—and lots more.

In the early 1920s Hubble worked for the University of Chicago, but was located at the Mount Wilson Observatory in California, with its 100-inch mirror telescope. He would take pictures of the heavens as viewed by the telescope, and he would examine them carefully. Hubble was frequently trying to measure the distance from Earth of the various stars that he was viewing.

To do this, Hubble needed 'a standard star'. To be a standard star in the Universe, Hubble would have to know (1) the star's *distance* from Earth, and (2) its *intrinsic* luminosity. If he had this data, he could then compare any other star with the "standard star's" intrinsic luminosity. He could then calculate its distance. Before describing how he proceeded, let us see what is meant by *intrinsic luminosity.*

Intrinsic luminosity means a star's brightness based on the measurement of the *actual* light it emits. You might say, the brightness as seen by a person located at the star itself—not as seen from its distance to Earth. Clearly, there is a big difference between the two. Prior to Hubble's general era, there was no knowledge of a star's *intrinsic* luminosity.

Any star, for anyone who would look from Earth, would appear significantly dimmer than what it would look like if you had been located much closer to that star. But, at that time, there was no means for Hubble to calculate a star's intrinsic luminosity. Now, the story of the discovery of intrinsic luminosity is a wonderful story in itself, as we will now see. (Each time I sit down to describe the following events—or reread what I wrote, I get a high'. So let us enjoy it, together.)

Hubble acquired this intrinsic luminosity data from work previously done by a Harvard Observatory astronomer, Henrietta S. Leavitt (1868-1921). Obviously, she was located remotely from Hubble, both in time and space. Let us take a side-trip from Hubble to discuss the work of Ms. Leavitt—and then we will get back to Hubble.

Ms. Leavitt first attended Oberlin College, a liberal arts college in Oberlin, Ohio, and then was later was a graduate of what is now known as Radcliff Institute For Advanced Study - Harvard University. She had a strong interest in astronomy and obtained a starting position as an assistant at the Harvard College Observatory. At the same, we should recognize that Ms. Leavitt's health was never very good, and she became deaf during the years that she worked at the Harvard Observatory.

Due to her high performance, she was later given the position of Chief of the Photographic Photometry Department—responsible for the care of telescopes. She was told, however, she could only do research on the observatory's photographic plates collection. She was essentially restricted from being a professional astronomer. (A case of male domination in Key positions?)

This was before the women's movement seeking the vote, professional recognition and other women's rights. In the U.S. and mostly elsewhere, women were rarely considered as professionals. Capable, professional women were essentially marginalized by most organizations, including, it appears, at Harvard,

Ms. Leavitt was first given access to the library of photographic plates, showing stars at various levels of brightness and distance. At this time there were no standards to measure a star's intrinsic brightness, or magnitude, but Ms. Leavitt developed a measuring system to do just that: She developed a standard measuring system and in 1913, her measuring system was adopted by the International Committee on Photographic Magnitudes.

Let us now review what she did. When she first went to work at the observatory, she had, what we would call today, 'a lot of drive.' And Ms. Leavitt had another interest—beyond photographic plates, *per se,* and that was to pursue work on variable stars, known as *Cepheids.* It was through the examination of Cepheids that her work became fundamentally important to word-wide astronomy. In 1912 Ms. Leavitt discovered a key, important feature concerning *Cepheid.* Ms. Leavitt did know, at that time, that a Cepheid is a young star, which is several solar masses (SM) heavier than our Sun, and is roughly 10,000 times brighter than our Sun's luminosity, or brightness.

She observed that the *luminosity of a Cepheid changes periodically and* that *the slower the luminosity change, the greater is their <u>intrinsic</u> luminosity.* (Again, *Intrinsic* means the basic luminosity of the star that is *independent* of distance from the viewer.) Ms. Leavitt thereby realized that the luminosity-changing time period of a Cepheid (an easy-to-observe number) was a *basic* indicator—or *standard, for determining the intrinsic luminosity of all the Cepheids in the sky.* This realization was her first step in discovering the *first standard for measuring distance and intrinsic luminosity in the Universe.*

(The steps described below may not necessarily be in the exact same order as the steps performed by Ms. Leavitt and other (male) astronomers, but they are correct *overall* and provide a clear understanding of how and what she accomplished.)

She knew she could use—and rely on, standard triangulation procedures for *close-in* stars to provide reasonably accurate data about the distance from Earth *to a close-in Cepheid.* Her next step was, therefore, to find and measure the distance to *a close-to-Earth Cepheid.* This step was the opening-key to her entire successful process.

After that, when she saw a close-to-Earth Cepheid, she would first observe its blinking period and from that, then determine its luminosity—using methods that she had developed earlier in her career. She then had the following data of a close-in Cepheid that would allow her to determine the distance to a far-off Cepheid (and only Cepheids) in the following way:

- The accurate distance from Earth of the close-in Cepheid, found by highly accurate triangulation *and* her estimate of its intrinsic luminosity *related* to *the time period* of its periodic change in luminosity.
- And next—and finally, the determination of *absolute luminosity* and its distance from Earth (according to Ms. Leavitt's now-standard setup) provided the information necessary for the measurement of *far-off* Cepheids.

She then used standard optical formulas to calculate the distances of farther-distant Cepheids that she later observed. She had her Luminosity Standards In Space, along with, most importantly, *any* distant-Cepheid's distance from Earth.

With this in hand, wherever she saw a Cepheid in space, she could calculate its distance from Earth and its intrinsic luminosity. *A remarkable discovery and accomplishment.* She was essentially putting 'light markers' out into space. This gave Edwin Hubble a new, important space-measuring tool. (Refer to Figure 15 below for the 'space-map' Ms. Leavitt used for her analysis.)

Albeit, she was never raised to the position of Astronomer—she was always an Assistant.

A sad and final note about Ms. Leavitt: In 1926 Professor Mittag-Leffler of the Swedish Academy of Sciences declared his intent to nominate Ms. Leavitt for the 1926 Nobel Prize in Physics for her role in the Period-Luminosity discovery. He had not been in contact with Ms. Leavitt and he had not learned that she died in 1921. He could not make that nomination for Ms. Leavitt since a rule that Alfred Nobel left as a rule for his prize is that the Nobel Prize cannot be awarded to a deceased person.

You can learn more about Henrietta S. Leavitt, by going to: http://www.womanastronomer.com/hleavit.htm and the many other web sites that describe Henrietta S. Leavitt and her significant contributions to astronomy.

Figure 15 How Henrietta Leavitt Defines a Standard Luminosity Star

(Enabling Astronomers to Calculate Distances of Other Cepheid stars—and to Other Stars in the Universe in General,)

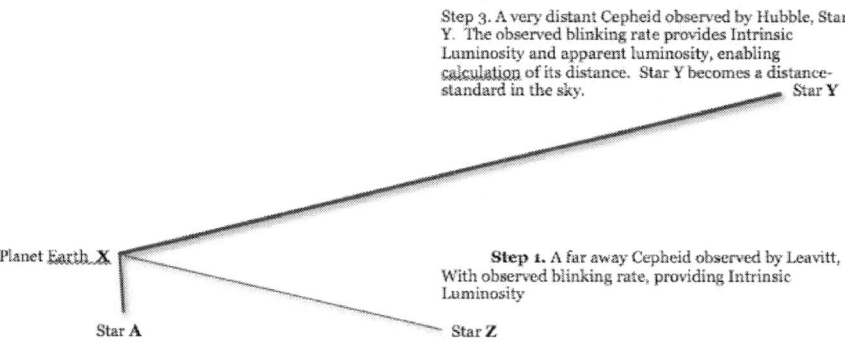

Step 3. A very distant Cepheid observed by Hubble, Star Y. The observed blinking rate provides Intrinsic Luminosity and apparent luminosity, enabling calculation of its distance. Star Y becomes a distance-standard in the sky.
Star **Y**

Planet Earth **X**

Step 1. A far away Cepheid observed by Leavitt, With observed blinking rate, providing Intrinsic Luminosity

Star **A**   Star **Z**

Step 2. A nearby Cepheid observed by Leavitt
With observed blinking rate, providing Intrinsic Luminosity
—and DISTANCE, enabling distance calibration of very distant Cepheid stars.

Note: All three stars are seen on Earth by what is called their 'Intensity at Earth'. This, of course, is different from Intrinsic Luminosity

106

Now, back to Edwin Hubble. In the early 1920s, armed with the data made available by Ms. Leavitt's findings, Hubble found a Cepheid-type star far out in the Universe and calculated that it was 2 million light-years away. (He arrived at this number, presumably, by calculating the intrinsic and observed luminosity, and using that data he could find it distance from earth —a reverse procedure of Ms. Leavitt's procedure.) He now realized that the Universe was much bigger than previously thought. (Once again, a light-year is the distance light travels in one year, or $6 * 10^{12}$ miles per year, or 6,000,000,000,000 miles.)

In 1929 Hubble went on to determine that the Universe was not a stable, tranquil system, but one in continual expansion. He did this also by observing the *actual* light from stars. He observed that the stars were red— not white, as would be in a non-expanding Universe, or blue, for a contracting one. This change in the frequency of light from a receding is known as 'the red shift', or, for an approaching object 'the blue shift'. This 'shift-phenomena' is explained below.

An explanation of the red shift and blue shift in light: Consider first an analogous phenomenon with sound waves, which is called the *Doppler effect*. As an example, when a police vehicle approaches a person with its sirens on, the sirens sound shriller to the person than it would be if the vehicle were standing still. That is, the approaching sound is at a higher frequency due to the approaching velocity of the vehicle.

When the same vehicle recedes from a person, its sound is lower on the sound-frequency scale than if would be if the vehicle were not moving. Similar effects occur when a railroad locomotive, or an airplane, approaches: higher pitch, as it approaches; lower pitch, as it recedes.

A similar effect occurs in light when objects approach us and recede from us. When the 'vehicle', or the planet, approaches, the light from the planet would have a bluish tint, which is seen back on figure 11, as being at the upper end of the visible light spectrum. That, is, a higher frequency than the standing-still case. Similarly, when planet recedes, its light has a reddish tint, the lower end of the visible light spectrum. Hence, the red shift occurs when objects move away from Earth.

Thus, Hubble would know the velocity rate that distant stars were moving from Earth. He did this by comparing the stars' red shift in their visible spectrum, with Earth's normal light spectrum. This also helped Hubble to calculate the speed that various parts of the Universe were moving away from Earth.

Given that he knew their speed-away-from-Earth—and distance, Hubble was then able to determine that they might have all been together at one time—let us say, 'when the Universe started.' At this point he postulated that the total Universe started out in one burst—in a Big Bang. He calculated that the Big Bang occurred 1.8 billion years ago. An amazing discovery—the event and the calculation of 1.8 billion years!

Hubble did not have the 'seeing' advantage of modern space satellites— with their telescopes out in space. They provide a much better picture of 'what is out there', which indicate that the Big Bang actually occurred about 13.7 billion years ago.

But Hubble did not stop there. He went on to develop a date file that tells us *how fast* one part of the Universe is moving away from another part. This file is still used today, and its data is frequently being improved upon. (Just put 'Hubble shift' in your web search mechanism for more information and insights into this fascinating, *new* phase in astronomy.)

The Size of the Universe

The following discusses the speed of light—that enables cosmologist to estimate the size of the Universe. *Light is our ruler,* or measuring stick, for measuring distances in the Universe. Light travels, as we know, at high speeds. It takes a certain time for it to go from its source, (whatever the source, a flashlight, a match or the Sun or a star) to another point. For example, it takes about 8 minutes for light emanating from the Sun to reach the Earth.

Light travels at the rate of 186,000 miles per second. The Sun, therefore, is about 93,000,000 miles from Earth. Most distances in the Universe that are being considered here are much further than our distance to the Sun, so we talk of 'light-years'. One light-year is the distance light travels in one earth-year. Hence, the distance light travels in one light-year is calculated by the

Allan Karson

following: 186,000 by 3600 seconds x 24 hours x 356 days = Six trillion miles in one year.

Cosmologists write six trillion miles in one year as 6×10^{12} miles/year. If you are unacquainted with, or forgot about numbers with exponents such as 10^{12}, it means that you write a numeral '1' and place 12 zeros after it: $10^{12} = 1,000,000,000,000$. And conversely, $10^{-12} = 1/1,000,000,000,000$.

The Universe is made up of billions of planets and stars and gas clouds—and they comprise billions of groupings that we call galaxies. And there are *clusters* of galaxies. And, as we know, the Universe is continually expanding. According to various web sites, cosmologists believe that the present diameter of the Universe is about 100 billion light-years.

Most cosmologists consider that the Universe to be located on the surface of a flat plane. Actually, it is not a truly 'flat' plane, but a plane that has rises and drops, similar to what you might see if you were to look out over land where no cities are located: Flat areas, crevices, valleys, riverbeds, rising hills and mountains. The fact that it is located on a "flat" plain had been a questionable issue for many years, but it is now an accepted concept among cosmologists that it is flat—with those many undulations.

The results from that 'all-revealing' space probe, WMAP, also indicate that the Universe is located on the surface of that 'flat' plain. Thus, that is the accepted conclusion—for now. So, happily, we do not have to deal with other geometries, such as hyperbolic geometry.

When we wish to know the speed that the different parts of the Universe are expanding from one another, we use, again, the work of Hubble. For example, we may want to calculate the speed that the Earth is moving away from a distant galaxy in the Universe. To calculate the speed for two galaxies in the Universe that are receding from one another, we use a mathematical equation that was developed by Hubble, appropriately named Hubble's Law.

Hubble's states $V = H*r$, where V represents the galaxy's receding (or recessional) velocity that the galaxy is moving away from the Earth, r is that galaxy's present distance from Earth, and 'H' is a special constant of proportionality, aptly named, *Hubble's Constant*. Hubble developed data tables that provide the 'H' for different galaxies, and you can find such

109

tables and descriptions on the web. (I advise you to first wait and read a few paragraphs below, for more information about where these tables are located on the web.)

Hubble's tables—that contain the values of 'H' according to distance from Earth, show that the farther away a distant galaxy is from Earth, the greater is the value of H. That is, the farther the distance of the galaxy is from Earth, the greater is the speed the galaxy is receding from us.

The following is an explanation of why this is so: Imagine two points on a stretched rubber sheet that are relatively close to another. Now, imagine another set of two points on the same rubber sheet, but these two points are far from one another. As the sheet is stretched, it can clearly be seen that both sets of points move apart, but at different speeds. The more-separated pair moves apart faster from one another than the closer-separated pair.

You can refer to the numerous web sites that provide the value of H and the associate speeds for numerous galaxies. You will see that most Hubble-related sites also have extensive background information concerning Hubble's constant. (An example of such a Hubble-related web site is titled, "Hubble's diagram and cosmic expansion" which has many graphs that are based on Hubble' tables, and that enable the reader to do some of Hubble's calculations.)

The discussion of 'how big the Universe is' would not be complete if we did not discuss the way cosmologists measure the enormous distances of the Universe. They use a measurement called "parsec". It is a unit of distance and is about 31 trillion kilometers or 19 trillion miles, or in exponential form, $19*10^{12}$ miles. To simplify matters, let us call it $20*10^{12}$ miles. (In comparison, another way to describe large distances is by using a light-year, which is $6*10^{12}$ miles.)

Our Universe is much bigger than that modest distance of a light year, which is (only) $6*10^{12}$ miles. When a cosmologist measures very large distances, such as between two galaxies, or the distance between Earth and a far-away galaxy, she refers to the distance in terms of *parsecs*, such as 100 parsecs, 200 parsecs or, in large distance, as far as 60,000 and more parsecs.

To calculate the distance of, say, 200 parsecs in miles, we would say $2*10^{13}$ miles x 200 parsecs, or $4*10^{15}$ miles.

Now, light itself (including the radio waves), moves at a speed of $6*10^{12}$ miles per year. So, for example, it takes light approximately .3 years to travel a distance of only, about, 1 parsec. (Actually .92 parsec.)

A brief summary of time-distance-units:

1 light-year = $3.2*10^{12}$ kilometers = $2*10^{13}$ miles

1 parsec = $31*10^{12}$ kilometers = $19*10^{12}$ miles = 3.26 light-years

The Milky Way galaxy, that contains the Earth's Solar System plus approximately 200 billion other stars (some with their *local* Solar Systems), is about 90,000 light-years across, or 27 parsecs.

Within our galaxy, the nearest star to Earth is *Proxima Centauri*, a distance of a 4.2 light years, or 1.3 parsecs. We might say, "It is right around the corner."

You can find distance calculators on the web, such as at: http://www.easysurf.cc/cnver15.htm#lyp6x

The Composition of the Universe

Recent findings of space satellites concerning 'what presently constitutes outer space' grossly differ from previous assumptions, which were based primarily on visual and radio telescopes located here, on Earth. These space probes went much further into outer space. (Out-in-space probes gain many benefits over Earth-based telescopic systems. One benefit is that their 'vision' is not obscured by Earth's congested, 'dirty' atmosphere)

In 2001 data was gathered by the Wilkinson Microwave Anisotropy Probe (WMAP) raised questions of our understanding of what matter is in outer space. For example, until then, we were taught in our chemistry classes that matter that we find on Earth—our chemicals of hydrogen, helium, iron, etc., were predominate throughout space. This appears not to be the case.

WMAP showed that the visible matter around us makes up only 4% of the total matter and energy of the Universe—and of that, only 4% is in the form of hydrogen and helium! And only .03% is made up of heavy elements such as iron.

Also, data from the WMAP indicated that about 96% of the Universe is made up of *unknown forms of matter and energy—both of which are totally invisible.* Twenty-three percent of the Universe is made up of matter, called *dark matter.* (When the gravity of *dark matter* is considered, it acts as 'normal' matter does—it attracts.)

More recent measurements in space *circa,* March 2013, indicate that 26.8% of the Universe is made up of *dark matter*—and is totally *invisible.* When the gravity of dark matter is considered, it acts as normal matter does—it attracts. And that isn't all: The 68% of the Universe is made up of an unknown energy (called *dark energy*! Dark energy acts with a *repulsive* force as opposed to gravity, which brings matter together. That is, its gravitationally repels. (An oxymoron?)

The presence of dark energy is believed to account for the *accelerating expansion* of the Universe, which appears to have started to take effect in the Universe five billion years ago.

A more recent space satellite, the Planck Satellite, was launched in 2009 with sensors that were more sensitive than those employed in the WMAP, and returned with more exact information than acquired by the WMAP. The summary data from that satellite is shown below. The earlier WMAP data (2001) is also shown, but in parentheses, following the new data. (The Planck data is from: http://io9.com/the-planck-satellite-discovers-extra-dark-matter-in-the-458250688)

Summary of the Findings of Planck Satellite (and earlier WMAP findings)

- Visible Matter around us makes up only 4.5% of the Universe.(4%)
- Invisible Dark matter makes up 26.8% of the Universe. (23%)
- Unknown Dark energy makes up 68.3% of the Universe. (73%)

I guess we should be pleased that they add up to 100%, since it appears we know all that is out there. (Actually, this data, taken from an 'unofficial' web site, had been rounded-out and adds up to 99.6%.) But that is not really the case. We know 100% *of what we know.* This means that there may be energies or matters or whatever that we may not know of. Types of, or variations of this statement indicate that science must continue looking into the unknown.

112

Man on Earth has a long way to go to understand our Universe—from the studies of Universe-size bodies (Cosmology) to the smallest (as will seen in later chapters, in Quantum Mechanics and String Theory).

Chapter 5. How the Universe Was Born

- The Big Bang—and How We Believe it Happened
- The Dilemma found in The Classic Big Bang Theory: Something Does Not Make Sense
- Five Theories (listed below) to Explain the Dilemma of the Big Bang
  - The Inflationary Theory, by Alan Guth
  - There was no Big Bang, by Stephen Hawking
  - The Speed of Light was Greater Than it is Now, by John W. Moffat and later, independently, by Joao Magueijo
  - The Cyclic Theory of the Universe, by Paul J. Steinhardt and Neil Turok

Before we discuss the event that constituted the Big Bang, we first ask, "Did the Big Bang actually occur?" Was there such an explosive event—such as a big bang? Or, for another less dramatic scenario, consider that we let the clock go backwards, and ask whether it emerged in a whimper—a Universe coming, gently, out of a shell, of some sort.

To answer that question, I introduce the Russian-born, George Gamow. George Gamow was a brilliant, farseeing cosmologist in the development of modern cosmology, and it was Gamow—along with Ralph Alpher, who conjectured the answer to those questions.

George Gamow, (1904–1968) was recognized by his peers in Russia as being a brilliant physicist. He had success and recognition in Russia, but due to increased oppression, he and his wife, also a physicist, defected from Russia in 1933. They used the 'reason to leave' that they wished to attend a Solvay conference in Belgium, which they did. They then emigrated to the U.S. in 1934, and it was there, that he added to his already-established international fame.

114

Note: Within the following, we will learn of Russia's leadership role in the study of Cosmology, during the period following the publication of Einstein's Theory of Relativity. End of note.

Gamow's first and long-time position in the U.S. was Professor of Physics at George Washington University in Washington D.C. In 1948 Gamow and his student, Ralph Alpher, provided inferential proof that the Big Bang occurred. (They could not show it actually happened, but showed, conclusively, that there was a residual, measurable, 'echo', that they called the Cosmic Microwave Background (CMB).)

It was prior to Gamow's coming to the U.S. that he had first acquired the idea for the Big Bang from his fellow-Russian mathematician, cosmologist and teacher-mentor, Alexander Friedman (1888-1925) and from the prescient writings of Georges Lemaître, a Belgium cosmologist-priest (1894-1966). (Note: Friedman's own contributions to the world of physics are discussed in this chapter.)

Now, about Alpher: Ralph Alpher (1921-2007) was a young, acknowledged-brilliant, graduate student working on his PhD at George Washington University, where Alpher's PhD advisor was Gamow. The impetus for Alpher writing his PhD on a 'formulation' or prediction of how the Universe started (a major undertaking for any PhD candidate) came from Gamow.

We will see that ideas leading up to Alpher's world-renown PhD thesis—and its aftermath, can be somewhat complicated and much of it is told at many varied web sites. I cite source book 47, *The Big Bang,* by Simon Singh, as providing a detailed description of the relationships and workings of the Gamow team. That is described mainly in the second half of the book. Singh goes through their working process at a pace that should be understandable to the person who is at least an 'intermediate-in-physics.'

Another physicist-scientist working with the team was Robert Herman (1914-1997). He worked closely with Alpher and contributed to the famous calculation of the (correct) residue temperature of the CBOE—the *echo* in far-off space of the Big Bang—and that still exists.

Gamow's work, even outside of cosmology, was *epic in proportion*. It covered a wide range of science topics, including, even genetics. For example, he discovered Quantum Tunneling (an aspect of the Weak Nuclear

Force) while at the University Göttingen in Germany and worked on radioactive decay of the nucleus, and Star formation. (This will be discussed when we consider Quantum Mechanics, in chapter 9.)

In 1948 Gamow, Alpher and Herman conjectured that the 'explosion' of the Big Bang would have produced electromagnetic waves in the microwave of the spectrum *that would still be present at the outer perimeters of the Universe.*

They reasoned that if the Big Bang had occurred, it would have initially radiated heat at *excessive* temperatures—and they considered that there *should still be some radiation or heat still present in the Universe.* The team made successive calculations of the temperature and they predicted that the present temperature of those waves would be in the 5 to 50 degree Kelvin-range.

These electromagnetic waves would have come from the hot bodies of matter in space, due to the heat of the bodies or elements. All bodies—of any kind of matter (stone, water, human flesh...) radiate electromagnetic waves of sorts. That is, when anything that is above zero degrees Kelvin (absolute zero) radiates.

Now, to a different science team, located in New Jersey. In 1965 at the then-world-renown Bell Telephone Laboratories, scientists, Arno Penzias and Robert Woodrow Wilson, were working on the laboratory's radio telescope and looking out at far-off space. They detected microwaves waves whose source they could not identify. There is a much written-about history of what then transpired, and we can read the tales of how Penzias and Wilson finally learned realized where the microwaves they observed came from—and were located at the "outside border" of the expanding Universe.

The Bell Labs team determined the temperature of the detected, mysterious waves to be 2.7 degrees K. The presence of the microwave and temperature all confirmed Gamow and his prediction—that there was a Big Bang!

The two scientists at Bell Labs, Penzias and Wilson, were awarded the Nobel Prize in 1978. No member of the Gamow-Alpher-Herman team— *The Predictors* of the still-present radiation and its temperature, were never equivalently awarded! (Physicists believe that Gamow and his team were

116

never adequately rewarded in prestige awards for their seminal contributions to our understanding about the beginning of the Universe.)

Digression. It appears we can make a list of persons who were similarly members of teams that won the Nobel Prize, but were not recognized by the Nobel Prize award committee. Such a list would include Lisa Meitner, whose former boss and colleague, Otto Hahn, won the prize in 1944, and Rosalind Franklin of the Crick-Watson DNA double helix team that won the prize in 1962. End of digression.

But, even without a Nobel Prize, there is some justice in the world. When the calculation of the radiation's temperature was announced to the scientific community, the entire scientific community, literally, roundly applauded the Gamow's team prediction—of the presence of the microwave radiation and their predicted temperature range (that was only 2.3 degrees away from the actual measurement of the temperature)—*all resulting from an event that occurred more than 14 billion years ago!*

About twenty years later, in 1989, the outer-space satellite probe, Cosmic Background Explorer (COBE) confirmed the Bell Labs data, showing that the waves are still in outer space, and with a refined temperature reading of 2.728 K.

You might ask the question, "What does the border, the outside part of the Universe look like? "Beyond all visible galaxies lies a dark shell of space, which appears to be totally devoid of stars of galaxies. In actuality, the dark shell is no different than any nearer region of space…" (From *The Cosmos,* Carl Sagan, Source 10.)

A digression about George Gamow's mentor, Alexander Friedmann. Alexander Friedmann (1888 – 1925) was a mathematician, physicist and a teacher of Gamow. In pre-Hubble days, when most physicists thought the Universe was stable, Friedmann was one of first world-physicists to make the major predictions that the Universe was expanding. He was also the first person to say that the Universe came into being through a 'Big Bang'. Friedmann's theory, announced in 1922, was based his analysis of Einstein's new Theory of Relativity.

Friedmann was also the first person to formulate the term 'Time-Space' and, in doing so, developed the 'physics-world famous' distance-equation, which encompasses all space-dimensions—and time: $(Distance)^2 = xt^2 + yt^2 + zt^2 - ct^2$. He also developed spatial-time drawings that clearly show the close relationship between time and space. Many scientists consider that Friedmann's contribution added fundamental ideas to Einstein's Theory of Relativity, and that without those additions from Friedman, the theory was incomplete.

In addition to being a brilliant academician, mathematician and physicist, Friedmann was also director of the Main Geophysical Observatory in Leningrad. He was also an active balloonist; setting records for attained heights, and was also a skilled airplane pilot. George Gamow had been a student of Friedmann's in Leningrad, until Friedmann's death in a balloon accident in 1925. End of digression.

The following describes the Big Bang, and, as the next section's title indicates, "What the Big Bang Looked Like"—and what were its after-effects.

If you are very interested in the history of the search that went on in the last half of the 20th Century—to understand how the Universe was initially formed, I recommend reading "The Big Bang" by Simon Singh, source 47. If you are primarily interested in the history of the Big Bang, the first part can be skipped by your starting on page 308, where George Gamow is first introduced. (According to Wikipedia, Simon Singh specializes in writing about mathematical and scientific topics in an *accessible* manner. I advise starting The Big Bang at page 308 because the first part of the book concentrates on technical issues that may be too difficult for the non-science reader.)

What the Big Bang Looked Like—and What Were its After Effects

The Big Bang occurred in three stages—and each stage is completely different from the other two stages:

Stage -1.   Before the Big Band occurred, or at T= -0.,

Stage 1.   The first stage is what happened during the first second,

Stage 2.   The second stage is a continuing one, starting at T=1 second, or T1 and continuing to the present, 14 billion years later.

Stage -1 is a Mystery, with a capitol M. There may be varied philosophical discussions and concrete ideas about this, but it is still a mystery. For example, there are some theories that describe the present Universe as being the *result of a succession of Universes*, with each one dying out and being reborn in a *new* Big Bang. And to make it more difficult, the Big Bang is not something that is easy to duplicate in a laboratory—at least, at the present time.

Due to the fact that stage -1 is far from being well defined by cosmologists at this time, we will not specifically consider its many possibilities. Certain aspects of this phase will be covered, however, later in this chapter, after we discuss stages 1 and 2.

Stage 1. There is what this book will call 'The standard model of the Universe' of what transpired in the first second of the Big Bang. The standard model is considered to be inadequate for being fully accurate in describing Stage 1 events—a similar dilemma as in stage -1. We will use this 'standard model', however, to introduce and start our discussion. We will later discuss the inadequacies of some of those theories about Stage 1, along with more modern and broadly accepted models of which there are many fascinating ones.

Another impediment till we get to Stage 1: Cosmologists normally have developed Stage 2 theories *before* developing theories for Stage 1. In that way, developers of Stage 1 theories knew what had to be 'output' of their Stage 1 theory. This makes sense, however, since stage 2 had to start out looking—in some way, as the *beginning* model of our present Universe. Another difficulty with Stage 1 is that while theories in physics are usually proven by observation or experimentation, Stage 1 is one event that probably will never be fully replicated.

In any case, what actually happened during that one-second will take some time to be fully understood—even by physicists. At the same time, it is natural that it should be pursued vigorously. The events that occurred at that brief interval—during that one second, had a significant impact on the details, the organization, the forces and the particles that were formed later, throughout this Universe. Given these caveats, let us still see 'what may have been' the general condition at Stage 1, according to the Standard Model of physics.

As a first step, assume a tiny bubble was formed, which the cosmologists call an *embryonic bubble*. (Remember, there is no widely accepted theory of what there was before this (probable) bubble. (As stated above, events before the bubble may be called the 'Stage -1 problem.')

The bubble was very, very small. The bubble's size is referred to as being on the order of what is known as a Planck length, which is 10^{-33} centimeters. It is much, much smaller than an atom or even an electron. Possibly, much smaller than even the Planck length. And it was very dense. Denser than *imaginable*. *VERY* Heavy!!!

Within this second, the small bubble expanded to an enormous size. (An understatement.) Its temperature became humongous—reaching unimaginably high values. Some theories say 'near infinite'!! This expansion in time, or explosion, is what is referred to as the Big Bang!!!

Stage 2. We now will look at what transpired next. It is a long stage, as we shall see. Later, we will come back to two theories that explain Stage 1. First, let us refer to Chart 16, shown below.

Chart 16 is an edited version of the chart that was on the web, provided by the European Space Agency (ESA) at: http://www.esa.int/esaSC/SEMC6TS1VED_index_0.html. A complete, detailed copy of the original web's chart is located in Appendix 6. The ESA chart assumes that the "Inflationary Theory" of Professor Alan Guth of MIT is the correct one for explaining the 'dilemma' posed to cosmologist for what occurred in phase 1. Those issues of 'Inflation' are described below.

For the most part, the chart is self-explanatory. You will see that these later sequences are divided into three phases:

- The first 300,000 years (approximate) where the complete Universe is opaque, or dark, (no light) and radiation rules the formation of everything.

- The second phase lasts for about 9 billion years and everything becomes organized throughout the Universe. Light reigns. Matter is 'King' and the key force is Gravity.

- The past 5 billion years, up to now: Dark energy takes over and we have to wait to see what happens. (Remember that dark energy makes up about 68% of the Universe.)

A note: The process of creating individual stars and planets is made by *successive shrinking and expansion* of the material making a star or planet, and was (is) accomplished over many, many, many years. A discussion of that shrinking and expansion process, known as the 'Virial Process', and the number of years it would take to make an individual star or planet, is presented in Chapter 6. End of note.

Chart 16 A Scenario of Events Following the Big Bang

Note: In order to increase clarity in this chart, the author has edited it by either adding parenthetical remarks or by underlining or italicizing important statements, or by adding short explanatory phrases.

During all of Phase 1 the Universe consist of Opaque Hot Gas Plasma, and Radiation (Electromagnetism) is the Primary Force.

Phase 1: The Radiation Phase

Starting at a millionth of a second after the Big Bang—and lasting for one second

- The Universe is roughly the size of our Solar System today. (Our Solar System is only a tiny, miniscule part of the Universe.)

- Overall, the Universe is extremely uniform in composition. But, at the same time, the Universe contains various dense portions and free-of-matter portions. (Note: Author changed this line from the original Chart.)

- It is a few-light years across.

- The Universe expands, become less dense and its temperature drops to 10^{12} degrees.

- Subatomic particles combine. They make the particles in the atomic nucleus, protons and neutrons.

- Dark matter and dark energy are present and invisible (as always). (Dark matter currently makes up about 23% of the Universe.)

Starting one second after the Big Bang, and lasting for three minutes

- Universe grows to about a thousand times the size of our Solar System today. (Our Sun and the planets that orbit about it.)

- Temperature drops to 10 thousand million degrees.

- Neutrons and protons combine to form the first nuclei: first deuterium, then helium and other light, simple elements. This is only the nucleus. No electrons are added.

Three minutes+ after the Big Bang

- Temperature is a thousand million degrees.

- It is too hot for the atomic nuclei to capture electrons and form real elements.

From three minutes to 300,000 years after the Big Bang

- Universe keeps expanding.

- Electrons wander around in space. They are too hot for electrons to be captured by any atomic nuclei that may be present.

Electrons, and any other matter, make up gaseous plasma.

- Electrons wander freely and interact with the photons (photons are light particles).

- Result: light is trapped and cannot propagate; hence, the Universe is opaque for 300,000 years.

Phase 2: the Universe Becomes Transparent and Matter (Gravity) Takes Charge.

After about 300,000 years following the Big Bang

- Universe has cooled to 3,000°C, allows protons (not photons) to capture electrons, and form the first element, hydrogen.

- Photons are now free—since protons took their electrons, and light (photons) can now form and propagate without hindrance.

Note: *The above process is a significant process*: It is the formation of hydrogen that removes the free electrons from Space and permits light propagation.

- The Universe suddenly becomes transparent. (This is the reason that our orbiting telescopes in space cannot see anything prior to about 300,000 years. It was previously opaque.

- The Cosmic Microwave Background radiation, (first predicted by George Gamow and his team of three) which we detect today, is that *'first light'*.

Soon after recombination (The Dark Age)

- The Universe is about one thirtieth of its current size.

- Matter becomes cool and luminous.

- 'Clots' or clumps of matter start to grow by gravitational attraction.

- Stars form when the lumps of matter grow to about 10 million times the mass of our Sun,

- Large groups of large objects in space, containing stars, coalesce to form galaxies and clusters of galaxies.

- First stars produce lots of ultraviolet radiation, thereby ending the so-called 'Dark age' of the Universe.

1,000 million years after the Big Bang

- The Universe is a fifth of its present size.

- Observations indicate that there are already fully formed galaxies.

- Later, when the Universe is half its present size, the nuclear reactions inside the stars produce most of the chemical elements that are needed to make Earth-like planets.

Phase 3: Five Billion Years Ago: No More Major Changes in the Universe and Dark Energy Takes over.

- 5 billion years ago, our Sun was formed from the collapse of a cloud of dust and gas, producing a very average-looking star.

- Remnants from the formation of the Sun, swirling in a disk around this infant star, gradually coalesce into planets that form part of our Solar System.

- 4500 million years ago, the Earth and the inner planets form with rocky mantles and molten interiors, while more distant planets become gaseous giants.

- 700 million years later, life begins on Earth. The oldest fossils of living organisms found on Earth, bacteria, are 3800 million years old.

A question: Immediately after the Big Bang, there was the *Radiation phase* (plasma), then the *Matter phase* (dark matter and ordinary matter), and now we live in the *Dark Energy phase*. What is next?

The Dilemma in The Classic Big Bang Theory for Stage 1: Something Does Not Make Sense!

In the first few lines of Chart 16 that describe what occurred *immediately* after the Big Bang, there is the following sentence:

"Overall, the Universe is extremely uniform in composition."

This implies that the matter that makes up the Universe has approximately the same density *throughout* the Universe, and that (also) energy is approximately the same density *throughout* the Universe. Given this information, the Universe can be considered to be homogenous at this early time. The Universe's being homogenous strongly implies that in some way, the entire Universe was set up and operated on in the same way by its laws —simultaneously. But Chart 16 shows that, at that time, the Universe is *already a few-light years across*!

One of the more important questions that cosmologists ask is, "How could all of the Universe have once been in *close* contact in order to come out to be homogenous, if the Universe is already a few light years across? Another way to look at it is to consider that the cosmologist are talking about an event that occurred in seconds, at most, but that would *require the time of at least a few light-years to propagate its laws throughout the, then, Universe!*

What might say that the 'standard model' for Stage 1 of the Big Bang does not provide an acceptable answer to this dilemma. The following presents *competing* theories that provide an explanation of how the Big Bang *might* have occurred. Each theory consists of a very different solution to the problem of defining what actually occurred during that utterly short time of Stage 1.

Each theory can be quite complex. What follows is meant to give us an idea of the many different concepts that have been brought to bear, to resolve this, still-present, enigma. Also, each of these theories is described in books written by the cosmologist/authors, and/or on various web sites. In the following discussion the name of the source is provided whenever the discussion is based on specific sources,.

You will now learn about a few of the many competitive theories among the players in the world of Cosmology. You will also see that such conflicting ideas are part of *the landscape of physics*.

Four Theories to Explain the Big Bang and What Came After

Theory #1: The Inflationary Theory: The Inflationary Theory was first proposed in 1981 by the MIT Professor, Alan Guth. It appears that his theory presently dominates the other theories and is the most-accepted one among the cosmological community. There are numerous web sites that contain a near-complete description of Professor Guth's Inflationary Theory.

The Inflationary Theory states that immediately after there was "something" in the Cosmos, consisting of smaller than particle-size matter, (such as the embryonic bubble) there was an abrupt and momentary highly *inflationary* stage in the development of the (ultra-small) Universe at that time. The new Universe expanded by a factor of 10^{50}! (Ten followed by 49 zeroes!!). And it is believed that this occurred quickly, *within something like 10^{-30} seconds!!!*

(.000,000,000,000,000,000,000,000,000,001 seconds.)

Inflation—starting at a point in time when everything was close to everything else, is a way to circumvent the problem of the Universe being too big for everything to have been in contact early in that beginning period, when everything is being formed.

This fast expansion also may have reduced any 'wrinkles' that may have been present in the Universe by 'ironing them out', much in the same way that a wrinkled balloon stretches out most rough spots as it inflates.

If that inflation did occur, other fascinating actions would have also occurred, such as that subatomic particles that made up the Universe at that time of inflation may have traveled apart from one another at abnormally high speeds (not as fast as the speed of light—but very fast). Also, there is the issue of the size of the Universe:

1. The Universe would be much larger than what we presently consider to be its size, implying:
2. There are parts of the Universe that can be so far away from our galaxy that we can never learn about them. That is, even if they were ever to

send us information that traveled to us at the speed of light, the information would never reach us.

The Inflationary Theory has received widespread recognition and is included in some textbooks. It presently appears to be the theory that is most-accepted. While there are many web sites that describe the theory and its founder, Professor Alan Guth of MIT, the advanced reader may be interested in Professor Brian Greene's book, *The Hidden Reality*, 2011, that relates the inflationary theory to a much broader theory, the *multiverse* theory.

Theory #2: The Speed of Light was (Momentarily) Greater than it is Now

First, some background about the two independent authors of this theory.

This theory, also referred to as Variable Speed of Light, or VSL, was developed in 1992 by John W. Moffat, Professor Emeritus in physics at the University of Toronto, Adjunct Professor in Physics at the University of Waterloo and a resident affiliate member of the Perimeter Institute for Theoretical Physics. In 1992 he submitted his paper, that describes the theory, to a leading physics journal in the U.S, *Physical Review D*, but, for some reason, the submission was rejected.

This may have been due to such 'heresies' that Moffat brought to the table, such as challenging a salient feature in Einstein's Special Theory of Relativity, that the speed of light was *different* from its now-given considered-to-be-constant value. Shortly later, he submitted his paper to the *International Journal on Modern Physics D* in Europe, where it appeared in 1993. No one, it appeared, noticed his paper or theory, and he returned to his work.

In 1995 he saw the title of a paper that seemed very similar to his, "Variable Speed of Light as a Solution to Initial Value Problems in Cosmology", by Andreas Albrecht and Joao Magueijo. Moffat also saw that their paper had been accepted for publication by *the Physical Review D*—and in neither case was there any reference to his earlier paper.

Well, the story goes on. Professor Moffat contacted Albrecht and Magueijo who truly did not know of Moffat's prior work—and they all became friends. As Magueijo—a cosmologist and Professor of Physics at Imperial College, London, says in his 2003 book, *Faster Than the Speed of Light*, source 42, "Imagine my shock when I discovered that another physicist had been there before us. (It as if) As we landed, a flag already flew over the Moon." (This history is described in John Moffat's book *Reinventing Gravity*, source 61, pages 102-103.)

We now examine what Professor Moffat says about VSL and what occurred in the first second after following the Big Bang. (Also from source 61: The time period in the chart above, corresponds to: "Starting one second after the Big Bang, and lasting for three minutes.")

"The speed of light was very large... about 10^{29} (or 100,000 trillion trillion) times the currently measured speed of light. This did away with the initial vale of problems of the big bang model—the horizon, homogeneity and flatness problems—just as effectively as inflation. ... A much faster speed... solved the horizon problem and therefore explained the overall smoothness of the temperature of the CMB radiation, because light now traveled extremely quickly between all pars of the expanding but *not inflating* universe."

A variable speed of light (VSL) also predicted that if the speed of light were initially huge, and then quickly slowed to become the present speed of light, it flattened out the universe.

(Professor Moffat also provides evidence that universal constants, of which the speed of light is one such constant, are not necessarily constant all of the time.)

Theory #3: There was no Big Bang: This 'non-Big Bang theory' is by the brilliant physicist, Stephen Hawking, formerly Lucasian Professor of Mathematics at Cambridge University.

The following is an extract from one paragraph on page 53 of Hawking's book, source 28, *A Brief History of Time*. The paragraph is divided into two parts here. In addition, some of the text has been italicized for the purpose of calling attention to key words or ideas.

"During the next few years I developed... The final result was a joint paper by (Roger) Penrose and myself in 1970, which proved at last that there must have been a big bang singularity... ...and nowadays nearly everyone assumes that the Universe started with a big bang singularity.

It is perhaps ironic that, having changed my mind, I am now trying to convince other physicists that there was in fact no singularity at the beginning of the Universe—as we shall see later, *it (the Big Bang) can disappear once quantum effects are taken into account.*"

Note: A singularity in nature is when something goes 'beyond standard thoughts' or 'out of sight.' This can be when calculations show that the temperature becomes infinite, or a piece of matter disappears, etc. It is usually not accepted by a scientist, to the point that one might say, it is *abhorrent* to the scientist. End of Note.

You may know of Professor Hawking through his other books. You then would know of his superior intelligence, as well as his being afflicted with the debilitating motor neuron disease, ALS (unofficially known as Lou Gehrig's disease). Amazingly, there is great deal of dry humor throughout Professor Hawking's book—something not in great evidence in most physics books. (Referring to the initial theme of *this* book, if science has not slept for 1600 years, his ALS condition might not have occurred or would be curable by now.)

A note about the name of the Lucasian Professorship (Chair) of Mathematics at Cambridge. The English philanthropist, The Reverend Henry Lucas, established the chair in 1663. According to Cambridge tradition, the 'ian' was added to the name, Lucas, to conform to a 'Latin style'. I wonder whether the name of the chair, Lucasian, would have the same romantic or erudite sounding name if Reverend Lucas had established the chair in a university in the United States, where it, probably, would have been called The Lucas Chair. End of note.

Theory #4: The Cyclic Theory of the Universe: The Cyclic Theory is described in *Endless Universe*, (2007), source 57, by Professors Paul J. Steinhardt of Princeton University and Neil Turk, who was at Cambridge University until 2008, and later became Director of the Perimeter Institute in Waterloo, Canada.

The authors of *Endless Universe* consider Professor Guth's Inflation Theory to be 'competitive' to their theory, so their book compares Guth's Inflationary Theory to their theory. Note: According to their book, *Endless Universe,* there is a strong collegial relationship between these three professors.

As the name of the theory implies, the theory postulates *that the Universe lives and dies periodically.* The Universe's cycle time is on the order of trillions of years. The idea that the Universe lives and dies has been thought of in the past by other physicists, but probably not with the depth described in this book. The book describes the details of the chain of events from a T-1 to the next T-1.

The Cyclic Theory adds a key ingredient that provides why or how the universe has that critical expansion—as differentiated from Professor Guth inflation, which does not specifically account for how or why inflation happens. Their theory relies on the unseen *Dark Energy* for the critical expansion of the Universe. The authors describe how Dark Energy is the prime reason for the Universe's earliest expansion, and how it reactivates this role about every five billion years. Note: Since NASA's satellites have only recently discovered Dark Energy, Dark Energy had not been given a role in the other, earlier theories, of the expansion of the Universe that are presented above.

Steinhardt and Turk propose that Dark Energy had a key role starting with the stage we call T=0. They provide a theory that essentially disagrees with the theories expressed above, all which include a *special* inflation period. Their theory, covering T-1, is essentially based on the idea that Dark Matter had a *major, preponderant* role in the T-1 period, obviating the need for any other methods or theories during that period.

According to the two authors, cosmologists were greatly surprised by NASA's finding of the presence of Dark Energy and Dark Matter in the Universe. We might ask why did it take so long to find this matter and energy—and the simplest response is that they both are invisible. (It might have been more straightforward if they were called invisible matter and invisible energy.)

Their presence can only be inferred: For Dark Matter, by detecting (not observing) when an attraction between bodies occurs, where one of the bodies is not visible; for Dark Energy, by observing an action that was observed first by Hubble, in which large celestial bodies were hurtling away from one another, with no observable gravity-repelling forces between bodies in space.

Digression: It is interesting to note that when Einstein was formulating his *General Theory of Relativity*, he added a factor into his equations that is called the *Cosmological Constant*. His reason for the inclusion of this constant was that 'it was needed to stop the Universe from continually expanding.' It now appears *that the* Cosmological Constant has been manifested in the physical form of Dark Matter.

Now, back to the Cyclic Theory. The question may be raised, "If the Universe is expanding under the force of gravity-repelling force of Dark Energy, how can it be a cyclic universe?" And this is where the originators of the Cyclic Theory 'bring life' to Dark Energy.

The Cyclic Theory advocates that the density of Dark Energy (that is presently a constant value throughout the Universe) eventually decays and allows the Gravity Force to regain it dominance in the Universe. And when the Universe has reduced the Universe to a "small size" (called 'the Crunch'), Dark Matter comes into play again, enabling a 'smooth crunch' to the Universe as it starts, again, it cycle of expansion.

In the Cyclic Theory, before the next Big Bang's Occurrence, or at T= -0, the previous Universe had a lifetime on the order of trillions of years. Space is flat and there is a smooth distribution of energy throughout. *Our* Dark Energy is from the decay of the *previous* Universe's Dark Energy.

The following are additional features of the Cyclic Theory:

The Big Bang: The Big Bang consists of the 'explosive' action caused by an event *attributed to String Theory*. String Theory will be described in Chapter 15. The action is two Branes (surfaces, defined in String Theory), of cosmological size, colliding together. (Note: One of the major attractions of String Theory in the world of physics is that it is presently the *only* theory that appears capable of joining the force of Gravity with the three other forces—thus completing the unification of the Four Forces.)

The Hot, Radiation-Dominated Period: This period is similar to the high temperature, radiation-dominated period that also occurs in Theory #1's inflationary period. This period lasts for approximately 75,000 years.

Dark Matter-Dominated Period: The temperature decreases, radiation decreases and particles and matter become the dominant form of energy. Matter, under the force of Gravity, becomes galaxies, planets and stars. This phase lasts about five billion years.

Dark Energy-Dominated Period: Dark energy, which is an anti-gravity force, becomes dominant and the Universe expands. This is the present situation in the Universe. (Note: The Dark Energy per centimeter throughout the Universe remains constant, so that even though the Universe is expanding, total Dark Energy increases.)

Contraction: Dark Energy decays, loses it expansionary force and changes to a mild pro-gravity force. As the Universe contracts, Dark Energy reverts to its anti-gravity force, and the normal gravity force and anti-gravity forces balance sufficiently to allow a smooth, steady contraction, or big crunch—whereupon the system starts off, once again, with its big bang.

A note: Obviously, I did not describe all the aspects of the Cyclic Theory that are presented in the book *Endless Universe*. Some points in the *Endless Universe* are straightforward; some are not. The book is one of the few source books that are listed as being appropriate (and interesting) for all three levels of readers: Beginner, Intermediate and Advanced. End of note.

Chapter 6. The Stars

This chapter discusses:

- The Lives of the Stars
- How a Star is made.
- Dying Stars and Black Holes
- Entropy and Black Holes

You may ask, "Why should we learn about stars? They are so far away and do not appear to affect us directly." There are three reasons (selected from many more):

Stars are active; planets, such as Earth, are passive. For that reason alone, stars are much more interesting than the passive planets such as Earth, Mars, Saturn and the rest of the planets.

Early in Earth's development, organisms on Earth responded to the light from the stars during the development of their optical system. This led to living creatures being sensitive to the Stars' radiation of visible light. It is estimated that this first occurred during the Cambrian Period, about 600 million years ago. Note: It is believed that all human and animal optical systems emanate from *one* source of light-system development.

The most IMPORTANT: All elements that are heavier than iron are made in the stars—and *only* in the stars, and are *thrown* to us by the stars' normal 'delivery' processes.

The Lives of Stars

In Chart 16 we saw that in the beginning of the Universe, only light elements were produced, such as hydrogen, helium and lithium. Production of all heavier heavy elements on Earth, starting with beryllium, the fourth heaviest element on earth, and going through iron and copper metals, and then to the even heavier elements such as uranium—*were all made in stars*.

These heavier elements required excessively high heat for their production —and such production factories are only located on stars

And they continue to be produced in the stars—through two types of nuclear fusion: 1. Nuclear furnaces within stars, and 2. An explosive action, known as Supernova Nucleosynthesis, which generates elements that are higher in atomic weight than those produce in the stars' fusion furnace. We will now examine the procedures of both nuclear fusion furnaces.

Nuclear furnaces became available when stars were formed, 1,000 million years after the Big Bang. These stars were large stars, and each star's gravity caused the stars to compress. The compression generated considerable heat—and each star was turned into a nuclear furnace and became very hot. The stars had not previously shined: Now they did.

Those new nuclear furnaces generated the heavy elements, those that go on the Periodic table from beryllium and on up to iron and nickel. Later in a star's life, it may have a change in its life cycle—and it would explode, throwing these newly formed elements out into the Universe, that could be attracted by a planet's gravity.

The exploding star could also become a Supernova star. The Supernova generates even higher temperatures than generated within the star's nuclear fusion furnace. Supernovas create fusion furnaces with much higher temperatures than the typical star can create, and these higher temperature fusion generate elements that are heavier than iron. This fusion process is called Supernova Nucleosynthesis. The entire fusion 'element production process', from the light to the heavier elements, is shown in Chart 17 "List of Elements in Order of Atomic Mass and How They Are Produced in Large Stars."

Later in this chapter, in the section titled, Dying Stars and Black Holes, we will discuss how only huge stars, those that are at least eight times greater in mass than our Sun, are the only stars that can provide this 'heavy element production role' to the Universe. Earth (and other planets) received these heavy elements, such as iron, from explosions of exploding stars. We might say, "They were sent by Space Express."

It is worthwhile to review the preceding sentence to understand its significance: Earth (and other planets) acquired those heavy elements, such as iron, from explosions of exploding stars. Human, animal and plant life include some of those heavy elements, such as carbon and iron, in the composition of their bodies and plants. Without stars, we would not exist as we know ourselves. As Carl Sagan, (1934-1996), the American cosmologist-astrophysicist and author said, "We are all made of stardust."

Chart 17. List of Elements in Order of Atomic Number and How They Are Produced in Large Stars.

Data in this chart was copied from

http://www.lenntech.com/periodic/mass/atomic-mass.htm.

This chart is presented in this book mainly for the purpose of completeness. That is, the author is not suggesting—in any way, that you try to learn the names of the elements or their atomic mass. It is presented here for completeness. On the other hand, the reader may find the items in the third column, Process and/or location, to be interesting.

Atomic Mass	Name of Element	Process and/or Location of Production of Some of the Elements
1.0079	Hydrogen	Only Hydrogen was produced
4.0026	Helium	during the Big Bang: Helium and
6.941	Lithium	Lithium came later.
9.0122	Beryllium	
10.811	Boron	Elements from Beryllium
12.0107	Carbon	to Nickel are produced in and
14.0067	Nitrogen	during a star's fusion process.
15.9994	Oxygen	
18.9984	Fluorine	
20.1797	Neon	
22.9897	Sodium	
24.305	Magnesium	

Atomic Mass	Name of Element	Process and/or Location of Production of Some of the Elements
26.9815	Aluminum	
28.0855	Silicon	
30.9738	Phosphorus	
32.065	Sulfur	
35.453	Chlorine	
39.0983	Potassium	
39.948	Argon	
40.078	Calcium	
44.9559	Scandium	
47.867	Titanium	
50.9415	Vanadium	
51.9961	Chromium	
54.938	Manganese	
55.845	Iron	Note: In this process Nickel decays
58.6934	Nickel	quickly to become Iron.
58.9332	Cobalt	
63.546	Copper	Elements heavier than Nickel are
65.39	Zinc	created by a process known as
69.723	Gallium	Nucleosynthesis. This process is
72.64	Germanium	known as the 'r' process.
74.9216	Arsenic	
78.96	Selenium	
79.904	Bromine	
83.8	Krypton	

Atomic Mass	Name of Element	Process and/or Location of Production of Some of the Elements
85.4678	Rubidium	Supernova Nucleosynthesis produ-
87.62	Strontium	ces much higher temperatures than
88.9059	Yttrium	are created in a star's fusion pro-
91.224	Zirconium	cess. They also produce variations
92.9064	Niobium	of the 'r process', and are
95.94	Molybdenum	known as the 'rp' and 'p' processes.
98	Technetium	
101.07	Ruthenium	Supernova Nucleosynthesis is a
102.9055	Rhodium	very high temperature fusion
106.42	Palladium	process that produces sulfur,
107.8682	Silver	potassium, calcium, scandium,
112.411	Cadmium	titanium and iron peak elements:
114.818	Indium	vanadium, chromium, manganese,
118.71	Tin	iron, cobalt, nickel. These elements
121.76	Antimony	are 'thrown out' into the
126.9045	Iodine	Universe—and are 'made available'
127.6	Tellurium	to Earth and other planets.
131.293	Xenon	
132.9055	Cesium	
137.327	Barium	
138.9055	Lanthanum	
140.116	Cerium	
140.9077	Praseodymium	

Atomic Mass	Name of Element	Process and/or Location of Production of Some of the Elements
144.24	Neodymium	
145	Promethium	
150.36	Samarium	
151.964	Europium	
157.25	Gadolinium	
158.9253	Terbium	
162.5	Dysprosium	
164.9303	Holmium	
167.259	Erbium	
168.9342	Thulium	
173.04	Ytterbium	
174.967	Lutetium	
178.49	Hafnium	
180.9479	Tantalum	
183.84	Tungsten	
186.207	Rhenium	
190.23	Osmium	
192.217	Iridium	
195.078	Platinum	
196.9665	Gold	
200.59	Mercury	
204.3833	Thallium	
207.2	Lead	
208.9804	Bismuth	

Atomic Mass	Name of Element	Process and/or Location of Production of Some of the Elements
209	Polonium	
210	Astatine	
222	Radon	
223	Francium	
226	Radium	
227	Actinium	
231.0359	Protactinium	
232.0381	Thorium	
237	Neptunium	
238.0289	Uranium	
243	Americium	
244	Plutonium	
247	Curium	
247	Berkelium	
251	Californium	
252	Einsteinium	
257	Fermium	
258	Mendelevium	
259	Nobelium	
261	Rutherfordium	
262	Lawrencium	
262	Dubnium	
264	Bohrium	
266	Seaborgium	

Atomic Mass	Name of Element	Process and/or Location of Production of Some of the Elements
268	Meitnerium	
272	Roentgenium	
277	Hassium	

The Universe is huge, billions of light years from one end to the other. It is still expanding and accelerating in its 'runaway mode'. It is estimated to expand 7% every billion years.

Stars 'manufacture' all the heavy elements and send them off to Earth—and to any other places in the Universe that will accept those Made-in-Star elements. Earth, along with the rest of the Universe, is the stars' junkyard. This topic is described in Chart 18. The chart summarizes the various ideas and insights presented in *The Accelerating Universe* by Mario Livio, pgs. 200 – 207, source 36.

I repeat below part the *battle* that is continually going on in most stars.

- The Nuclear reaction *within* the stars, acts as an anti-gravity force, *pushing outward.*

- The Force of Gravity *within* the stars, acts as an anti-nuclear force, *pushing inward.*

Thus, Stars spend their entire lifetime battling between the force of gravity and the nuclear force.

In that chart you can see all your favorite stars, ranging from our favorite one that we call the Sun, to the much discussed special class of stars, called Black Holes. We will delay our discussion on Black Holes by first examining the lives of the various sized star, and then examining how stars are made—which is, once again, a fascinating topic.

Chart 18: The Secret Lives of Stars

- Stars are born via the gravitational collapse of large, dense gas clouds.

- Stars spend most of their lifetime transforming hydrogen into helium at their centers—performed by a Nuclear Fusion Reaction.

- Stars are hot, resulting from their nuclear reactions. Our Sun, which is like many of the other stars, maintains a temperature at its center that is approximately 15 million degrees Kelvin.

- The nuclear reaction converts the star's mass to energy, according to Einstein's equation, $E=Mc^2$, (Example: 1 ounce of matter contains approximately 600 kilotons of equivalent-TNT energy.)

- The nuclear reaction acts as an anti-gravity force, pushing outward.

- Stars spend their whole lifetime battling against the force of gravity that is pushing inward.

- The outward high pressure, generated by the nuclear reaction, is normally in balance with the star's force of gravity.

- When the star's energy source is exhausted, gravity wins and the star starts contracting.

- While contracting, a new nuclear reaction may start if the temperature gets high enough—and a star is 'reborn'.

The most important factor in determining a star's brightness and evolution is its mass.

"The brightness of a star is proportional to the cube of its mass. For example:

- If Star A has five times the mass of Star B, Star A will be 125 times brighter than Star B.
- If Star A has 10 times the mass of Star B, Star A will be 1000 times brighter than Star B."

The more massive a star is, the shorter its life.

Note: In the chart below, our Sun's mass is referred to as SM. The chart uses our SM as a basis for comparison with other stars.

The following describes a star's life, according to its size:

Small stars

- Stars that are lighter than about 8% of our Sun's mass (known as solar mass, or SM) never become hot enough to ignite a nuclear reaction.

- Such stars become, what are called, 'brown dwarfs.'

Modest-sized stars

- Stars that are more massive than the 8% and less than 80% of SM have attained, what might be called, 'a long life.'

- They continually burn and transform hydrogen into helium.

Big stars

Stars that are more massive than 80% SM and up to 8*SM, end their lives with a whimper.

Huge stars

- Stars that are more massive than 8*SM live a short but very-exciting life.

- They, too, continually burn and transform hydrogen into helium.

- They get so hot that they transform the fuel in different stages—going from light hydrogen to heavier helium to carbon and oxygen to heavier iron.

- The core eventually becomes iron—plus a shell of 'ashes' resulting from previous nuclear reactions.

- Gravity starts winning, the star compresses and temperature reaches a few billion degrees.

- Then all hell breaks out and there is a Supernova. A BIG explosion—much more powerful than any seen of Earth.

- The explosion, itself, generates still heavier elements, such as iron, lead and uranium.

- The explosion disperses the varied star-made elements (such as iron, lead and uranium) into outer space—and sends it to us and all the other nearby planets.

Huge stars can have two different end-of-life scenarios, as follows:

If the Huge Star is smaller than 20*SM

- The supernovae explosion leaves behind a core that is called a *Neutron star*. Neutron stars are very, very dense.

- For example, a Neutron star has a mass that is about 40 percent greater than the Sun's mass—*but with a radius of only ten kilometers* (six miles). Neutron stars, therefore, are *so dense that one cubic inch has a mass of about a billion tons* (or 2000 billion pounds). (From source 36, *The Accelerating Universe*.)

Note: By converting the preceding density from the U.S. measurement system to the MKS system, the density is 55 million tons (metric) per cubic centimeter.

If the Huge Star is larger then 20*SM

- Gravity usually wins.

- There may or not be a supernovae explosion.

- Most become Black Holes.

How a Star is Born.

The process of how a star is born is a long and involved process. Chart 19, below, provides information about the overall process that transpires in our Universe during the making of a star. Star-making is a total-system process.

In terms of systems engineering, it involves inputs, outputs, transfers, etc. Various active elements contribute to its survival.

There is also what engineers call 'feedback' in this system. Carbon is sent (via star-express) to other, non-local regions, providing the basic body for making of other, future stars. Thus, a form of feedback, but within the total system. And there are other mystifying aspects of this operation that are not fully understood by cosmologists that could lead to their hypothesizing additional mechanisms in space.

I suggest a source-book that made me very aware of fundamental, cyclic activities that are always going on, or present in stars. *Extrasolar Planets and Astrobiology* by Professor Caleb Scharf of Columbia University, source 63. This book describes the many activities presently underway in Man's search for life in the Universe outside of planet Earth. Professor Scharf's book provides the basis for the information presented in Chart 19 that describes the amount of time involved in the (amazing, wondrous) star-making process. (The book is for the intermediate and advanced reader.)

An important section of that book is its description of the *Virial Theorem*. We will first briefly describe the Virial Process and that will be followed by Chart 19, which describe the overall and the many operations and processes that go into making a new star, that is called, a proto-star.

The Virial Theorem is the fundamental process that accounts for the life cycle of the stars—in its growth from "infancy to maturity". For our purposes, we will refer to it as the Virial Process. The Virial Process begins when star-to-be is at first, a relatively small body in space. Gravity 'comes along' and takes over and reduces the size of a star to where the protons inside the protons in the star-to-be build up great pressure—which causes the star-to-be's heat to increase, then the heat causes star-to-be increase its volumetric size, which is then followed, again, by Gravity taking over again, reducing the size and, again, there is proton pressure causing heat to rise—etc., etc., a *long*, continuous, cyclic process. This cyclic process is known as The Virial Process

During the early millenniums of the Universe, when matter was first beginning to form, matter was formed in constantly changing clumps, and these clumps eventually became the stars and planets in the Universe. These clumps grew in size and shrunk in size, grew in size and shrunk in size, and on and on, according to the detailed process just described—according to the *Virial Process.*

Note: During the years 2012 to 2014, there was a web site that provided a simulated visual and sound model of a growing star, growing, demonstrating the Virial Process—getting smaller, growing, then smaller, etc., presented on the web by a Professor of Astronomy. Sadly, it is no longer present on the web; it was a creative, unforgettable, and informative presentation. End of note.

Chart 19 So You Want To Be A Star

Follow These Steps, to Plan, to Be a 'Star in the Making', called a Proto-Star—and Finally, Become a Bright, Shining Star.

First, Plan: Think about gathering together a lot of dense gas that you will see in gas clouds in a galaxy. Watch out, however, since most clouds of gas in galaxies are not dense enough to enable gravity to bring all the gas molecules together. Also, make sure that the clouds of gas are not too warm (too active) to prevent gravity from helping you force, or bring, all those gas molecules together. As a matter of fact, the cloud of gas must be pretty cool, about 10 degrees Kelvin!

A conundrum and a potential problem: When you gather gas and the gas becomes dense, it gets hot—and at that time, the force of gravity may not be strong enough to help you. So you may ask, "Where cans we find cold, dense clouds of gas? That is, most things in the universe depend on location. (Once again, that key factor on earth *and* in the heavens, 'location, location and location'.)

Step 2: You must find a Spiral Galaxy and look in the discs of Spiral Galaxies. (There are about 10^{11} galaxies in the Universe, consisting of both spiral and non-spiral galaxies.) By the way, you will have many other stars to share your galaxy with, since a galaxy may have about 10^{11} stars within its vast spatial domain.

Step 3: Now, search for a special cloud of gas (and dust) in the Spiral Galaxy that contains carbon and other organic elements! "Why carbon and dust?" you might ask. Because carbon and dust help keep the temperature of the gas cool, by shielding the gas from other hot stars. (Carbon and the other elements (and the dust) were originally made in other stars.) "Where do they come from?", you may ask. As described in Chart 18, The Secret Lives of the Stars, HUGE stars can explode into what is called a Supernova, dispersing carbon (and other elements) into the space. Thus, your selected galaxy must have previously caught some of that "space junk".

Step 4: Now, the hard work. Once you get your Cool Gas Act together, you begin the process of forming a solid mass. That process consists of amassing enough gas, and pressuring to its minimum size, so that a Virial-type action can start. (That is, when Gravity causes the contraction and Proton/Nuclear energy causes the expansion). While there is no definite time period for this pre-proto-star stage, it can last between 100,000 and three million years. At the completion of this stage in your early star's life, you are now referred to as a *proto-star*.

Now, your question, "How long will those expansions and contractions take until the proto-star finally becomes a full-fledged brilliant star?" For a star ten times as massive as the Sun or larger, the time scale is 30 to 40 million years. Admittedly, this is a very difficult subject to be able to provide any exact numbers, but those provided here are meant to give you an idea of the *immensity of time* involved in your star-building process. Remember, you must be patient.

Dying Stars and Black Holes

In the above title, the phrase 'Black Holes' is placed after Dying Stars, since a Black Hole is the final stage for a (big) dying star. (Refer to Huge Stars in Chart 18, above)

Black Holes have received considerable attention and notoriety in the popular science literature due to the uniqueness of the characteristics of the Black Hole, itself. Much is due to the initial popularization of Black Holes in the books by Professors Stephen Hawking and Kip Thorne, and other leading physicists who are interested in this compelling topic. (Note: For those truly interested in this topic, I recommend the book by Kip Thorne, *Black Holes & Time Warps.)*

A brief description of a Black Hole: An object in space that is *very* compact. One cubic centimeter could weigh as much as 100 million tons! DENSE. In other words, a Black Hole has a large amount of mass in a *very small* volume.

A Black Hole's gravitational force is so strong that it prevents just about anything from escaping from the black hole. (The term 'just about' is an important term and is explained later.) It is so heavy that its gravitation field (which is a function of its mass) prevents even weightless photon particles (that are normally not effected by Gravity) from leaving the Black Hole.

But there is a more important reason for understanding and studying Black Holes than its unique, science fiction features. Black holes appear to be a *complete* manifestation of the force of Gravity. Matter comes to Black Holes due to the force of Gravity. The force of Gravity exercises its force over the complete local environment outside the Black Hole, and objects entering the inside of the black hole are also acted upon by the Black Hole's 'own' force of Gravity.

First, a bit of history of Black Holes: The first person to predict black holes was John Mitchell, a Cambridge University don (or teacher). Mitchell wrote a paper in 1783, that appeared in the Philosophical Transactions of the Royal Society of London, in which he pointed out that a star that was sufficiently massive and compact would have such a strong gravitational field that light could not escape. Amazingly prescient.

Black Holes and Karl Schwarzschild (1873 – 1916) Next—and to add a touch of humanity to Black Holes, think of the impressive analysis and discovery performed by the German physicist and astronomer, Karl Schwarzschild. Schwarzschild was the second person to predict the presence of Black Holes in space. He had a firm base from which he made his prediction, Einstein's "Theory of General Relativity," which had just been published in 1916.

Schwarzschild was born in Potsdam, a suburb of Berlin, Germany, where he was considered to be a child prodigy. He studied celestial orbits and wrote his PhD thesis on the work of the renowned French scientist, Poincaré. He became a professor at the Institute of Göttingen, and later became director of the highly respected Astrophysical Observatory at Potsdam.

Aside from his work on astronomy and his future work on Black Holes, Schwarzschild was a man of many scientific skills. For example, he had contributed to the study of electromagnetic radiation at the level of electrons. In 1914 Schwarzschild enlisted in the German Army with the rank of lieutenant, where he was placed in the artillery arm of army and was quickly placed on active duty. He first served in Belgium and France, and starting in 1915, he was in the combat zone on the Russian front.

Two points are made here: He joined the army even though he was 40 years old and was having a very successful career. And the rank of lieutenant in WWI was, presumably, a much more significant rank than it was in WWII —for most nations' armies.

As noted above, it was in 1916 that Einstein published his "General Theory of Relativity". This paper described how the Gravitational force works in the Universe. (New, heady and difficult material, to say the least.) Schwarzschild obtained a copy of the new publication and immediately studied and analyzed it. Within only a period of one month after obtaining the document, Schwarzschild prepared two papers on Einstein's new theory. In one of these papers he wrote and developed a prediction and wrote about his new revelation: How huge, dying stars can become, what we call today, Black Holes in space.

Schwarzschild predicted that the nuclear furnace in a huge star would eventually dies down. This would lead to the end of the star's brightness and life. And given that the star is HUGE, gravity would take over. The star would then become VERY dense and compact. While in the star's beginning, this type of situation—a huge, compact mass, initiated a nuclear furnace, now all the necessary elements to rekindle that furnace have been burnt out. Just no more fuel exists, so there was no nuclear or heat pressure to fight against gravity.

Schwarzschild then predicted that, due to the very strong gravitation field that the star possesses—resulting from this density, nothing can get away from the star. NOTHING—including light. In the dark, black sky, they would appear black to any observer. Hence, the later-given, descriptive name, Black Hole.

Schwarzschild sent his analysis to Einstein, and Einstein immediately presented the Schwarzschild papers at the next meeting of the Prussian Academy. Thus, Schwarzschild rose to everlasting fame. But...

While still in Russia, Schwarzschild contracted a rare, painful skin disease and was sent home from the front. He died shortly thereafter, in 1916. He never learned of his acknowledged accomplishment and future worldwide notoriety. Especially, since it was Einstein, his discoverer and chief supporter. In 1959 The German Astronomical Society established a special lectureship in honor of Schwarzschild and a Karl Schwarzschild medal. The first recipient was, appropriately, Martin Schwarzschild, his son.

For more on Schwarzschild, I suggest visiting the numerous web sites describing more on this brilliant person, such as at: http://www.groups.dcs.stand.ac.uk/~history/Biographies/Schwarzschild.html

A significant feature of Black Holes, which Schwarzschild identified and is named for him, is referred to as the *Schwarzschild Metric*, or, equally, the *Schwarzschild Radius*, or *Event Horizon*. The latter term will be used in this text. (There are also other equivalent terms, such as the 'Gravitational Radius'.)

The *Event Horizon* is always larger than the radius of the Black Hole's radius, and it can be considered to be an invisible sphere that surrounds the Black Hole. The size of the radius depends on the mass of the Black Hole itself. The critical importance of that 'outer sphere' is that the *escape velocity* from the surface of the sphere would necessarily have to be equal to, or be greater than the speed of light. Thus, normally, the gravitational force inside that radius is so powerful that any body entering that sphere is forced to undergo *irreversible* gravitational collapse by being captured by the Black Hole.

Black Holes come in various sizes: Big Black Holes, mini-Black Holes and as small as to micro-Black holes. The following presentation will refer only to the 'regular' big Black Hole; the reader can learn about the special features of the other, smaller Black Holes at their other, numerous web sites describing Black Holes. The Schwarzschild radius of a currently hypothesized, supermassive Black Hole at the center of our galaxy is approximately 4.8 million miles!!! Or 7.7 million kilometers!!!

An important point: Light, consisting of weightless photon particles, once it is within the Event Horizon, cannot escape the hold, or the clutches, of a Black Hole. Thus, special optical data processing systems have to be developed to observe this 'exotic' former star.

To astronomers who may be using 'standard telescopes' and who are looking for Black Holes in the night sky in space, the area appears as a dark sphere that is surrounded by illuminated stars. Figure 20 shows what happens in the surrounding volume of the Universe when a huge star becomes a Black Hole. The viewer essentially sees nothing near the Black Hole.

The formula, for the Event Horizon is: $2*G*M)/c^2$

Where G is the Universal Gravitational Constant, M is the mass of the black body and c is the speed of light. (The value of G in the MKS system is: 6.674×10^{-11}Newtons meters^2/ kg^2).

If we apply this formula to the size of typical adult, who we assume became very heavy and very dense—becoming a Black Hole, his/her Event Horizon is 10^{-23} centimeters. Note that this gets close to the smallest distance physicists presently consider, the Planck Length, which is 10^{-33} centimeters. Physicists may consider experiments with ultra-small Black Holes in order to gather information about the smallest entities.

Entering into the realm of bodies in space that we are acquainted with, the Sun's Event Horizon would be about 3 kilometers (1.86 miles) and Earth's would be about 9 millimeters (.35 feet). And then there are small, mini-micro Black Holes.

That theory also says that clocks in different inertial systems can show different times. Such is the case here. It shows the enormous difference in the observation that can occur between two different, separate groups that are both making what they believe to be a similar observation.

A rocket ship falls through the Event Horizon of a Black Hole. That activity is observed by an observer who is hovering close to, but not too near, the horizon. The action, as felt/experienced by the occupants in the rocket ship, occurs in a fraction of second—as would be shown on the ship's clocks. Gulped and lost in the Black hole.

Figure 20 Detecting the Presence of a Black Hole

1. Black Holes are detected (observed) when a particular volume of outer space does not radiate any light or detectabel radiation.

2. Observers cannot see any matter within the space of the Schwarzschild Radius.

3. Some light, however, may escape from small Black Hole. This has never been observed.

4. Stars surround the Black Hole.

5. The Event Horizon: The unseen circle of the surrounding stars. It is only perceived.

Chart 21 Some Interesting Facts About Black Holes

Population of Black Holes in Universe:

- A typical galaxy, such as the one that contains our Solar System, could well contain tens or even hundreds of millions of Black Holes.
- Many galaxies, if not most, including our own, seem to have an enormous Black Hole at their center, with a mass millions of times that of our Sun.
- There appear to be two types of Black Holes: one type is where a star dies and implodes, referred to as a stellar Black Hole; the second is where there is a galactic Black Hole at the center of huge galaxies and these galactic Black Holes can weigh millions to billions of solar masses.
- Only a small fraction of stars end up their lives as Black Holes.
- Some Cosmologists/Astronomers estimate there are at least a billion billon (10^{18}) Black Holes in the Universe.

Some Characteristics of Black Holes:

- All the Black Holes found in space are rotating rapidly.
- Light that is just inside the Horizon never reaches us; light photons just outside the horizon leaves the horizon very slowly—taking a LONG time. The Gravity, surrounding the excessively weighted Black Hole, essentially *stretches* out any light that is trying to leave that mass of Gravity.
- The area bounded the Event Horizon (as seen by an observer) is proportional to the information that 'has been trapped' in the Black Hole. Since it is 'lost information' to all outsiders of the Black Hole, it is referred to as the Entropy of the Black Hole.**

**Jacob Bekenstein formulated this relationship when he was a PhD student of Professor John Wheeler's at Princeton in the early 1970s. Stephen Hawking had initially opposed this idea, but later accepted and added to Bekenstein's ideas. The concept of area of the Horizon is proportional to the information, or entropy, lost in the Black Hole is now fundamental to the relationship between information, entropy and Black Holes.

Professor Wheeler (1911 – 2008) was the person to apply the name, "Black Holes" to these masses in space. He was at Princeton from 1938 to 1976 and returned there in the year of his death, 2008. He was known to be a wonderful developer of physics talent, and his graduate student list includes such eminent physicists such as Richard Feynman and Hugh Everett of 'Many-Worlds' fame, whom we will discuss in a later chapter.

Entropy and Black Holes

The following describes some of work of Professor of Physics Stephen Hawking, who held the position of Lucasian Chair of Mathematics from 1979 until his retirement in 2009. His book, source 28, *A Brief History of Time*, was first published in 1988 and immediately became a worldwide best seller.

Originally, Hawking and other cosmologists considered that nothing could escape from Black Hole. Nothing! In about 1970, prior to the book's publication, Hawking was 'prodded' on the subject by the Israeli, U.S.-trained physicist, Jacob Bekenstein (1947-2015), who questioned Hawking about the entropy of Black Holes.

Entropy was introduced in Chapter 2, where it indicated that entropy 'rears it head' in many parts of cosmology'. Here, once again, is proof of entropy's ubiquitous nature. Physicists and cosmologists still do not have a clear idea of what that information represents. That is a key issue that was being sought in their overall investigations of Black Holes. Based on Hawking's subsequent investigation, that considered entropy along with Black Holes, in 1974 Hawking announced they are not entirely black! That is, some radiation can escape. This would, however, would depend on the size of the Black Hole.

Hawking theorized that radiation, which would manifest itself in a form of heat, or radiation, could escape—but only for small Black Holes—which in itself should be a rarity since *all* Black Holes are considered to be massive to begin with. We should not expect, however, that this temperature will be easily detected. The mathematics shows that the temperature associated with the expected radiation is on the order of one ten-millionth of a degree above absolute zero.

For those who already have an understanding of black holes, I suggest they read Chapter 12 (pp. 324 to 341) in Source 53, *The Cosmic Landscape* by Leonard Susskind. (This book has a rating of Intermediate.) This chapter contains no introductory material and immediately throws us into the Battle of the Black Hole War between Susskind and the arch, *Defending Knight*, Professor Richard Hawking. The ultimate ending of the battle (chapter) is complicated and may be considered bloody.

An assortment of information culled from the Web:

http://en.wikipedia.org/wiki/Supermassive_black_hole

As of November 2008, another binary pair, in OJ 287, contains the most massive Black Hole known, with a mass estimated at 18 billion solar masses.[21]

On 4 January 2010, at the meeting of the American Astronomical Society, ... 33 merged galaxies with pairs of supermassive Black Holes orbiting around a common center at speeds exceeding 4,600,000 kilometers per hour (2,860,000 mph).

Currently, there appears to be a gap in the observed mass distribution of Black Holes. There are stellar-mass (star-size Black Holes), generated from collapsing stars, which range up to perhaps thirty-three solar masses. The minimal, supermassive Black Hole is in the range of a hundred thousand solar masses. Between these regimes there appears to be a dearth of Black Holes. (Source: http://en.wikipedia.org/wiki/Micro_black_hole)

It is possible that such quantum primordial Black Holes were created in the high-density environment of the early Universe (or Big Bang), or possibly through subsequent phase transitions. Astrophysicists might observe them in the near future, through the particles they are expected to emit by Hawking radiation.

(http://www.sciencemagnews.com/how-often-do-giant-black-holes-become-hyperactive.html)

Most galaxies—if not all, and including our own, are said to contain supermassive Black Holes at their centers, with masses ranging from millions to billions of times the mass of the Sun. For reasons not entirely understood, astronomers have found that these Black Holes exhibit a wide variety of activity levels: from dormant to just lethargic to practically hyper.

The most-lively, supermassive Black Holes produce what are called "active galactic nuclei," or AGN, by pulling in large quantities of gas. This gas is heated as it falls in and glows brightly in X-ray light.

"We've found that only about one percent of galaxies with masses similar to the Milky Way contain supermassive Black Holes in their most active phase," said Daryl Haggard of the University of Washington in Seattle, WA, and Northwestern University in Evanston, IL, who led the study. "Trying to figure out how many of these Black Holes are active at any time is important for understanding how Black Holes grow within galaxies and how this growth is affected by their environment."

An Afterward note: A relatively recent book by Columbia University Professor Caleb Scharf, *Gravity's Engines*, 2012, indicates "how voracious Black Holes profoundly alter the way in which the universe assembles itself. They (BHs) temper the production of stars and restrict and oversee the build of entire galaxies." That is, they are an active element in the Cosmos, not the traditional passive, invisible has-been star.

Chapter 7. The Particles in the Universe

This chapter is in two parts:

- Part 1: How Particles in the Universe Are Found/Detected in a Particle Collider
- Part 2: A List of the Known Particles and their 'Relative Locations'

Part 1

Up to now, we have been discussing 'BIG' things, such as forces, heat and temperature, the Universe and its large constituent parts (stars, planets, suns...). Now we look at the smallest of all these—particles. Particles make up the matter we observe in the Universe, and they are also basic to the four forces.

These particles are known by a large variety of names—electrons, neutrons, protons, gravitons, quarks, pions, and muons, to name only a few. All are smaller than what are called 'microscopic'. They are usually referred to as *subatomic-sized* particles.

Some particles, such as gravity particles that are called gravitons, have never been seen. This is may be due to the fact that our present-day instrumentation is not sufficiently sensitive to detect a graviton. They can, however, just not exist, but knowledgeable physicists who have performed related experiments, have predicted them, in spite of their present invisibility.

To be able to detect the particles, physicists 'shoot' particles at other, target-particles. They plan 'shooting experiments' in which a smaller particle will be 'chipped' off from the larger, target particle. When the collision occurs, the physicists take pictures and records data of the collision to observe and measure the sub-particles that result from the collision.

The planning, 'shooting' experiments and after-experiment data analysis, can be very time-consuming and manpower consuming, innately difficult, and of course, very expensive. Also, the energy levels required just to 'shoot' is enormous. In addition to shooting at a high speed, they must focus the collision in order to get the correct angle at the collision. Focusing increases the system's requirements. The machine or system that does the shooting is called a Particle Collider or Particle Accelerator. They are huge. Let us now take a brief look at one of the colliders, known as, Large Hadron Collider (LHC).

Note: This one is specifically chosen since it is relatively new and is mentioned in the popular press and on TV from time to time. Two other particle colliders are in the U.S., and they are described later in this chapter.

The Large Hadron Collider (LHC) is the largest collider ever built. It was built by the European organization, CERN, (English: Centre European for Nuclear Research). CERN is headquartered near Geneva, Switzerland.

HLC is in the shape of a circle that is 16.7 miles in circumference and is located 320 feet under the ground. Particles are shot down the tube and huge magnets bend and keep the particles focused on their circular path. The particles' speed is increased successively, more and more, by 'boosts' provided by other giant magnets. After attaining the desired speed, the particles are then focused on their target particles—and they are guided to hit the target-particles. Thousands of sensitive, high-speed photos are taken and analyzed for each shooting. This is a huge industrial-level operation.

HLC makes electrons or protons go faster and faster in the circle. For example, it takes about 90 microseconds for an individual proton to travel once around the new collider, a trip of 27 kilometers, or 16.7 miles. (16.7 miles divided by (90 x 10^{-6} seconds = 185,000 miles/second—almost the speed of light!)

These numbers are significant: The moving particles' speed is faster than 99% of the speed of light (186,000 miles/sec)—and, according to Einstein's Special Theory of Relativity, the mass of the individual particles correspondingly *increases* significantly. Thus, the amount of energy to make those particles go that fast can be enormous.

(Chapter 11 presents Einstein's equation that shows how objects increase their mass significantly when they move near the speed of light. This increase in mass requires a significant increase in the amount of energy that is needed to attain and maintain those very high speeds.)

In order to get a 'bigger bang' than has been achieved in earlier particle colliders, HLC has two sets of particles going around the tube in opposite directions. After each has attained its maximum speed, they will both be focused on one another, causing a much bigger bang than if only one set hit a stationary target. (Chapter 17 describes what colliders do along with the information that colliders provide to physicists. The chapter also contains details about the construction and complexity of the Large Hadron Collider.)

The Large Hadron Collider is the largest and most powerful collider ever built. It can investigate particles whose size can be as small as 10^{-20} centimeters. These astoundingly small sizes are associated with Quantum Theory and the Standard Model, which will to be discussed, respectively, in Chapters 9 and 13.

Physicists anticipate a problem, however, when they will seek to perform experiments that consider much smaller particles or smaller shapes than normally associated with point particles. This fact is important for future experiments concerning String Theory, a theory that involves sizes approaching the order of 10^{-33} centimeters—a size approaching what is called the 'Planck Length', presumably, the *Ultimate Small Lengths*. (Some of the theoretical features of String Theory, whose theoretical development was started toward the end of the 20th Century, are presented in Chapter 15.)

Investigations at 10^{-33} centimeters would require energies in a collider about ten orders of magnitude beyond those presently built, including the LHC. Unless physicists come up with ideas on how to circumvent that energy limitation of 21st Century colliders, they may have to wait for a new generation of colliders for experimenters to catch up to theories.

Or, possibly, or hopefully, some new, brilliant physicists will come into the field to come up with different experiment-settings that will solve this problem.

Part 2 The Basic Particles That Comprise an Atom

In 1913 the Danish physicist, Niels Bohr, postulated that within the atom, electron particles revolve around the nucleus. He also said that each electron stays in a *different* orbit or different circular path. Also, each element, or unique substance in nature, would have a different number of electrons circling the nucleus. (We will be coming back to Niels Bohr in Chapter 10, where we will learn of his leadership role in the development of Quantum Theory.)

In Bohr's model, each element has a different number of electrons. And that was the basic concept or picture of the atom—until later in the 20th Century.

Bohr's neat, comprehensive picture did not explain everything—or even much, about the atom. But it was a giant, first step. In the second half of the 20th Century, numerous experiments were performed at the then-largest, most-powerful U.S. colliders, and every year new particles were discovered in these colliders.

The two most powerful, at this time in the United States, was the Tevatron at the Fermi National Accelerator Laboratory, located near Batavia, Illinois, and the SLAC at the National Accelerator Laboratory, which is managed by Stanford University and University of California, Berkeley.

The Tevatron has a circular ring of 3.9-mile length and the SLAC has a two-mile linear (non-circular) accelerator. With CERN's HLC now on the scene, for the first time, superiority in colliders is now is located in Europe.

(During the early 1990s a very powerful collider, comparable to Europe's later-built Large Hadron Collider, was proposed to be built in Waxahachie, Texas, but the U.S. Congress canceled the massive project in October 1993. By 1993 the cost projection for the U.S. collider had risen to over $12 billion, and with limited financial resources, the U.S. government was forced to choose between funding the International Space Station (ISS) or the super-particle collider.)

As will be seen, there are many different particles and, as important as quantity may be, many of the particles are linked or closely associated with one or more of the forces. And these particles may be grouped together, so there is another set of names to learn—*the names of the groups of particles.*

The list below contains only *some* of particles, along with only *some* of their characteristics.

Electrons, protons, neutrons, photons, quarks, anti-quarks, leptons, gluons, neutrinos, muons, and so on. Some of these particle names represent a sub-group of particles. For example, within the quark-group there are what are called three 'colors' of quarks *and* there are six 'flavors' of quarks. And there are categories to group the basic particles in, such as the category of particles being either classified as a boson-type or a fermion-type.

Most, if not all of the particles, can have different weights. One of them, the photon, has no weight. (The photon, presumably, can also have infinite life.) Some have an electric charge and the charge can be positive charge or negative charge—or no charge.

And some of these particles spin to the right, some to the left, and at different rates of spin. Some have no spin. (Actually, the particles do not spin, but the results of their actions show that *they act* as if they were spinning.

Chart 22 is an introduction to the names, structure and vocabulary of the world of subatomic particles that comprise the atom. In the following, all particles and subatomic particles will be referred to only as 'particles'.

A digression: The discipline within the world of physics that investigates the particles and their interrelationships is known as 'solid-state physics'. This is a relatively new field of physics; it was initiated by Albert Einstein, as follows: After Albert Einstein published his work on the characteristics of light in 1905 and on General Relativity in 1916, he followed these two Nobel-level works into a new and different direction— identifying the specific heats of the elements, and that study led to the establishment of what is currently known as science of *Solid State Physics*. End of digression

The particle world starts at particles that can be considered to be large, such as the (multi-particle) nucleus of an atom, and it proceeds down to particles that might be called small, such as the electron. The electron has a radius of almost 3×10^{-13} centimeters.

To get an idea of the relative sizes of the particles, imagine the atom as the size of a professional baseball stadium. The size of the nucleus would be about the size of a baseball. Ants would be far too big to fit into this ball field. (From http://www.sciencebyjones.com/subatomic_particles.htm)

Learning the names of all, if not most, of the particles can be a brutal, time-consuming, exercise. Enrico Fermi, (1901 – 1954) the great Italian-American physicist, for whom the general classification of 'matter particles' is named and called the fermion, once said to a colleague, "If I could remember the names of all these particles, I'd be a botanist." And Fermi was just talking of the names that were identified in his time—and not the plethora-of-particles identified since 1954.

Note: In preparing the layout of this book, I debated where to place the following two-page chart, No. 22. One of the main purposes of Chart 22 is to make the reader aware of the many, many particles that the physicist must deal with in his and her experimental work, ranging from the one-room laboratory—to the LHC—to the Cosmos, and to their analytic and theoretical work in developing equations to represent these particles. End of note.

Chart 22 Subatomic Particles That Make Up the Nucleus of an Element's Atom

- A list of subatomic particles that make up larger particles —which, in turn, are the primary occupants of the Nucleus—i.e., the protons and neutrons. (The ubiquitous electron is not normally inside the nucleus.)
- The particles in this chart are presented from the largest to smallest.
- Certain names or terms below are in italics. These items are considered to be 'more important classifications—which the reader may read more about, than the other items.

1.   Composite Particles

1.1   Hadrons

1.1.1   Baryon

1.1.2   Meson

2   Subatomic Particles, consisting of two categories—Fermions and Bosons

Allan Karson

2.1   *Fermions*

2.1.1   *Quark* group

2.1.2   Leptons

2.2   *Boson* (The Boson is the last on this list, but Bosons 'make the world go round.')

Composite Particles are not strictly defined (an understatement). They cover a wide range of masses and sizes. They range from atoms and down to the nuclei and hadrons (to be described below in this section, 1.1). Composite particles can also consist of the smaller, defined classes, bosons and fermions.

1.1 Hadron. Varied composite particles, consisting of Baryons and Mesons.

1.1.1 Baryon is a composite particle made of three *quarks* (which are fermions). Baryons are the constituents of protons and neutrons that make up most of the mass of the visible matter in the universe,

1.1.2   Mesons are made of one quark and one antiquark. Both baryons and mesons belong to the hadron family, which are the particles made 'by assembling' quarks.

2 Subatomic Particles, of which there are two categories—*Fermions* and *Bosons*

2.1 *Fermions*

Fermions are usually associated with *matter* particles. The Standard Model contains two types of elementary fermions, Quarks and Leptons. These two are presently considered to be *basic* building particles

2.1.1 Quark group consists of six quarks. Each one of these has a corresponding antiparticle.

U up   Charm c   t top   d down s strange   b bottom

163

Quarks make up protons and neutrons, and composite fermions and composite bosons. (When quarks are grouped together to make up the protons and neutrons, the groups of quarks are called Hadrons. Leptons and Quarks are the basic building blocks of matter. (Remember what Enrico Fermi said.))

2.1.2 Leptons are a sub-group of Fermions. Leptons are subatomic particles that respond only to the electromagnetic, weak and gravitational forces—and not to the strong force. They are considered:

- To be elementary particles, i.e., not to be made up of smaller, subatomic particles
- Can carry one unit of electric charge or can be neutral (no charge).
- There are six leptons and each one of these has a corresponding antiparticle.
- Charged leptons are the electrons, muons, and taus. These all have a negative charge.
- Uncharged leptons are the three neutrinos (not listed above) and have no mass.
- Charged leptons can combine with other particles to form composite particles.
- Each lepton has angular momentum, or spin.

2.2 Boson

Bosons are often force carrier particles. That is, they are particles that transmit interactions between other particles, such as the fermion class of particles (matter particles). Bosons are the constituents of radiation, similar to the photon being the force-carrier in the electromagnetic force. They are not/never the *end-carrier* of the field; its constituents perform that function.

The Boson, the Force carrier particle, operates in all four Force Fields:

- Electromagnet field
- Weak nuclear field
- Strong nuclear field
- Gravity field

Examples of Bosons:

- Photon, active in electromagnetism
- Gluon, active in strong nuclear field
- Z_0 W+ W- active in weak nuclear field
- Graviton (postulated for activity in gravity, but has never been detected)
- Higgs (Higgs particles are postulated as providing/attributing mass to all other particles. In July 2012 a particle was 'detected' for the first time, whose behavior so far has been "consistent with" the expected characteristics of a Higgs boson.

The study of these particles has enabled physicist to place these particles and their interrelationships within what is known as *The Standard Model of subatomic particles*. The overall Standard Model is described in Chapter 13.

A final note about the world of subatomic particles: Actually, the note is not about real particles, but about Nobel Prize winner Leon Lederman and a book he wrote about particles—with some of it dealing with reality and of it some dealing with fictional particles.

But first, in 1962 Dr. Lederman and two colleagues, Melvin Schwartz and Jack Steinberger, earned the Nobel Prize for identifying the second particle belong to the class called (ghost-like) neutrino, the muon neutrino. Lederman went on to be Director of the Fermi National Accelerator Laboratory in Batavia, Illinois.

Dr. Lederman's book, *The God Particle*, source 19, is wonderful reading for anyone who wants to know more about subatomic particles. It tells the history of such particles, starting with Dr. Lederman's 'conversation' with Democritus, of Thrace, Greece (*circa* 460 BC – 370 BC) who was the first to identify the 'concept' of an atomic theory.

Later in the book, Dr. Lederman describes, in detail, the history and technical details of many of the world's major collider/detection systems. His book includes descriptions of the various and numerous experiments that have been made to detect that *plethora* of subatomic particles. He does all of this with humor. For example, after describing many of the particles, he compares that world of particles to an extended family of parents, children, grandparents, in-laws, cousins, distant cousins, etc., all living together in one house—that has only one bathroom.

We now will describe how many famous physicists first discovered these particles and physical laws. After that, we discuss the physicists who laid the groundwork for those who, shortly later, developed the Crown Jewel of the 20th Century—Quantum Theory.

Chapter 8. Leaders at the Birth of the Atomic Age (1885 to 1915)

According to the general descriptions of the major discoveries in physics, starting in the years of 1885 to 1905, scientific advancement was smooth, clear and predicable. As Voltaire might have expressed in a modern version of his satire, *Candide*, "It just had to happen that way." Well, it did not happen that way. It was far from easy. The following describe only some of the obstacles to any smooth progression of events and discoveries.

The first problem: For all new theories, there was usually a person or persons of high stature who did not agree with the theory. It was also difficult to contact other people who might help you verify your position on the theory. So circumventing the negative person(s) would be difficult.

The means for travel was not well developed in those early years, especially between Europe and the United States. Hence there was a dearth of international conferences that would enable comparisons of efforts and ideas. The first such conference was the Solvay Conference held in Brussels, 1911, sponsored by the Belgium industrialist, Ernest Solvay.

The world of manufacturing, at that time, was not yet devoted to the development, or manufacture of test equipment for testing physical theories. This was also a major impediment.

We will see in Chapter 17 that the principle method physicists used to examine and determine the various elements was by analyzing the spectrum of the element's radiation. Such spectrum analyzers had to be developed by the individual laboratory itself—by its physicists, chemists and their small staffs.

The presence of good recording photographic film—to record at least visible parts of the spectrum, would have helped a great deal. But film (at least by Eastman Kodak in its Rochester Labs) started to be considered a serious scientific tool only in the early 1920s. Examples of equipment that had to come along were the Geiger counter, cloud chambers, and the 'refractive grating' that possessed high optical resolving power that enabled detection of the frequency spectrum of the element under observation.

In addition, there were no easy-to-access centralized libraries, as there are now, where a physicist could review the current literature on a subject—and in so doing get an idea of where he/she stood in the investigations being done on the topic.

And, of course, there were no computers to store data and information, and to make calculations. And need we say there were no ways to enable scientists, who were not co-located, to collaborate in group-telephone sessions.

The quantity of physicists and mathematician was significantly fewer than it is today, so many experiments could not even be attempted. According to the physicist-writer, Abraham Pais, in *Inward Bound*, (page 5) "... in 1900, the worldwide number of academic physicists of all ranks was about 1000. Among these, the number of senior faculty in theoretical physics was eight in Germany, two in the United States (Gibbs at Yale and Pupin at Columbia), one in Holland (Lorentz).

In 1899 the American Physical Society was founded by 38 physicists. Compare that number to the membership of that society's membership in 2016—which is over 48,000 members.

I leave this topic-of-difficulties, however, with two now-positive observations: The physicists of the time appeared to have a strong interpersonal, collegial feeling and excellent working relationships—albeit, mainly locally. And, except in rare instances, they appear to have commented freely and in positive manner with their colleagues, independent of national origin.

But before considering the pre-quantum physicists and their discoveries— made mainly in the beginning of the 20th Century, we should acknowledge the importance—and fundamental role of:

- J.J. Thomson and Ernst Rutherford, British Scientists, who provided the scientific guidance to the scientists of this early 20th Century period

- Laboratory System Developers who provided laboratory equipment to the scientists of late 19th Century - early 20th Century that were fundamentally necessary for these investigations.

And in addition, the earlier contributors. Who cast a long, positive shadow over the entire 20th Century:

- Isaac Newton (1642-1727): First to recognize that light consists of a spectrum of colors.
- Fox Talbot (1800-1877): Developed the photographic process.
- Robert Bunsen (1811-1899): Developed a non-luminous burner, freeing analysis of matter of the disturbing spectrum of a luminous flame—enabling analysis of (only) the spectrum of the matter being observed.
- Anders Jonas Angstrom (1814-1874): In 1853 performed spectral analysis on various gases and was the first to observe spectral lines—four: red, blue-green and two in violet.
- Johann Geissler (1825-1898): Developed the Geissler tube, made of glass and used as a low-pressure gas-discharge tube.
- Jacob Balmer (1825-1898): Developed an empirical formula for the visible spectral lines of the hydrogen atom—which provide guidance to the determination of other elements' spectrums.
- Clerk Maxwell (1831-1879): Predicted that the light from remote locates, such as the stars and sun, could be spectrum-analyzed—providing their constituent chemical makeup.
- Gustav Kirchhoff (1824-1887)
  - First to identify the important role of Spectrum Analysis in the analysis of the numerous, varied constituents of (1) matter on earth, and (2) light coming from the Sun and the Stars.
  - Since Spectrum Analysis was crucial, later, in Max Planck's identification of Quantum Theory—and since Planck is known as the father of Quantum Theory, Kirchhoff may be called the Grandfather of Quantum Science.
- Henry Augustus Rowland (1848-1901): Developed high-quality diffraction gratings used in spectral analysis.

169

The Pre-Quantum Physicists and Their Discoveries: 1895 to 1917

I start this section of the chapter with the remarks made by the person who, indirectly, provided me with much of the following information, Professor Abraham Pais. In his book, *Inward Bound*, source 11, he describes the findings made in the late 19th Century by various experimenters, and he then describes the accelerated rate of discovery that occurred in the early 20th Century. He introduces that early 20th Century era with the following:

> *The great divide.* Progress in science depends vitally on a backlog of experimental data in need of interpretation. The whole purposes of this chapter (9) up to this point can be encapsulated in one phrase: to demonstrate that in all of the twentieth century (to date) *the experimental backlog in physics was never greater than during its opening years.*

Wilhelm Conrad Röntgen (1845-1932), Professor of Physics and Director of Physics Institute of the Pleicher Ring, Wurzberg, Germany: Discovered X-Rays, 1895. Röntgen has the honor of being the recipient of the first Nobel Prize awarded in physics. A plaque was placed in that Physics Institute in 1905 to commemorate the anniversary of the finding of X-Rays. Leaders in physics, such as Ludwig Boltzmann, Hendrick Lorentz and Max Planck attended the placing of the plaque.

Antoine Henri Becquerel (1852-1908), Professor at the Museum of Natural History, Paris: Discovered radioactivity, 1896. This was a fortuitous discovery: He found photographic plates, which had been stored near uranium. The plates had been blackened in the same way that X-ray plates are blackened. It is considered that this was the first time 'nuclear energy' was ever observed. Becquerel was elected to the Academy of Sciences, France, and in 1908 he became the Academy's president—a position that both his grandfather and father had also previously held.

Marie Sklodowska Curie (1867-1934) and husband, **Pierre Curie (1859-1906),** Professors at the Sorbonne, Paris, France: Discovered radium in 1898, a new element, and started the study of radioactivity. In April 1898 she wrote, "Radioactivity is an intrinsic property of the atom." This was the first time that radioactivity was shown to be an intrinsic property of the atom. She also introduced another property of radioactivity: "Radioactive properties are a diagnostic (tool) for the discovery of new substances."

Pierre and Marie Curie shared the Nobel Prize with Antoine Henri Becquerel in 1903.

Marie Curie was born in Poland and came to France in 1891, at the age of 24, where she continued to be an excellent student as she had been in Poland—and achieved many 'firsts' in her life. The list, to say the least, is impressive. She was the first woman to be:

- Awarded a PhD in France,
- A professor at the Sorbonne in the Sorbonne's 600-year history,
- Awarded the Nobel Prize, in physics in 1903, for her discovery, with her husband, of "Radioactivity In Uranium Work On Radioactivity Based On Becquerel's Discovery"
- Awarded a second Nobel Prize—first time, for a woman or a man, in chemistry in 1911, for her "discovery of the elements radium and polonium". Note that the second award was in a different scientific discipline from her first award—(which was also a first).

Note: The following is a list of the three other persons who have won two Nobel Prizes:

- Linus Pauling, U.S., for Chemistry, 1954, and for Peace, 1962
- Frederick Sanger, U.S., for Chemistry, 1958 and in 1980
- John Bardeen, U.S., for Physics, 1952 and 1972

We will now examine an investigation of the key relationship between matter and electromagnetism.

Hendrick Lorentz (1853-1928) and Peter Zeeman (1865-1943), Dutch physicist.

First, a bit of history. Faraday, in the 1860s, had attempted to determine if there was any relationship between electromagnetism and the flame that is given off by heating a 'piece of matter', (where matter is the elements on the Periodic Chart, such as hydrogen, helium… cobalt, etc.). This is a process known as *spectral analysis*, first described in chapters 3 and in greater detail in chapter 18.

Maxwell had also made similar, unsuccessful, investigations. We will 'forgive' both of them for different reasons—Faraday did not have the benefit of a Rowland Grating and Maxwell is not known for being an experimentalist. Enter Professor Hendrick Lorentz of the University of Leiden and Peter Zeeman, Lorentz's assistant. During the period around 1896, they, too, were investigating the effect of magnetism on matter. Zeeman knew of the prior 'unsuccesses' of both Faraday and Maxwell in their respective attempts, but he felt that if they thought there was a relationship between magnetism on matter, there must be something there.

In order to do this, Lorentz and Zeeman observed—over time, the radiated spectral lines of heated samples of the chemical, sodium, which they had placed in a magnetic field. They were trying to see if there were any changes to the spectrum of the radiated rays from what happens when not-so-heated sodium lines similarly go through a magnetic field.

At first, they, too, observed no changes in the sodium's spectral line's rays on spectral gratings. Similar to Faraday, Lorentz and Zeeman had no success. They must have been very disappointed, took a few days off, and came back and learned that the Leiden Laboratory had just acquired a more precise, Rowland Grating! And yes, they tried again.

And within a month and to their surprise, they observed the sought-after effect—that magnetism *does* have an effect (an indirect effect) on the sodium's electromagnetic radiation. The magnetic field changes the *energy level* of the electrons making up the sodium, and, thus, affects the spectrum of the sodium. Also, the magnetic field does not interact with light. For making these fundamental finding about the relation of matter, its spectrum and magnetism, Lorentz and Zeeman shared the Nobel Prize in 1902.

Digression. The Rowland Grating was designed and made by the **Physics Professor Henry A. Rowland (1848 –1901)** of John Hopkins University in Baltimore, Md.) We should recognize that Lorentz and Zeeman were using an early model of Rowland's grating (discussed later in Chapter 17) to make their spectral line measurements. We can also assume that if Faraday, in the early 1800s, had a Rowland Grating, as Lorentz and Zeeman had, he might have been the first to make the discovery. The Rowland Grating had a significant impact at that time—and later models are still used to observe spectral rays. End of digression

(During this period, Lorentz and Zeeman also developed numerous equations that were not necessarily related to the work described above. One of these equations, about Light, is referred to as the Lorentz Transformation. We will be discussing that equation in chapter 11, when we shall discuss Einstein's Theory of Special Relativity.)

J. J. (Joseph John) **Thomson (1856-1940)** English physicist at the Cavendish Laboratory, Cambridge University. Colleagues and students consider that it was Thomson's intentness-of-purpose that bought him, at the age of twenty-eight, to be named Professorship of Experimental Physics at Cambridge University, the same position once held by Clerk Maxwell. Thomson later became president of the Royal Society and Mastership of Trinity College in Cambridge. He is mainly remembered for his experimental discoveries of the electron and of the first stable isotopes.

In 1897 Thomson discovered 1. The electron possessed mass, and 2. The ratio of its charge to its mass. He achieved this by using a cathode ray tube that was similar to our non-flat TV tube—but not as elegant or simple in design as ours are today. Electrons are shot from its cathode, its 'electron gun', and go down the 'long' path to the face of the tube. By using a magnetic field to deflect the charge going to the face of the tube, Thomson was able to separate the particles according to their individual mass. (This was nearly a half-century before the home TV set's cathode ray tube was produced and sold to the world's population.)

Thomson was awarded the Nobel Prize in 1906 for this discovery. The numeric value of the charge (alone) of the electron was determined in 1909 by the American physicist, R. A. Millikan. Millikan's famous experiment is called the *Oil-Drop Experiment.*

If you are interested in learning more about the specific setup and mathematics that Thomson (and many others) used in experiments, I suggested you read source 11, *Inward Bound*, by Abraham Pais, pgs. 84-86. In addition, *Inward Bound* describes many of the early experiments and findings of numerous other physicists.

For example, in the discussion in *Inward Bound*, Thomson goes on to identify atoms that are called an 'isotope' of a 'basic' atom. Isotopes are atoms whose nucleus contains a *different* number of neutrons (higher or lower) from the element's standard number. All isotopes of atom usually have the same name. This change in the number of neutrons *changes the atomic weight* of an atom from its 'standard' weight to a higher (or lower) weight. Thomson identified that neon has one isotope by using similar techniques, some of which are described in *Inward Bound.*

Inward Bound also describes how an additional eighteen isotopes were detected; but there are probably hundred of undetected isotopes. For a complete list of the presently known isotopes, go to:

https://en.wikipedia.org/wiki/List_of_radioactive_isotopes_by_half-life

We will later discuss the fascinating topic of 'duality' when we discuss the contributions of Duc Louis de Broglie. In 1924 de Broglie was the first to hypothesize that all particles could also possess the duality of being a particle or a wave. You might say it is a reappearance of Einstein's finding concerning the 'duality' of light in 1905, i.e., both quanta and waves. We will hold that discussion for later, however, in order to stay with the chronology.

But before leaving the subject of duality, we note here that J. J. Thomson's son won the Nobel Prize in 1937 for proving the *wavelike* properties of electrons, known as 'duality', first postulated by Duc Louis de Broglie.

Ernest Rutherford, New Zealand-born (1831-1937) Research Student at the Cavendish Labs, Cambridge; University (1895); Professor at McGill `University, Quebec, Canada defined the constituent parts of Becquerel's rays and was awarded the Nobel Prize in chemistry, 1908.

After Rutherford received the Nobel Prize, he went on to make significant additional contributions in identifying the nucleus of the atom (via an experiment described in Pais's book *Inward Bound*) and. as will be seeing in the next chapter, we might consider that Rutherford was also a mentor of Niels Bohr, or at least a strong, open and thoughtful guide to the future Impresario of Quantum Theory.

Rutherford is one of the few Nobel Prize winners to have made greater contributions to science *after* receiving the award. In 1912, Rutherford made a second major finding—he discovered the Nucleus within the atom.

174

(Both Rutherford's and Thompson's ash-remains reside in Westminster Abbey, near the tombs of Newton, Kelvin and Darwin.)

At this period in time, when momentous observations had been made by J. J. Thomson in Cambridge University, England and Ernest Rutherford in Manchester, England and McGill University in Canada, *circa* 1900 to 1915, *for the first time*, physicists had a basic (but incomplete) idea of the structure of an atom.

They both identified that electrons circle the nucleus—but in ill-defined patterns. This was an important input for the young, recent PhD., Niels Bohr, who was doing post-doctorate studies at their laboratories, Cambridge during autumn, 1911, and Manchester, during late spring, 1912. (Bohr's doctoral thesis was on the electron theory of metal.)

In comparing what Bohr learned during those two visits, the visit with Rutherford's team was more fruitful. Also, as we shall soon see in the next chapter, Niels Bohr defined those all-important electron-actions—and showed their significance in *both* Atomic Theory and (later) in Quantum Theory. Both those topics are discussed in the next chapter.

Chapter 9. Quantum Theory, The Crown Jewel of the 20th Century Physics

The following is a quotation of Abraham Pais, a 20th century physicist, technical author and associate of Albert Einstein: "Were I asked to designate just one single discovery in the twentieth-century as revolutionary, I would unhesitatingly nominate Planck's of December 1900 when he identified the Quantum Phenomena."

A definition of Quantum Theory

(from http://whatis.techtarget.com/definition/quantum-theory and amended by author): Quantum theory is the theoretical basis of modern physics that explains the nature and behavior of matter, waves and energy on the *atomic* and *subatomic* level. The nature and behavior of matter and energy at that level is sometimes referred to as quantum physics and quantum mechanics. Note: Its rules and methodologies frequently do not conform to what is known as *rational thought* or *common sense—and to certain standard laws of classical physics*.

During the 20th Century, three new revolutionary, "mind-boggling" theories were introduced in the world of physics. They are: *Quantum Theory*, introduced by Max Planck and followed up by Niels Bohr and others, and *Special Relativity Theory (dealing with Light) and General Relativity Theory, (dealing with the space-coordinates of the Universe)*, the latter-two introduced by Albert Einstein. This chapter describes Quantum Theory.

Most topics, concepts and ideas in physics can be introduced to the-novice-in-physics in one, sequential history: First, its theories and the physicists who developed them, and then, where it is now. That type of formula-for-presentation is what was done in the preceding chapters on topics such as Thermodynamics, the Cosmos, Forces and the Big Bang. One might say, "a normal, sequential story."

Quantum Theory is different, however. Since it contains many non-intuitive concepts, it has a history that contains many controversial issues and theories. To make this exciting and powerful part of the world of physics intelligible, Quantum Theory is presented in the following sequence:

Part 1 An Introduction to Quantum Theory

Part 2 A Closer Look at the Discovers and Formulators of Quantum Theories

Part 3 Quantum Theory—A Perplexing Concept

Part 4 Quantum's Mysteries

Note: Within the world of physics, Quantum is usually referred to as Quantum Theory or Quantum Mechanics, depending on which part is being discussed. In cases where it is *not clearly one* of those two cases, *theory* or *mechanics*, this book does not follow that rule and uses the word 'quantum' alone. End of note.

Part 1 An Introduction to Quantum Theory

- The Difference between Classical Physics and Quantum Physics.
- An Introductory History of Quantum Theory and Quantum Mechanics.

You will find that Quantum poses many dilemmas. Let us immediately clear up one of them: Both names, Quantum Theory and Quantum Mechanics represent, overall, the same theory/concept. Quantum Theory is the term usually used to describe Quantum's studies/activities/findings/meetings *before* 1925, and Quantum Mechanics is the term used to describe (most of) the studies/activities/findings and meetings *after* 1925.

One of the reasons for this change in nomenclature is that up to about 1925 to1926, Quantum Theory consisted of what might be called 'a group of scattered findings and ideas'. In about 1925, 'the group coalesced into one 'system'. Probably, the first 'new system concept' was due to the seminal contribution made by the German physicist, Werner Heisenberg (1901 – 1978) in which provided Heisenberg provided 'A complete mathematical system' for defining a particle in space. Hence, part of it went from a theory to "mechanics". But the two parallel paths of the study of Quantum—theory and mechanics, continue on.

177

There are many concepts, theories, and results of experiments in Quantum that defy our natural senses, defy our commonly held beliefs—and appear to be 'just wrong'. Well, they usually are not 'just wrong.' They are, however, counterintuitive. Repeat: counterintuitive. And we should learn to accept those concepts, theories, and results of experiments.

We should remember that people of earlier times (and probably now, too) also experienced certain dilemmas that went against their reason and their every day experience. Some examples were when Newton identified an unseen force, gravity; when Faraday and Maxwell also identified an unseen force—plus waves, that they called electromagnetism; and when the highly logical mathematician, Raphael Bombelli (1526 – 1572) introduced an imaginary number, 'i', into the system of real numbers.

We live and we learn to live with illogical, mystical concepts. Such counterintuitive concepts in science are frequently the way of advancement in science.

This chapter, as are some of the others, is written for the non-technical person. However, it may include technical ideas and words that may be difficult the first time they appear. If you have difficulty in understanding what is said, relax. Concentrate on the name of the physicist, the years, and what they did. You will also see that much of what is presented is repetitive. (In many instances, it is purposely, repetitive.) Therefore, if at first it may not be clear, it may become clear on the second, or third repetition.

The Difference Between Classical Physics and Quantum Physics

First, Quantum Theory is a break from many of the theories in classical physics. You might now ask, "What is the difference?" Well, here goes:

In classical physics, most things are *continuous*. You might refer to Chart 11 again, the Electromagnetic Spectrum. It is continuous—there are no breaks in the spectrum. Similarly, when we look at the ocean, we see one big mass of water. It is continuous. It is not in little pieces of water. So the ocean, or a glass of water, is an example of most theories in classical physics—continuous. Now, what about the word 'Quantum'. The dictionary says that the word 'quantum' means 'particular amount' or 'package' or 'discrete packet'. Let us see what that means in the world around us.

178

We go on an elevator; it stops at each floor. We can say that each floor is located at a different level or at a different 'quantum'. We listen to the radio. We move the dial across the tuning band and hear radio programs at only certain locations on the radio dial. That is similar to a quantum idea. And if we are standing on a sand beach, the overall sand looks as if it is smooth, but when we look closer we see that is consists of small particles. We might call the particles 'quantum'. But 'true' Quantum Theory looks at even smaller pieces of matter.

Quantum Theory *is the study of small sub-atomic-sized particles and their interactions.* We are talking of particles whose size or distance across the particle can be on the order of 10^{-6} centimeters for a small chemical molecule (A molecule is not considered a quantum particle. It is listed here for comparison purposes only.)

10^{-6} cm. for molecules

10^{-7} cm. for atoms

10^{-12} cm. for nuclei

10^{-13} cm. for neutrons and protons

10^{-17} cm. for electrons

10^{-33} cm. for Planck Length, representing the minimum length met in physics

In learning about Quantum's history, it is important to appreciate that the physicists who first observed the physical phenomena felt they, too, were observing something that was strange and unexplainable. It was very different from their training and their daily experiments in classical physics. That early stage can be thought of as Phase 1 of Quantum Theory (1900 to about 1920-1925).

Part 2 The *Very First Step* in The History of Quantum Theory and Quantum Mechanics

We will now see how the study of one discipline of physics, Thermodynamics, led to the discovery of the Quantum Theory. A case of an 'everyday' science, involving everyday, observable heat and light radiation, led to this new discipline in the 20th Century—and a complete change in how we view the Universe.

Max Planck (1858 – 1947), a highly respected German physicist, introduced the beginning concepts of Quantum Theory to the world in 1900. Planck was a Professor at the University of Berlin and a respected physicist and person during all his life. As a teacher, his lectures at the university were held in high regard, which may be considered a rarity at that time.

Planck was then seeking to answer the question, first asked in 1857 by Professor Gustav Robert Kirchhoff, (1824-1887) of the University of Heidelberg. "What is the relationship between the *frequency* of the heat radiated and the *energy* of that heat?" Planck's interest in this topic was due to a commission he received from the German electric companies to create the maximum light from light bulbs—with minimum energy. (It is interesting to recognize that a whole, new, discipline in physics (Quantum Theory) was first started by the light bulb—which we might call 'a prosaic product.')

The German government's future Bureau of Standards in Berlin, then named *Physickalisch-Technische* Reichsanstalt, was also very much interested in measuring the equilibrium distribution of heat radiation within a cavity. (The term and object, 'blackbody' will be used in place of cavity.) Planck had available the data that had been gathered in experiments conducted by the government bureau, where the experiments had been conducted using a heated box, known as a 'black-box', which is described in detail, later below.

Planck and the Bureau, were (separately) seeking to answer *the* question posed by a earlier predecessor to his position at the university, Gustav Robert Kirchhoff (1884-1887). Basically, the question was, "What is the relationship between the *frequency* of the radiated heat and its *content of heat*?" In mid-October 1900 Planck considered he found the answer to Kirchhoff's question and was able to announce the final version of his findings in December 1900.

In the course of his working on the problem, Planck was the first to identify (or to say) that very small things (small amounts of heat, for example) operate in 'packets', and should be measured in 'steps', or in quanta. *This was completely different from what physicists thought previously.* Thus, a revolutionary idea! Let us now look over the shoulder of Planck as he proceeds with his experiment.

Planck observed (or, to be closer to the reality of what Planck did) Planck *hypothesized,* at first, that heat is transmitted in small packets (or 'quanta', as those small steps were shortly referred to) within all and every *frequency band.* (A repeat: *Heat's radiation energy is transferred to matter in very small quanta, or, in very small packets.*) And each heat quanta is at one specific frequency. Planck observed this (or, as the author prefers to say) *hypothesized this* while measuring heat emanating from a box whose insides had been preheated.

The box, that had been made warm inside, also had a small hole on one side to allow heat to come out (or, let us say, *'radiate'* out). That is, it is a typical small-to-medium size box, with warm (or hot), facing inner surfaces. A box like any box, but here, referred to as a 'black-box.'

(The word 'black', when appearing before closed or semi-closed devices. is also meant to imply that that they are excellent *radiators* and *absorbers* of heat. Historical note: It was Gustav Kirchhoff who first defined what constitutes a blackbody or black-box.)

The following is a quote from Planck's *Scientific Autobiography,* source 1. Page 34-35), with italics added.

"...Thus, this so-called Normal Spectral (frequency) Distribution represents something absolute, and since I always regarded the search for the absolute as the loftiest goal of all scientific activity, I eagerly set to work. I found a direct method for solving the problem in the

application of Maxwell's Electromagnetic Theory of Light. Namely, I *assumed* the cavity (inside of box) to be filled *with simple linear* (frequency*) oscillators or resonators,* ...and I *expected* the exchange of energy caused by the reciprocal radiation of the oscillators (measureable quantities) to result, in time, to a stationary state of the normal *frequency energy* distribution corresponding to Kirchhoff's Law."

Planck—a conservative, well-established physicist, *made the assumption* that the radiant heat (or energy) is transmitted in what are small individual 'packets'—generated by those individual (nonexistent, but hypothesized) oscillators. He called these heated packages, *quanta*. He also assumed that *each quanta had one, specific frequency*. (This is opposed to light's continuous spectrum that is shown earlier in figure 11.) Planck first informed colleagues of this assessment of how heat is made up of quanta in October, 1900, which is considered the year of the birth of Quantum Theory.

Planck continued to refine his work and observations and presented his final, *completely-correct* equation to his colleagues in December of that year, 1900. Note: In this instance the equation's results were *in complete agreement* with actual laboratory data. The author, Pais, calls the October to December 1900 period, The *Heroic Period* of Planck's life.

Note: I have read a few different reasons for Planck making this (extraordinary?) assumption about identifying that heat radiates-in-quanta. For example, we see that Planck was working on an exhausting, important, experimental problem—providing a realistic answer to Kirchhoff's question —which no one could answer. Thus, the pressure he was under may have been one of the reasons for his making this basic, lasting, far-reaching assumption—which may—or probably, have simplified his task. End of note.

To enable Planck to calculate the *heat* of *each of these quanta of heat*, he developed a constant, '*h*', that he called "the elementary quantum of action" which is the *same* for all Quanta. That is, if a particular measured quanta-height (or strength of its heat) has a certain value, which, in general would be called '*e*', (An amplitude value that Planck measured for *each* quanta or oscillator), the product of *h* times *e* is the **energy** *of that individual quanta*. Thus, *The answer* to Kirchhoff's question.

At the time of this discovery (or experiment), which occurred toward the end of year, 1900, Max Planck was already an experienced and recognized physicist. But he had no idea of the overall significance that his discovery would have on the future of physics—and the worlds of electronics, electronic devices, computers, communications—and many other related scientific disciplines.

Many years later, Planck wrote in his autobiography about the numeric constant, *h*, which is now known as the *Planck constant*. He wrote that he

182

developed and introduced that constant in his mathematical equations dealing with the relationship between the Frequency and Energy of a quanta's radiation, to make them both correspond to his observations.

The idea of h, made it evident that the elementary *quantum of action* plays a fundamental part in atomic physics, and that its introduction opened up a *new era* in natural science. It heralded the advent of something entirely unprecedented, *the idea of quantum*, itself. It was destined to basically remodel the physical outlook and thinking of mankind for *very small items*, which, ever since Leibniz and Newton lay the groundwork for infinitesimal calculus, *was founded on the assumption that all casual interactions are continuous*. We might say, 'Planck started a revolution,' or, 'Planck-*The Revolutionary.*'

Part 3 Further Developments in Quantum Theory

The next major development in Quantum Theory was made in 1905 by Albert Einstein (1879– 1955) when he explained how light interacts with matter. This was known as the photoelectric effect. (Note: The famous German physicist, Heinrich Rudolf Hertz, was the first to observe the photoelectric effect in 1887, and it was he who brought it to the attention of the scientific world. (The phrase, 'cycles per second' for frequency, is also frequently referred to as 'hertz', e.g., 5,542 hertz.)

Einstein later postulated that weightless particles, later named photons, should be considered to constitute light, and he attributed the photoelectric effect to these particles. A Quantum of Light! Einstein also said that light should also be considered to be wave, acting according to Maxwell's classic electromagnetic equations. A duality, quanta and waves.

The concept that light could be a particle was, at the time, a revolutionary concept. (This was even though Newton had *first* considered that light was (only) a particle.). The physics community of Einstein's era did not accept the concept of light also being a particle—a duality, until about eighteen years later. Arthur Compton, at Washington University in St. Louis, performed experiments in the 1923 that proved that duality. And even then, there continued to be doubters—including Compton.

Note: In the previous chapter it was mentioned that Niels Bohr, with a recent PhD, visited both Thompson's lab in Cambridge and Rutherford's in

Manchester during 1911 and 1912 respectively, and during those visits he learn a considerable amount about the latest ideas concerning the atom. In all probability, Planck's recent work probably was discussed, and the information gained during those visits can be considered to be a prime contribution to Bohr's soon, initial identification of 'quantum' activity within the atom. That is, the abrupt rise and fall of the rotating electrons in the rim of the atom, which is discussed below. End of note.

In 1913 the Danish physicist, **Niels Bohr (1885 – 1962)** introduced the idea of "Quantum" to the structure of the atom—and thereby provided a large impetus to the activities investigating the new Quantum Theory. Bohr hypothesized the structure of the tiny, tiny atom. He essentially provided a three-dimensional 'dynamic picture' of the atom, as follows:

> Bohr hypothesized that electrons go around the nucleus, with each going at different, specific (quantum) level orbits, or, at different, specific *energy* levels. It is these electron energy levels, plus the quantity of electrons going around the nucleus, that are the major factors in defining and differentiating the numerous, different elements. (The composition of the nucleus is the other major factor.)

Quantum arrived at in its next stage in the 1925-time-frame when the German physicist, **Werner Heisenberg (1901 –1976)** and the Austrian physicist, **Erwin Schrödinger (1887 – 1961)** each provided a different, but basically equivalent, mathematical model of Quantum Theory.

Heisenberg described a particle by using *matrix* mathematics; Schrödinger described a particle as being a *wave* in space and time, with numbers describing the particle's amplitude, or strength, at its positions in space. Since physicist are usually 'more comfortable' with differential equations than matrices, Schrödinger's system has had a broader use. Heisenberg was awarded the Nobel Prize in 1932, and Schrödinger was awarded it in 1933.

This new way of describing the "quantum system"—through an *overall* mathematical system such as those by Heisenberg and Schrödinger, led to most of Quantum Theory to then be called *Quantum Mechanics*.

Shortly later, **Max Born (1882 – 1970)**, a professor of mathematic and physics at the German University of Göttingen, added an important 'probability feature' to Schrödinger's Quantum model. Schrödinger's equations calculated *where* a particle *might* be in space. Born's new feature said that the square of the overall numeric value in a Schrödinger wave equation is *the probability* that the particle—described by a wave, is located *where* the equation say it is. Born was awarded the Nobel Prize in 1954 for this contribution.

(The introduction of probability in Quantum Theory by Born, as a key fixture of Schrödinger's theory, essentially established a distance between Einstein and the Quantum Theory community. We shall learn more about Einstein's disagreements in Chapter 10. As Einstein has been often been quoted as saying, "God does not play dice," implying that God does not deal in probabilities.

There were many other physicists who contributed at these early stages and during the later stage and their contributions are shown in Time-Chart 23. The chart is titled 'Quantum Fathers …' Chart 23 denotes who had attained the title or eminence of a professor. In Europe, at the time of Quantum Theory development, the title of Professor was much more significant than it has ever been in the U.S. university system.

End-note: Teachers at the universities had a wide variety of titles, such as of professor, professor extraordinaire and privatedozent. In the European university education system, there was, however, only one 'Full' professor. Any other person with the title of professor, such as Professor Extraordinaire, was considered 'down the ladder'. And these 'other' professors did not command much salary or prestige.

Professional teachers who had obtained their PhD usually started their teaching career as a *privatedozent*. This is almost equivalent to an assistant professor in the U.S. system. There was a major difference, however, since university did not pay them, and it was expected that their students would pay the *privatedozent*. End of note.

Chart 23 Quantum Fathers: *Circa* 1895 to *circa* 1960+,

Chart 23 consists of five 8.5" x 11" pages across and three pages down.

The chart is on the web at **http://bit.ly/akchart3**

The pages in the chart should be printed out on a color printer and taped together by masking tape to obtain one complete, overall chart of the Quantum Fathers.

The chart shows the most important physicists who made the direct and indirect contributions to Quantum Theory and Quantum Mechanics.

Fourteen physicists are listed on the top horizontal axis: Planck, Sommerfeld, Einstein, Hahn, Meitner, Born, Bohr, Rutherford, Schrödinger, de Broglie, Pauli, Heisenberg, Dirac and Fermi.

The vertical axis goes from 1900 to *circa* 1960.

These are physicists who were instrumental, or close to, the development of Quantum Theory and Quantum Mechanics.

References to physicists within the chart are color-coded according to the physicist's contributions, the physicist's university position, award of Nobel Prizes and other important issues and matters.

Part 2 A Closer Look at the Early Discovers and Formulators of Quantum Theory and Quantum Mechanics—and *What* is the Difference Between Quantum Theory and Quantum Mechanics

Let us now look, once again, but with more detail, at those historical steps in Quantum's development. We will see that during the period 1900 to 1940, Quantum went through the following two phases:

Phase 1. Basic physical concepts were discovered about the parts of an atom, and how these 'parts' interact. Hence, Quantum Theory is about subatomic particles. This phase consisted of deriving quantum's 'building blocks'. During this phase, investigations are considered, made and their results added to Quantum Theory.

Phase 2. Starting in the mid-1920s, mathematical formulas were developed to describe and understand the *overall* and intricate actions (and mysteries) of quanta—and probability was introduced later, in the 1940s, to define an important parameter for the quantum wave, Location, in quantum. During this phase, investigations are considered made and added to Quantum Mechanics.

The following is a repeat of what Max Planck did to arrive at the actual heat, or radiant energy, associated with *each* Quantum's contribution of energy: Planck multiplied the quanta-amplitude of each observed frequency term by a certain number that he had derived—the constant 'h', the "quantum of action", which would be that same for *all* Quantum/frequency components.

Planck—in his derivation-process, determined the value of h to be 6.55×10^{-34} meters^2 kg seconds, a unit of energy. (Or 6.55×10^{-34} joule-seconds.) Today the value is considered to be 6.626068×10^{-34} meters^2 kg seconds and is known as the 'Planck constant.' All the following might be *components* of the total radiant energy in the black-box, *whose dimensions determine the fundamental frequency*, f_1. All other frequencies that are multiples of that fundamental frequency, f_1, and are actually present in the black box and are constituents of the total energy in the box: $1 f_1 h$, $2f_1 h$, $3f1h$, $4f_1 h$,... $11f_1 h$, $12f_1 h$, $13f_1 h$,, $14f_1 h$,.... $100f_1 h$, $101f_1 h$, $102f_1 h$, $103f_1 h$,..

The picture below displays Standing Waves as they actually occur in heated-in-the inside boxes, such as the black-box that Planck used in his identification of Quanta:

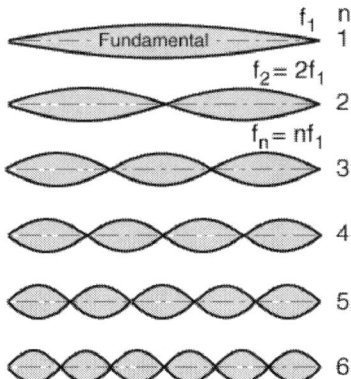

We can stop now and take a breath and review the situation. Planck was really a conservative person. He was not attracted to new, novel ideas. But he had opened a new theory or discipline in physics. During the years

187

following his discovery of the quantum effect, he spent many years trying to incorporate this quantum theory into classical physics—so it would not be a new science. But he failed to accomplish that; he *really* did discover something new.

But physicists who learned of the work of Planck—as well as Planck himself, did not appreciate the significance of the new finding. This situation continued for many years. When Planck was finally awarded the Nobel Prize in Physics in 1918, it was cited for his discovery of 'energy quanta'. I imagine it would have been impossible—at that time, to award him for being the first to enter into 'the *forthcoming* Quantum Age.'

The delay in making the award, from 1900 to 1918, may be attributed to the fact that many in the physics community only, *slowly,* accepted Quantum Theory. As will be seen with the next major contributor to Quantum Theory, Albert Einstein, he also waited a comparable amount of time to be awarded the Nobel Prize for his discovery of the photon, which will be shown, in the next section, to be another 'constituent', pressing for the belief and relevancies of the new Quantum Theory.

With that introduction, we meet *The* second major contributor to Quantum Theory, Albert Einstein. In 1905 Albert Einstein wrote three articles—each monumental, in three different domains of physics. All three were published in one issue of the highly respected German physics journal, *Annalen der Physik*, 1905. One of the articles described Einstein's theoretical analysis of the wave/particle duality of light, that was briefly mentioned above.

Einstein's defining that light could be a particle or a wave was the first inkling that matter, in the form of light or particles, could have a 'dual life.' This duality was later referred to as 'complementarity'—and became a key term during Niels Bohr's early leadership of Quantum Physics.

(Chapter 11 presents Planck's "unique" analysis of blackbody radiation and Einstein's contributions in more detail. Schrödinger's contributions—and his dislike of the German Nazi movement, are also discussed in that chapter.) But first, let us build on the then-incompletely-known atom, with a review of some basics:

Joseph John (J.J.) Thomson discovered the *electron*, a negatively charged particle more than two thousand times lighter than a hydrogen atom, and

Ernest Rutherford identified a key and vital part of the atom—its *nucleus*. And it was, again, Ernst Rutherford, working at McGill University in Canada where he discovered the effect of radioactivity that lead to the half-life of certain elements.

I now note from *Niels Bohr's Times*, source 35, quotes of the writings of J.J. Thomson's thoughts about what he (and possibly, also, the physics world) thought the atom consisted of, as science entered the 20th Century, the pre-Quantum era. In 1897 Thomson wrote:

"These primordial atoms have constituents which we shall for brevity call 'corpuscles'. I regard the atom as containing a large number of ... corpuscles."

Note: Thomson continued to refer to electrons as corpuscles for some years. They were actually electrons rotating around the atom in a certain prescribed orbit. Their *reason for being*, at that time was completely unknown. End of note.

Again, Pais, who describes how Rutherford treated his new-found object, the nucleus.

"Rutherford and his team, "one might say, treated the atom as 'naked', as a nucleus without surrounding electrons.... Thus up to that time, June 1912, the nucleus and the atomic electrons had been considered separately.

Some physicists of the time may have also considered the atom to be constructed, as we might picture a fruitcake, with the electrons being distributed within the cake, and that if a particle were shot at, it would just enter fruitcake, loose it energy in that cake-mass—and stop."

The central question of what the structure of an atom was like, given that it actually contains a nucleus surrounded by electrons, had not yet been systematically addressed. This changed with the arrival of Niels Bohr on the scene in Manchester.

Starting in 1912, Niels Bohr (1885-1962) worked as a postdoc (A person who just received a doctorate) with Rutherford's experimental physics team at the University of Manchester. This experience enabled him to learn about

Rutherford's concept of the makeup of the atom and how radioactivity affects the atom. After he returned to Copenhagen in 1913, he prepared a paper the he sent to Rutherford a paper that introduced a new concept of how the components of the atom work together.

This paper is referred to as the "Rutherford Memorandum." It was written in June and July 1912 and its title is "On the Constitution of Atoms and Molecules." Bohr used the same title later, for his three definitive papers of 1913.

The following is from http://www.faqs.org/docs/qp/chap04.html (some (not all) italics added.)

Bohr's Explanation:

Bohr proposed that while circling the nucleus of the atom, electrons could only occupy certain discrete (thus, quantum) orbits (or rings}. That is equivalent to saying that *each* electron had to have certain *discrete energy levels*.

Further, Bohr said that electrons give or take energy only when they change their energy levels. If they move up, they take energy (say from outside light), and if they move down, they release energy to the outside of the atom in the form of light. This light-energy itself is released in discrete packets, called photons. (Einstein's indirect contribution).

Furthermore, Bohr also said that an electron that is not in its native energy level (in other words, which has been excited to a higher energy level) always has to fall back to its original, stable level.

This is the physical picture of the electrons in the atom that the world has lived with since 1913, the time of Niels Bohr. It is frequently referred to as the Bohr-Rutherford model.

Bohr was awarded the Nobel Prize in 1922 for providing this (quantum) conceptual picture of the atom—and he went on to be the world leader, (or, as I refer to him later in the text, *The Impresario*), of the new branch of physics that became known as Quantum Mechanics.

Allan Karson

The 'pictorial concept' developed by Bohr has been replaced by more mathematical systems, but it still is adequate for a general understanding of the actions that go on in the atom.

We can ask, "Where did Bohr get inspiration for that idea of applying the concept of Quantum Theory to the atom?" If we review notes that he wrote in a draft document while he still was in working with Rutherford's team in Manchester, we can clearly see the sources for his 'quantum inspiration.' The sources were also from earlier papers by Planck and Einstein—plus (indirectly) an idea from a member of Rutherford's team, Charles Galton Darwin, grandson of the naturalist, Charles Darwin.

Let us now see the process of how Bohr ultimately introduced his new ideas about the World of Quantum:

1. Bohr's doctorate was about the theory of metals, which he received before his stay at Manchester. In that doctorate paper he had written a very perceptive, thoughtful statement: "The assumption of mechanical forces (by author: as in classical physics) is not *a priori* self-evident, *for one must assume that there are forces in nature of a kind completely different from the usual mechanical sort...*"

2. Bohr was aware of a paper written by the physicist, Charles Galton Darwin, which described the loss of energy in alpha particles due to collisions between and inter-atomic electrons, with the nucleus playing a negligible role. Darwin did not speculate how the electrons move inside the atom; he also assumed they were free of the atom when they collided with an alpha particle. (Obviously, Bohr drew a completely different assumption, a correct one—that electrons were in the atom, not free and not tied to the nucleus.)

According to Pais, "Bohr knew very well that his ...examples had called for the introduction of a new and as yet mysterious kind of physics, "Quantum Physics." Continuing with Pais:

"In his Rutherford Memorandum Bohr wrote that his new hypothesis 'is chosen as the only one which seems to offer a possibility of an explanation of the whole group of experimental results, (results of Rutherford's team) which gather about and seems to confirm the

191

conceptions of the mechanisms of the radiations as the ones proposed by Planck and Einstein."

And continuing with source 35 "In the course of reminiscing about what he did next, Bohr once remarked: "It was in the air to try to use Planck's ideas in connection with such things." (From an interview of Bohr, by physicists, Nov 7, 1962, bottom of page 156, referred to in source 35)

Spectroscopy and Bohr: And Bohr did more during this period. In order to appreciate his overall impact to the world of physics—in addition to quantum, it is helpful to look at the science known as, *spectroscopy,* and see how Bohr's new and brilliant model of the atom provided a deeper understanding of what spectroscopy represented in the overall understanding of the atom.

Physicists and chemists had long been using (visual) spectrometers to look at the elements and complex bodies (such as the sun, stars and chemicals on earth) to analyze them and identify the various light frequencies that the elements and bodies emit. (Chapter 17 describes the science of spectrometers.)

In 1885 The Swiss mathematician physicist, Johann Jacob Balmer (1825-1898) developed the mathematical formulas to interrelate the frequencies of the spectroscopic lines given off by hydrogen. Bohr had not known of the Balmer formulas when he made his first hypotheses of the electrons' energy rings, but when he did learn of them, the picture became clearer to him.

Bohr postulated that the frequencies detected in spectroscopy—the study of the spectrum of individual atoms, are indicators in the *change* (Italics added) in quantum states among the electrons in the atom. *Bohr now provided a quantum reason for spectroscopy.* And the merger of Balmer's mathematical formulas and Quantum Theory assisted in giving birth to the Quantum-atomic world.

Bohr also showed *why* there was a difference in elements. He described that the difference between elements is mainly due to the *number* of electrons in the atom and the *energy levels* of the electrons. (Scientists have identified 118 elements on Earth. A list of *all* of them, making up the periodic charts, is at http://www.webelements.com/.)

A final, additional note about Bohr's model of the circular path of electrons: During the early years of Bohr's development of the Bohr model, Bohr realized that the electrons *could* be moving in an ellipses, instead of circle, but he was not sure how to characterize the motion. He discussed this observation with the senior German physicist, Arnold Sommerfeld (1868 – 1951) who was Professor of Physics and Director of new Theoretical Institute at University of Munich (1906 -1939). Somerfield volunteered to examine that 'atom situation, ' and eventually showed that the electrons follow an elliptical path—not a circular one. Bohr was so pleased with this innovation and his model, that for a time, the model became known as the *Bohr-Sommerfeld model*.

Digression: Does this sound familiar? It reminds me of Kepler's determination that the planets go around their stars in an elliptical path, and not in a circular one. End of digression. An additional note about Sommerfeld: While he may not have been a great innovator, he is best known for his teaching and mentoring. His students who were awarded the Nobel Prize are Werner Heisenberg, Wolfgang Pauli, Peter Debye and Hans Bethe. His post-graduate students who were awarded the Nobel Prize are Linus Pauling and Isidor I. Rabi, who both completed their careers at U.S. universities.

Phase 2: Quantum Mechanics

In the short time-period, 1924–1927, physics entered a new and challenging era, referred to since then as the *Quantum Mechanics era*. During this era, new and astounding revolutionary ideas, or insights, were provided concerning the makeup and behavior of what makes up the inside of the atom, that may be referred to as 'analysis of particles'—and the overall conception of the atom's behavior with other atoms and particles.

And the pace of discovery quickened. Physicists and mathematicians such as Niels Bohr, Max Born, Louis de Broglie, Paul Dirac, Werner Heisenberg, Wolfgang Pauli and Ernest Schrödinger were leaders in this new era—and they all were awarded the Nobel Prize!

Also, as a result of the new theories and discoveries at the particle level, most physicists realized that physics would never be the same. Some, not all, considered they were experiencing a complete break from classical physics, which had ruled physics since the time of Isaac Newton.

The following are some of those new breakthroughs that changed the character of their work from a theory, Quantum Theory, to this new discipline of physics, Quantum Mechanics. It might be said that Quantum Mechanics invoked a 'systems theory' in quantum, if we can borrow a term used by engineers. Note that there are still many physicists who continue to work on the theoretical aspects quantum, Quantum Theory.

Among the early contributors was **Louis de Broglie, (1892-1987),** a French physicist. In 1924 he hypothesized that if light could have a duality in being either a particle or a wave—as previously hypothesized by Einstein for light, all particles could also possess this duality. De Broglie was awarded the Nobel Prize in 1929 for this revolutionary concept—which probably 'gave a jolt' to the world of physics.

When Einstein first learned of de Broglie's work, he thought highly of it and circulated the idea in the physics community. Einstein is also said to say, ironically, "I believe it is a first feeble ray of light on this worst of our physics enigmas."

Historical note: Even Einstein, who was among the first to develop areas of Quantum Theory, had certain reservations about its potentially breaking with Classical Physics. We will see his cautionary attitude in Chapter 10. End of note.

But the REAL leaders were the German physicist, **Werner Heisenberg (1887-1961)** and the Austrian physicist, **Erwin Schrödinger (1901-1976)** who made *fundamental, long-lasting* contributions by providing mathematical models of the atom—according to Quantum Theory.

Probably one of the reasons for the change in name from Quantum theory to Quantum mechanics is the result of the accomplishments of these two physicists, Werner Heisenberg and Erwin Schrödinger. These physicists developed different *systems* of mathematical equations from one another, while both providing firm, mathematical 'platforms' that are universally used by physicists in their studies and experiments in the field of Quantum Mechanics.

The mathematical systems of these two leading physicists are very different from one another. Heisenberg expressed his work using *matrix algebra*, and Schrödinger expressed his work using *differential equations*. In this application of the two systems, the overall results may be expressed differently, but physics-wise they are considered equivalent. For various technical and familiarity reasons, physicists typically prefer and use Schrödinger's system of differential equations. Heisenberg was awarded the Nobel Prize in 1932; Schrödinger was awarded the Prize in 1933, which he shared with Paul Dirac, an English physicist, whose contributions are described later below

Max Born (1882-1970), a German mathematician-physicist, had helped both Heisenberg and Schrödinger develop their mathematical systems. It was Born, who, in 1926, independently introduced the concept and role of probability into the interpretation of their equations—and into Quantum Theory at large. (Born's introduction of probability will be discussed again, later in this chapter.)

The introduction of probability into physics—was a major change from the world of classical physics, wherein probability had previously rarely played the key role that it acquired in the world of Quantum—with the important exception of Thermodynamics.

Born was awarded the Nobel Prize in 1954. The inclusion of the probability —along with most of the other fundamental ingredients of Quantum Mechanics, became known as the *Copenhagen Interpretation*. (We have not discussed all the other parts of the Copenhagen Interpretation, but it is more than sufficient to have, at least, an idea of what the Copenhagen Interpretation signifies.) Two other future Nobel Prize winners, Wolfgang Pauli and Paul Dirac, also made important, additional contributions to the definition of the atom.

In 1925, **Wolfgang Pauli, (1900-1958)** a young Austrian-born, American physicist, made a contribution that amended Bohr's concept of electron rings. Pauli postulated that two electrons in an atomic structure that have the *same* (quantum) properties can never be in the *same* quantum state. This Quantum Mechanics law is known as the Pauli *Exclusion Principle*; Pauli was awarded the Nobel Prize in 1945 for this contribution.

A note about the Exclusion Principle: The Protons, electrons, and neutrons, neutrinos and quarks (the constituent particles of protons and neutrons)— are all subject to it, and the structure and chemical behavior of atoms is due to it. It also explains why the rings in the Bohr model of an atom have particular limits to the number of electrons in each ring.

The Pauli Principle helps account for *why* the Periodic Chart of Elements has its orderly, precise set of successive rows and columns. In some way, the Pauli Principle also accounts for the chemical properties of the elements, as well as why "solid" objects are solid. (The periodic chart was developed in 1869 by the Russian chemist and inventor, Dmitri Mendeleyev (1834-1907) —long before Quantum Theory was even thought about.)

Also, you may observe in other books or studies about physics, this principle is a way to *verify*, or check-out, a concept dealing with atomic structure—much in the same way that entropy is a way to verify a system concept in thermodynamics or in matters dealing with energy. End of note.

Paul Dirac (1902-1984), an English mathematician physicist, held the Lucasian Chair of Mathematics at Cambridge University from 1932 to 1969. According to Dirac's biographer, Graham Formelo, of source 64, Dirac was the first person to submit a paper for a PhD, *anywhere*, in which the topic dealt with Quantum Mechanics—and, in addition, Dirac, at the age of thirty-one, was the youngest physicist ever to receive the Nobel Prize in physics. He was awarded the Prize in 19933 and shared it with Erwin Schrödinger.

If you never heard of Paul Dirac, you are not alone. The following are the words of Stephen Hawking, who delivered the Dirac Memorial Address in Westminster Abbey,

"… But he was never well known to the public. His death in 1984 drew a short obituary in the *Times*, but otherwise it went almost unnoticed. It has taken eleven years for the nation to recognize that he was probably the greatest British theoretical physicist since Newton."

Paul Dirac was involved in a wide range of studies in Quantum. For example, it was he who proved that the system equations of Heisenberg and Schrödinger were equivalent in output. This work by Dirac was called *The Transformation Theory of Quantum Mechanics*. He also predicted the existence of the positron, (the 'positive electron'), spin theory and a wealth of other issues and problems in quantum mechanics such as finding 'the relativistic quantum mechanical treatment' of spin. That work led to the development of his famous (inscrutable for most of us) Dirac Equation.

One of Dirac's most important universe-wide discovery—and concept, was his analysis of what happened immediately following the Big Bang, which was: Any elementary particles that formed at that time—and that had an electric charge, plus or minus, *also had a twin partner* that had a charge opposite to its partner.

That is, at that very early time in the life of the Universe, an electron would have a negative twin, which would be a positively charged electron, now called, a *positron*. And so on for other (presently one-sided charged particles, such as the electron). Thus the original, very early Universe had a balance of plus and negative particles. But, for reasons not fully understood, it appears that nearly all of the 'other-half' was destroyed during the Big Bang in the time period called, T-2. (T-2 is known as the Lepton Epoch, the period from 1 second to 3 minutes after the Big-Bang in which: the majority of electrons and positrons collided and annihilated each other, leaving the 'surplus' of electrons.)

Dirac also believed that any mathematical formula that is meant to represent a physical process *must first be beautiful*. It should also be simple, but simplicity should be subordinate to beauty. Paul Dirac was also taciturn—to say the least, he was also precise in his use of the English language. In his lectures, if a student raised a question, Dirac would respond with the same words that the student originally used to describe the question.

Digression: A possible/probable reason for his being (unbearably) taciturn: Dirac's father came from Switzerland, settled in Bristol, England, and was a successful, teacher of the French language at Bristol's Merchant Venturers' Technical College. But he was a very domineering person to members of his family. Dirac once told a friend that his father forced him to speak only in French when the two would sit together. During Paul's early years, he was not fluent in French so he did not speak much or well. And he said that he always felt browbeaten by his father. Thus Paul's colleagues are said to attribute his extreme reticence to talk to his father's insistence about speaking French. End of Digression.

Part 4 Quantum Mechanic's *System* Innovators

- Werner Heisenberg
- Erwin Schrödinger
- Max Born and Probability

Werner Heisenberg: Heisenberg's system definition of the Quantum System was completed in 1925. The system's consists of mathematical matrices equations that express how each subatomic particle is defined by its three parameters: A particle's Energy, its Momentum (Mass * velocity) and its Position. Heisenberg new system of equations was one of the key factors that promoted Quantum Theory to transition to a different 'state of being' and to become known as Quantum Mechanics. That is, it was no longer only a theory, and became an integral part of the world of physics.

It would have been 'nicer' if a number could represent each of these parameters, but that is not the case. A matrix defines each of the three parameters; thus, there were three matrices per each parameter. Before describing what Heisenberg accomplished in the development of his matrix system-based model of Quantum, I would first like to describe Heisenberg's recent activities, which led to this accomplishment.

Heisenberg, his university associations and his previous work at Bohr's Institute in Copenhagen, September 1924 to May 1925:

It can be said that it was Werner Heisenberg who introduced the analytic methods into the study of Quantum Mechanics. His system changed the path from what is known in physics as a 'theory' to what is traditionally known in physics as a "system", and for which the term 'mechanics' is frequently applied—no matter *what the system consists* of.

(That is, the term 'system' can be applied to a liquid system, an electronic system, a cosmic system, etc. And even in the situation where mathematical statistics is being applied to analyze a system, mathematical statistics is called *statistical mechanics*.)

During the early 1920s, Heisenberg studied physics and mathematics at the University of Munich under the guidance of Professor Arnold Sommerfeld —who recognized Heisenberg's superior technical abilities, and at Göttingen, studying physics under the guidance of Max Born and

198

mathematics under David Hilbert. (In the U.S., we might consider that a 'home run.')

Heisenberg had already created his professional career path. At the age of twenty-three years old, he had already authored (or co-authored, with Sommerfeld and with Born) twelve papers on Quantum Theory.

In 1922 Sommerfeld expanded Heisenberg's network of physicists-of-notoriety by introducing him to Niels Bohr at the *Bohr Festival* held in Göttingen, June 1922. Bohr, too, immediately recognized the qualities of Heisenberg and from then on there was some communication between Bohr, the distant mentor, and Heisenberg, the disciple.

Heisenberg received his doctorate in 1923 under Sommerfeld at Munich. He completed his Habilitation (I.e., being qualified to teach at a German university) in 1924, at Göttingen, under Born. However, Bohr invited Heisenberg to do research at Bohr's Institute of Theoretical Physics in Copenhagen, where he stayed for almost a year, from September 1924 to May 1925. Heisenberg returned to Göttingen to work with Max Born, a mathematician-physicist, and Max Born's fellow physicist, Pascual Jordan.

Let us now examine what Heisenberg did while working within Bohr's team. It had been recognized by most of the Copenhagen team that they had a 'theory', but not what is called in physics, a *well-defined system*. Such a system would include analytic equations, and, most importantly, parameters that could be *measured* in a laboratory— all in addition to its already in-place, descriptive material. (They did have a *qualitative* description of how electrons orbit (go around) a nucleus and they did have three (later expanded to four) parameters that (grossly) described those paths (with letters m, n, m).)

Only recently, Bohr's "right-hand physicist", Dutch-born Hans Kramers (1894–1952) was trying to determine an analytic way to measure what was going on in an associated field of interest—radiation of light. He was attempting to express the characteristics of light that is emitted by an atom when the atom is exposed to, or excited by, a light beam. (Note that Kramer had developed his system using the mathematics system of matrix algebra.)

When Heisenberg arrived at Copenhagen, Bohr was engaged in a study of the characteristics of fluorescent light, its polarization, and he handed it over to Heisenberg to work out the analytics of the problem. Given the nature of a team spirit at the Institute, Heisenberg worked closely with Kramers to work on the problem, and they produced a joint paper in December, that year. But, as Pais says, "from his research into what went on, "Its redaction was entirely due to Kramers."

(Later, Heisenberg wrote about his own work, in what became knows as Quantum Mechanics, which was based on the use of matrix algebra, originally introduced to him by Kramers. Heisenberg would write, "This new scheme (matrix mechanics) was a continuation of what I had done with Kramers…" (The author always has regretted that Kramers was never awarded the Nobel Prize for this and for his other contributions that are described in Source 33, *Quantum Generations*.)

On Heisenberg's return to Göttingen he considered it was not necessary to describe clearly the explicit actions and activities of the electrons going around the nucleus. Rather, since he observed that the electron acted as an oscillator, his equations describe how an oscillator acts—and not specifically the way an electron may physically work. And an oscillator's actions can be described in matrix form—*shades of Kramers*.

Let us stop to think about this step. Prior to Heisenberg, physicists thought in terms of explicit, measurable parameters, such as mass, velocity, acceleration, position. And more importantly, they were usually observing those physical parameters in the physical systems they were modeling—which they were attempting to describe the models in their math. Obviously, since Heisenberg was 'modeling an electron', he could not see or measure the electron's *actual* specific motion.

So, we will see what are the parameters *that represent the action of the electron*, within Heisenberg's matrices. To be very precise on this issue, I turn to source 64, page 84, a wonderful book by Graham Farmelo, which is about the ultimate 'conqueror' of the atom, … *Paul Dirac, Mystic of the Atom*. (Italics and following parentheses were introduced by Farmelo in his text.)

> Each number in this array (matrix of Heisenberg) is a property of a pair of the electron's energy levels and represents the *likelihood* that the electron will jump between that pair of energy levels. So, each number

200

can be deduced from (measurable) observations of the light given out by the electron when it jumps between (those pairs). In this way, Heisenberg demonstrated how to build an entirely new atomic theory solely in terms of *measurable* quantities.

Hence, Heisenberg created a new way of looking at particles. Rather than considering the specifics of an action, Heisenberg described accurately its *overall* effect. (When the physics community learned of this technique, many considered it was an ingenious advance in methodology.)

(To this author, it appears to be reminiscent of what Max Planck did, two decades earlier, in his six years of examining the radiation of a Black Body. The difference being that, in Planck's work, he measured radiation in the non-visible range; in Heisenberg's, he measured radiation in the visible range.)

I would call this an *Einsteinium Moment*. And Einstein did acknowledge to Heisenberg the genius of his novel approach. In 1928, Albert Einstein nominated Heisenberg, Max Born and Pascal Jordan, who had both worked with Heisenberg on the formulation of Heisenberg's matrix system, for the Nobel Prize in Physics. This award was delayed until 1933—but it was awarded only to Heisenberg.

Born eventually received the Nobel Prize in 1945 for a later contribution he made, his statistical interpretation of the wave function that was developed by Irwin Schrödinger, which is also a presentation of Quantum Mechanics —but not for his contribution to Heisenberg's matrix algebra presentation of Quantum Mechanics. Thus, Born had his hand in the two foremost interpretations of Quantum Mechanics. (Schrödinger's presentation will be discussed in the next chapter.)

Heisenberg represented these parameters with what are called *Fourier wave equations* (oscillators), and placed these equations in matrices—and operated on these matrices. He created a large, complicated, mathematical structure. A brilliant achievement!

A matrix can be a square, or rectangular in shape, with many numbers in it, depending upon the particular item being measured. A '4 by 4' matrix representing an electron's energy could look like this, where each letter can stand for a number or another mathematical equation or another matrix. Yes, it can become a very complex matrix—and difficult to work with.

A11	B12	C13	D14
E21	F22	G23	H24
I31	H32	J33	K34
L41	M42	N43	O44

To help you understand what a matrix is, think of describing a sweater with a 3 by 3 matrix.

Wool11	Sleeveless12	Patterns13
Stripes21	Blue22	Yellow23
Collar31	Size32	Buttons33

Heisenberg's matrices had only a few rows and columns, but matrices, in general, can be very complicated and large. Many of the matrices used to define items being investigated, today, in research centers in physics, such as at CERN, are sometimes too large even for today's supercomputers.

Erwin Schrödinger's system of equations was completed in 1926. He followed a different mathematical path from Heisenberg's; he defined the actions of an electron, mathematically. He started off using classic differential equations to represent the activities goings on in the atom.

Differential equations may be called a 'subset' of Calculus, or a 'derivative' from Calculus. Engineers and scientists had long used differential equations to describe physical phenomena, such as liquid flow (hydrodynamics). (We will see in Chapter 16 that some of the earlier contributors to formulizing this "total" system of mathematics used by physicists included Daniel Bernoulli, Johann Euler, Isaac Newton and Gottfried Leibniz—and many others.)

Schrödinger represented the electron as a *waveform* and wrote what are called 'quantum laws of motion'. His theory is called Wave Mechanics and the equations he developed describe how the quantum state of the physical system changes over time. Schrödinger later showed that the different sets of equations produced by Heisenberg and himself *produce the same results*. He actually showed that he could derive Heisenberg's from his own set of equations, the Schrödinger equations. They both explain, at the detailed level, how quantum matter works.

Most physicists felt more comfortable dealing with Schrödinger's representation of the quantum, as opposed to Heisenberg's. One reason, as mention previously, is that physicists prefer to work with differential equations, rather than with matrices.

Max Born and Heisenberg: Max Born worked closely with Heisenberg in formulating his equations. Actually, it may be said that it was Max Born who actually formulated the Heisenberg system into its final and accepted form. As Born has said, "Before he (Heisenberg) came to me, Heisenberg did not know anything about matrices." Born was always resentful that he and Jordan did not share in the Nobel Prize that was awarded to Heisenberg in 1933.

But Born went on to work with Schrödinger on his wave equations. And Born's analysis of those equations introduced that then 'revolutionary' concept of *probability* into Schrödinger's equations—and into the quantum world at-large, further differentiating Quantum Physics from Classical Physics.

Born showed that the *square* of the final values in Schrödinger's waves, should be interpreted as a statistical predictions of the wave's position, i.e., where it could be, instead of representing an exact location. For those who find the concept of a wave in space—that is ill-defined in position, to be difficult to think of, or to visualize, try thinking of the wave as being a modest-sized *mist in space.*

Max Born's applying probability to Quantum, at this time, was revolutionary, since it was one of the first major introductions of probability into the world of Quantum. As mentioned above, Born received the Nobel Prize in 1954—mainly for making this new, fundamental contribution to physics, that events can have a probabilistic basis.

This concept became a clause célèbre in the physics world—particularly between Bohr and Born one side, and Einstein on the other. They handled the argument in a very professional and gentlemanly way throughout their lives. It is partially due to Born's introducing probability in the system equation, that Einstein is said to have disagreed and said, "God does not play dice."

But in the book, *Subtle is the Lord,* source 8, *The Science and Life of Albert Einstein*, the author quotes from Chapter 26, *Einstein Vision*, "In 1936 he (Einstein) wrote, "It seems clear…that the Born statistical interpretation of the quantum theory is the only possible one' and in 1949 Einstein declared "The statistical quantum theory (is) the most successful theory of our period."

That chapter is mainly devoted to Einstein's 'inner' and later' thoughts about the issue of probability, and how it dominates much of Quantum. (We will come back to this issue, 'Probability in Quantum', later in this chapter.) While Paris's book is for advanced students of physics, Chapter 26 is a more-general explanation of Einstein's thoughts, rather than being a *heavy* technical discussion. Even if you are a 'beginner or intermediate', I believe you will appreciate what Einstein has to say in that chapter, and you might consider reading that chapter at a local library.

Other physicists also developed ways to look at and analyze Quantum matter. For example, in the 1940s, the U.S. physicist, Richard Feynman (1918 – 1988), developed a new way to explain Quantum operations. Feynman introduced *vector diagrams* so it could be done *pictorially*. (He also introduced new, but-heavy, mathematical equations.) The Feynman diagrams are used mainly for a small number of particles. They show basic interactions of subatomic particles. When the number of particles becomes large, the diagrams become too complicated, and the physicist uses his mathematical approach.

The theories of Feynman also explain, in terms that are meaningful to physicists, how Quantum Theory works—with each having a different way to show and analyze quantum. At the same time, remember, they do not explain *why* it works the ways it does. You can see what Feynman drawings look like on many web sites. One specifically for beginners is at: http://www.egglescliffe.org.uk/physics/particles/parts/parts1.html

Part 5 Quantum's Special Phenomena

- Observe How a Particle Can Be a Particle or a Wave
- Heisenberg's Uncertainty Principle
- The Phenomenon of "Tunneling Out"
- Virtual Particles

Part 5 describes four special phenomena of Quantum—but there are many more. Some of us may find this material to be difficult to comprehend. You should not feel bad since there is much in quantum theory that is difficult to comprehend—or even to believe.

No matter what you learn about Quantum; be assured that wherever Quantum Theory is applied in our world, in a far off planet—no matter where in the Universe, *it works*. But it defies what we refer to as *intuition*. It is *non-intuitive*. Or, it may be said that it defies our 'Earth-bound' concepts. In order to live with Quantum Theory, or adapt to it, it is necessary to abandon our personal rules or thoughts about how everything works, from subatomic particles—to the Universe.

You will not be alone in facing these dilemmas. To give an idea of what leading physicists have said about them, here is what three Nobel Prize physicists have said about it:

Murray Gell-Mann, "Quantum Theory, that mysterious, confusing discipline, which none of us understand but which we know how to use".

Richard Feynman, "...I think I can safely say that nobody understands quantum mechanics." Quantum mechanics "describes nature as absurd... from common sense. And it fully agrees with experiment."

Stephen Weinberg, "I admit to some discomfort in working all my life in a theoretical framework that no one fully understands."

Observe How a Particle Can Be a Particle or a Wave

In most physics book about Quantum Theory, you will see that they will introduce a key dilemma by showing pictures of how a particle can be either a particle or a wave. Some of those books present many such examples, and the fullness of the story can sometimes (most of the time) lead to more confusion on our part. We will next examine, one example:

Shooting an electron through narrow slits and registering the outcome on a photographic plate, described in Figure 24, below.

Fig 24 Two-Slit Experiment Shooting an Electron Through Narrow Slits and Registering The Outcome on A Photographic Plate.

This figure is a edited copy of the web site titled, The Tom Bearden Website: http://www.cheniere.org/books/excalibur/2slit.htm,

TWO-SLIT EXPERIMENT

This experiment is fundamental to all of modern physics. Feynman, Nobel Prize winner in physics, has stated that no physicist understands this experiment, and that it cannot be explained by any classical means.

In the experiment, electrons are emitted from a source and travel past a double-slit wall region on their way to a screen. The apparatus is shielded against light. The emitted electron is a little particle. Just like a tiny baseball.

It should go through one of the slits and not the other. It would then hit the screen at one of the two spots indicated as the Expected Distribution. There should be a little scatter from those electrons that chip the edge of the slit a bit. Electrons that do not hit the holes but strike the wall are absorbed in the wall. But we do not get this expected pattern!

Instead, the pattern is essentially the same as the one a person would get if each electron were a wave front passing through both slits at once. However, each electron still strikes the screen in only one point; the distribution of these points fits the Actual Distribution pattern shown. This stuns the physicists.

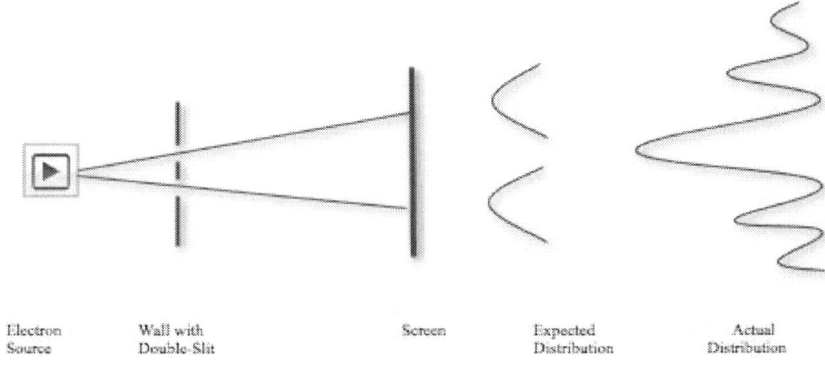

| Electron Source | Wall with Double-Slit | Screen | Expected Distribution | Actual Distribution |

I

207

Heisenberg's Uncertainty Principle

If Heisenberg had stopped his discoveries with his contribution to Quantum Theory with his Quantum Mechanics, the Quantum-world might have been a simpler place to understand. Instead, he introduced one of the most famous dilemmas in physics. Heisenberg said:

"We can never know simultaneously the exact momentum and position of a subatomic particle."

This is called the 'Heisenberg Uncertainty Principle'. It forms the basis of many of the major controversies that once raged in the world of physicists. *This principle has never been disproved.*

The principle states: The *product* of the uncertainty of both the *position*, x, and the *momentum* of a particle, p, (p = mass * velocity) is:

$$\Delta x * \Delta p \geq h/4\Pi$$

where Δx is the uncertainty in position and Δp is the uncertainty in momentum.

Thus, there is *always a minimum* uncertainty and the value is equal to h, the Planck constant (6.62068×10^{-34}) divided by 4pi, which equals .527 $\times 10^{-34}$ joule-seconds—a minute number in any system of measurement. (Note: One joule-second is equivalent to one watt).

Note 1: The measurement of the indeterminacy is in joule-seconds. This is a measurement of energy-time. Joule-seconds can be directly associated with product of the parameters being measured, uncertainty in position and in momentum.

Note 2. The Planck constant is a *universal* number. These are other numbers that are believed to be the *universal* throughout the Universe and do not change with the passing of time.

(Some of those other numbers, that are also considered to be a universal numbers, are: 'c', the speed of light; G, the gravitational constant and 'e', the value of elementary charge carried by a single electron, and 'q', the electric charge carried by a single proton.)

208

Heisenberg's Uncertainty Principle has had a far-reaching impact on showing the *limitations* of measurements in physical science. These limitation arise when a scientist wants to (or needs to) make measurements to a greater accuracy than that allowed by the Uncertainty Principle.

The Uncertainty Principle means that if you measure an atomic particle (even using high quality, precise, highly accurate measuring instruments), the experimenter (the person) could never know both the exact location and exact velocity of the particle, as shown below, in Figure 25.

Starting from the time the theories was accepted in about 1925, there has not been a deviation from the rules or theories of Quantum Mechanics—including the Uncertainty Principle.

Figure 25 The Essence of Heisenberg's Uncertainty Principle

An Electron

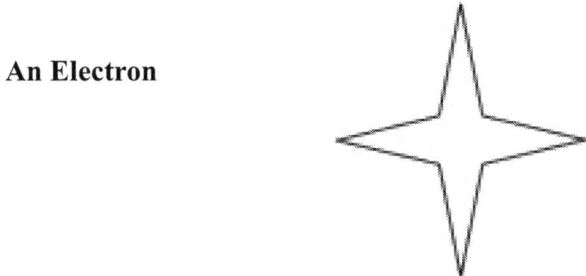

Example 1. (Below)

Position of Electron is completely defined

Therefore

Velocity of Electron cannot be defined beyond a certain accuracy

Example 2. (Below)

Velocity of Electron is completely defined

Therefore

Position of Electron cannot be defined beyond a certain accuracy

I close the discussion of indeterminacy with a brief discussion of measurement situations in physics that do not require the highly accurate measurements involving the Uncertainty Principle.

This is the case for most of the measurements where the size of an object is being measured. This measurement is for small things, but not as small as atomic size. In such measurement we might find an explanation for an indeterminacy being present. This is when a light source is used as the measuring system.

•	For the light source to be effective and provide an accurate measurement, the wavelength of the light must be much smaller than the length object being measured. In order to attain short wavelengths, the frequency must be very high—or ultra high.
•	Conversely, the higher the frequency of the light wave, the higher the energy of each light quanta. Thus high definition, high frequency light beams have high energy that can bounce off (or hit) the particle being measured—thereby altering the particles position.

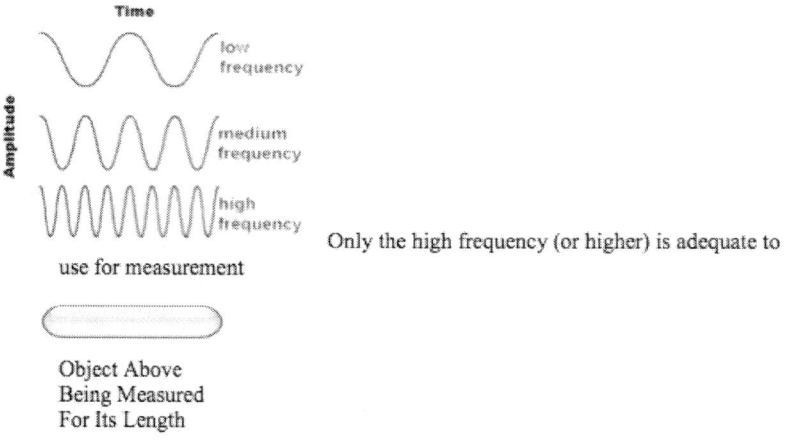

Only the high frequency (or higher) is adequate to use for measurement

Object Above
Being Measured
For Its Length

But in considering 'what is the real limitation in the measurement process', the real limitation is the Uncertainty Principle—hands down.

Allan Karson

The Phenomenon of "Tunneling Out"

Heisenberg opened an exotic problem with his Uncertainty Principle—at least for people who wake up normally in the morning, put their feet on the ground, look into the mirror and see a reflection of themselves—and live the lives we are supposed to—work, play sports, eat normal meals, go to school and have kids. (That is, all of us.)

You might ask: What would be the lives of people (subatomic people, of course) in "Heisenberg Land". Perhaps some answers or ideas could be drawn of the 'Heisenberg Life' from the following examples.

First, everyone who has ever opened a book that just mentions Quantum Theory and Heisenberg's Uncertainty Principle must have read about the following situation (or similar situation) about an electron, or another subatomic particle that 'tunnels out of its immediate environment. An (edited) quote from the Britannica Concise Encyclopedia. With italics-added:

> "In physics, the passage of a particle (such as an electron, quark..) through a seemingly impassable energy barrier. Though a particle's energy may be too low to surmount a barrier in classical physics, the particle may still cross the barrier as a consequence of its *quantum-mechanical wave properties*."

This statement makes no 'buts or ifs'; such as 'if the environment is made out of lead or some other material, it may not work'. *Unequivocally*, it works —for *all* materials. (Of course, with a lead box, you might have to wait a long time, but the particle will eventually get out.)

In 'classical physics', a particle is confined in the nucleus because of the high-energy requirement that would be necessary for it to escape the strong attraction of the nucleus. (Remember that the Nuclear Force is the strongest of the four forces at short distances.) But physicists have observed some particles escaping from (or tunneling out from) the Nuclear Force. At first, there was no explanation for this particle escaping or decaying.

It was the versatile George Gamow, in 1928, who first solved what was then called the problem of the alpha decay of a nucleus. He found that the decay was attributed to 'tunneling out', or 'escaping' of the particle.

Later, the German physicist, Max Born, recognized the generality of quantum-mechanical tunneling. He realized that the tunneling phenomenon was not restricted to nuclear physics, but was a general result of quantum mechanics that applies to many different systems—*even to studies of cosmology.*

The impact of the tunneling effect can be appreciates by noting that the Nobel Prize in Physics in 1973 was awarded to three persons—*All* for their efforts at proving that the tunneling effect exists: Leo Esaki (Japanese-born American, with IBM), Ivor Giaever (Norwegian-born American, with GE) and Brian David Josephson (Welsh, Cambridge University).

Tunneling, today, exists in our products in varied ways: One is a negative aspect of Tunneling, where it is a source of 'major current leakage' in Very-large-scale-integration (VLSI) electronics that results in the substantial power drain and heating effects that plague high-speed and cell phone technology.

But tunneling can also have positive uses. For example, a tunnel diode, or Esaki diode, was developed in the 1950s and was named after its discoverer, Nobel Prize winner physicist, Leo Esaki. It is a semiconductor diode capable of very fast operation, well into the microwave Giga Hertz region.

(One GHz represents 1 billion cycles per second.) In the 1950s and 1960s the tunnel diode was used in many microwave applications because semiconductor devices of the day could not reach the high frequencies that are necessary to generate microwaves.

Another application is in electron-tunneling microscopes, which can resolve objects that are too small to see using conventional microscopes. Electron tunneling microscopes overcome the limiting effects of conventional microscopes (optical aberrations, wavelength limitations) by scanning the surface of an object with *tunneling* electrons.

There remain different aspects of Quantum Theory that are so different from *anything* that a person would find in classic physics. This concept has been used many times to explain certain phenomena occurring in outer space—i.e., in the Cosmos. As an example, physicists sought to find what happens when the Uncertainty Principle is applied to that vast, far-out region.

The results of the investigations show that in that outer vacuum space of the Universe, subatomic *virtual* particles are *momentarily* present—and they are called *Virtual Particles*.

In Chapter 6, in the section titled 'Dying Stars and Black Holes', certain new insights into Black Holes are described. It was Professor Stephen Hawking who developed these new insights. His analysis showed that Black Hole *could* emit radiation. To arrive at his proof, Hawking introduced 'virtual particles' in his analysis.

Some of the features of virtual particles are:

- Subatomic virtual particles can exist for a very short time. (They exist closer in time to a theoretical Planck Time, which is 5×10^{-44} seconds.)

- It is believed that virtual particles are continuously being created in the vacuums of outer space.

- A virtual particle's energy and momentum do not have to obey the usual relationships expected of them.

And, while virtual particles exhibit some of the phenomena that real particles do,

- A single virtual particle cannot be detected because that would result in its becoming real.

- Their kinetic energy may not have the usual relationship to velocity. That is, a virtual particle's energy can be negative.

- Virtual particle can be considered a manifestation of quantum tunneling.

- A 'real particle' can join up—momentarily, with a virtual particle to cause momentarily a 'real' inter-particle action.

Finally, we end the present discussion on virtual particles and quantum tunneling with text that was obtained from the two following sources:

140:Parikh, Maulik K.; Wilczek, F (2000). "Hawking Radiation As Tunneling". Physical Review Letters 85 (24): 5042–5045. doi:10.1103/PhysRevLett.85.5042. PMID 11102182.

Hawking, S. W. (1974-03-01). "Black hole explosions?". Nature 248: 30–31. doi:10.1038/248030a0.

"At the end of its lifetime, a star that is heavier than more than about 20 solar masses can undergo gravitational collapse to form a black hole. According to classical physics, these massive stellar objects exert a gravitational attraction that is strong enough to prevent anything, even electromagnetic radiation, from escaping past the Schwarzschild radius. However, it is believed that quantum mechanical effects may allow Hawking radiation to be emitted at this distance. Electrons (and positrons) are thought to be created at the event horizon of these stellar remnants.

When pairs of virtual particles (such as an electron and positron) are created in the vicinity of the event horizon, the random spatial distribution of these particles may permit one of them to appear on the exterior; this process is called *quantum tunneling*. The gravitational potential of the black hole can then supply the energy that transforms this virtual particle into a real particle, allowing it to radiate away into space. In exchange, the other member of the pair is given negative energy, which results in a net loss of mass-energy by the black hole. The rate of Hawking radiation increases with decreasing mass, eventually causing the black hole to evaporate away until, finally, it explodes."

Chapter 10. Niels Bohr, The Impresario in the Development of Quantum Theory and Quantum Mechanics

In 1963 Werner Heisenberg, the German physicist, whose contributions to Quantum were discussed in the previous chapter, wrote an obituary of Bohr: "Bohr's influence on the physics and the physicists of our century was stronger than that of anyone else, even than that of Albert Einstein. (Source 35)

In all of physics, there has not been as strong an identification of *one scientist* within the worldwide scientific effort as there was between Niels Bohr and the world of Quantum Theory and, later, Quantum Mechanics. They were *One*. It is true that there were scientists who made significant contributions to Quantum Theory earlier than Bohr did, such as Max Planck and Albert Einstein, but the tie between them and Quantum is not as strong as that between Bohr and "Quantum Overall."

I believe that Einstein was a Genius—and that Bohr was a "Mystic Genius"

To add credence to that claim and to show his innate feeling for physics and its mechanical forces, the following is from the introduction to his 1911 PhD thesis. And this is before he was exposed to *anything* that could be construed to relate to Quantum Theory. (Pais in source 70—and his italics):

The assumption of (mechanical forces) is not a priori self-evident, for one must assume that there are forces in nature of a kind completely different from the usual mechanical sort; for while on the one hand the kinetic theory of gases has produced... mechanical, there are on the other hand many properties *impossible to explain if one assumes that the forces which act within the individual molecules...are mechanical also.*

That is, *impossible to explain,* the essence of Niels Bohr's (future) Quantum Theory.

We now describe the contributions of Niels Bohr. This chapter presents:

- An Introduction to the physicist-author, Abraham Pais, to whom I will be relying on for this part about Niels Bohr
- Part 1 Niels Bohr's Personal Development in Physics and in Quantum Theory
- Part 2 Niels Bohr's Leadership, the Theoretical Physics Institute and the Copenhagen Interpretation
- Part 3 Einstein's Questions/Challenges to Bohr—and EPR
- Part 4 The Aftermath of EPR
- Part 5 The meeting during World War II between Werner Heisenberg, head of the Nazi's atomic bomb project, and Niels Bohr, a citizen of conquered Denmark

An introduction to the Physicist-Author, Abraham Pais

I will be turning to the physicist/writer Abraham Pais for guidance and information concerning Bohr, and this will based on Abraham Pais's scientific biography of Niels Bohr, source 35, *Niels Bohr's Times, In Physics, Philosophy, and Polity.*

I frequently refer to Pais since he knew both Bohr and Einstein very well. As he says in his book on Bohr, he (Pais) was probably the only (senior) physicist who knew both Bohr and Einstein well. Pais also says, 'He loved Bohr'. Only a person who knew Bohr well, such as Pais, could describe what follows below.

I will also be relying on the source 25 by Abraham Pais—that is essentially his autobiography, *A Tale of Two Continents, A Physicist's Life in a Turbulent World.* I may be copying complete sentences or paragraphs, or 'rephrasing' Pais's sentences. (I have not asked his publisher for permission to do any of this. I do this under the 'fair use' concept in the copyright laws. I use the 'fair use' concept in many part of the book.)

First, perhaps, it is appropriate to provide a brief introduction to Abraham Pais. In the next chapter, Chapter 11, where I describe his relationship with Einstein, there is additional information about him. In Pais's personal biography, *A Tale of Two Continents*, he describes the history of his family, who were descended from Sephardic Jews who were driven from their Spanish homeland in 1492 and emigrated to Friesland, the name of the area of northern Holland. Pais's forebears called this area the "Jerusalem of the West." Pais was in born in Amsterdam, and he describes how he and his family survived the Nazi occupation of Holland that started in 1940 and ended in 1945.

Pais had gained his doctorate in physics in 1941 at the University of Utrecht —only a few days before the Nazi cutoff date of 14 July 1941 for any Jewish person to receive a doctorate. In 1943 Pais went into hiding to avoid deportation, but was later arrested and released. The release was possibly due to the pleas for help made by senior Dutch physicists to Werner Heisenberg, the German physicist in charge of the (unsuccessful) Nazi atomic bomb project.

When the war ended in 1945, Pais received offers to work in Niels Bohr's organization in Copenhagen, Denmark and, later, to work at the Institute for Advanced Study (IAS) in Princeton, New Jersey. Pais first went to work at Niels Bohr's Institute in Denmark, in January 1946.

Pais stayed with Bohr for less than a year, and in September 1946 he went to the Institute for Advanced Study (IAS) in Princeton, New Jersey, where Einstein had been living since he left Germany in 1933. Bohr wrote to Pais in early 1947, asking him to return to Copenhagen—and he might have, but Pais learned that the situation at IAS was going to change for him in a positive way—as we shall now learn.

In April 1946 Robert Oppenheimer, the world-famous physicist of the University of California, Berkeley and Los Alamos/A-bomb fame, had been (secretly) offered the Directorship of IAS. Before making the decision to accept the position, Oppenheimer visited IAS in order to understand IAS's environment and scholars.

In January 1947, while he was still considering the IAS offer, Oppenheimer had been invited to make a presentation to the American Physical Society's (APS) meeting at Columbia University in New York City. Pais happened to be attending the APS's meeting at Columbia, and at one point he went over to introduce himself to Oppenheimer—a normal procedure among physicists. They chatted briefly and Oppenheimer said he wanted to talk further, after his presentation to be given that day.

They met a little later, went out from the university and walked down Broadway and found a bar, where Oppenheimer told Pais of his still-secret Directorship offer at IAS. He asked Pais not to leave the Institute without first informing him, since he wanted Pais to stay there to help him develop the Institute. Can anyone imagine what Pais must have felt about this meeting? A twenty-nine year old recent-PhD with a minimum of experience, being asked such a request by one of the most world-renown physicists at that time!

Pais agreed to Oppenheimer's request and in February Pais wrote to Bohr— while not telling him about Oppenheimer, said he probably would stay at IAS. Pais shortly became a professor at IAS, and over the years Einstein and Pais developed a distant, but friendly, professional relationship. After all, there was a generation gap of thirty-nine years between the two.

In the meantime, Robert Oppenheimer (1904 – 1967) did accept the offer of Director of IAS at Princeton in 1947 and held that position until 1966—the longest tenure of any director of AIS. He died of throat cancer in 1967.

(I would also like to tell how Pais *first* went to IAS in 1946, before his trip to New York City that is described above. In 1946 he first went to IAS, at Princeton, New Jersey, and had planned to work with the brilliant Wolfgang Pauli, whom Pais had been told, had recently been at IAS. On arrival he learned, however, that Pauli had recently left to work at an institution in Zurich. Thus, I say, "Pais was stuck with Einstein.")

Pais left IAS in 1963 to be professor and head of the physics group at the new, also-prestigious Rockefeller University in New York City. Another offer he could not refuse.

In 1972, Pais had started a parallel career, writing many substantive books about physics and the physicists he knew well, such as scientific biographies of Bohr, Einstein and Oppenheimer. And over the years, he frequently met with Niels Bohr, both at IAS and in Copenhagen. Among his many books is *Niels Bohr's Times, In Physics, Philosophy and Polity.*

All of Pais's books were all heavily researched, and relied on a tremendous amount of original source material. For example, in the mid-1980s, Pais wrote *Inward Bound*, a detailed history of the many projects that were performed over the years to define what is in the atom. That book, and most of his other books, contain references to hundreds of source documents. Abraham Pais died in the year 2000, and even though I never met him, I felt I lost a close friend. End of introduction about Pais.

Part 1 Niels Bohr's Personal Development in Physics

According to Pais—and as previously described in chapter 8:

"Bohr received his PhD in physics from the University of Copenhagen in May 1911 and the first place he worked at was the Cavendish Laboratory at Cambridge University. He was to work under the direction of the Nobel Prize winner, J. J. Thomson, but that was not a good and lasting relationship. It was neither Bohr's or Thomson's fault; it was just a case of two very different personalities."

"Also, it was Rutherford's discovery of the nucleus that eventually led to the most important discover by Bohr of the structure of the atom as a whole." (Source 35)

Pais also provides a vignette in source 35 that shows the superior insights of Bohr—even over an experienced person such as Rutherford. At this time, Bohr had only recently obtained his PhD, and was the low man on the Rutherford team working at Manchester as a postdoc. Note that Manchester was an important center for performing experimental studies on radioactivity.

It was probably in 1911, at a time when it was known for a fact that there exist several groups of chemically *identical* substances with distinct (different) atomic numbers, A, defined as the atomic weight of a substance. For example, the five substances at one time called thorium, radiothorium, radioactinium, ionium and uranium-X are isotopes of thorium with A=232, 228, 227, 230 and 234.

219

Bohr has recalled how after listening to Hevesy, a member of the Rutherford team, the idea immediately struck ... the immediate conclusion was that by *radioactive* decay the element ... would shift its place in the periodic table...corresponding to the decrease or increase in the nuclear charge accompanying the emission of alpha or beta rays respectively.

Bohr then presented his idea to Rutherford—five times, according to Bohr. According to Pais, "Rutherford must either not have listened or not understood or not have believed what Bohr was trying to tell him"—which, it turned out, was the correct description of what was happening that produced the isotopes.

The end of this story is that Rutherford convinced Bohr not to publish this idea—which, if he did publish it, would have made Bohr the discoverer of isotopes, and (an earlier) recognized player in the world of physics.

Note: The following is a brief description of isotopes:

Isotopes of an element all have the same name (Hydrogen, Helium, Carbon...), but with a slight addition to the name. For example, Carbon-12 and Carbon-13. Isotopes are one or two or more atoms, or elements, with the same atomic number. The atom's *atomic number* is based on the quantity of protons that are in the nucleus, but that remains constant. Similarly, the quantity of *protons* is the *same* in all the isotopes of the same element.

The important (sole) difference *among isotopes of the same element* is that the quantity of *nucleons* (protons plus neutrons) in the nucleus is *different*. In Carbon-12 there are 6 protons and 6 neutrons; in Carbon-13 there are 6 protons and 7 neutrons. End of description of isotopes.

Returning to Bohr's travels, Bohr returned home from Manchester to Copenhagen in July 1912 and he was ready to take on a new, modest teaching assistant position at a Copenhagen technical middle school—his first professional position. A more senior physicist, Professor Martin Knudsen, at the University of Copenhagen had arranged this position for him. Also, and probably more important to Bohr than the teaching opportunity, is that when he returned, his fiancé, Margarethe, was waiting for him and they were married that August.

The newly married couple honeymooned in England and Scotland. A point to keep in mind: While in England, Bohr met with Rutherford in Manchester and gave him a draft version of what Pais calls "a manuscript." The document contained a preliminary analysis of what Bohr had observed and learned while doing his postdoc working under Rutherford in Manchester. That is, the previously mentioned, Rutherford document.

On the newly married couple's return to Copenhagen, it appears his position had been upgraded, since he was now an assistant to Professor Knudsen— but was *also* a *privatdocent*, a recognition that he could teach independently at university level—and can teach and supervise PhD students independently. Bohr lectured on the mechanical foundations of thermodynamics from October to mid-December, 1912. In January 1913 he applied for a leave of absence for the fall semester of that year.

During this entire recent period Bohr had been thinking deeply-and-often of what he observed and learned while working and learning in Cambridge and Manchester. In all probability, the subject of quantum was also discussed at the Manchester labs. Bohr had recognized that the scientists/physicists recognized there were electrons that were going in various orbits, and there was a nucleus where the atom's main mass was concentrated, and there were alpha rays. But, to Bohr, it appeared that *no one was looking at—or trying to develop, an integrated, systems approach to the 'new' Quantum atom.*

(An attempt at levity: This situation is similar to that in Mark Twain's short story, *The Stolen White Elephant*, where a swarm of detectives try to identify what they had before them (A white elephant), by examining the elephant with their individual magnifying glasses.)

On 6 March 1913 Bohr sent to Rutherford a letter and the first chapter on "The Constitution of Atoms and Molecules", asking him to forward that manuscript to English-language *Philosophical Magazine* for publication, where it was published in three parts in the summer and fall of 1913. (In the first decade of the 20th Century the *Philosophical Magazine* was a rival to the German-language magazine, *Annalen der Physik.)*

This paper also included many of the ideas that had been defined by Planck and others who had any thoughts and ideas about quantum—since 'quantum was in the air.' But it was Bohr who *first* applied such ideas to the *configuration and dynamics* of he atom. This new paper of Bohr's was to make him a world figure in science.

Note: The following are only two of the many important points made in Bohr's memorandum to Rutherford—and which later became known as *The Manchester Memorandum*. They both have the concept of Quantum Physics as their *fundamental* mechanism—the first time Quantum enters into Atomic physics:

- Electrons move in set, discrete, stationary orbits inside the atoms, and go in one of two paths, or states, depending on the energy of the electron.
- Radiation (to the outside of the atom) arises because of the electron's jumps between these states (different sized orbits)

Other important points to be made concerning that 'memorandum':

- The quantitative confirmation of these ideas by Bohr, confirmed by treatment of hydrogen and ionized helium, mark a turning point in the physics of the 20th Century and are a high point in Bohr's creative career.
- The insistence on the role of the outer-most ring of electrons as the seat of most chemical properties of the elements, in particular their valences, constitutes the first step toward Quantum Chemistry.
- The sharp distinction between atomic/molecular and nuclear physics begins with Bohr's realization that Beta rays emanate from the nucleus.
- Gave the first *firm* and *lasting* direction toward an understanding of atomic structure and atomic dynamics. In that sense he may be considered the father of the new 'Quantum atom.'

Bohr's Acumen and World-Personality (According to Pais in source 35.)

To those of us who knew him then, Bohr had become the principle consolidator of one of the greatest developments in the history of science...Quantum. Bohr was one of the most open minded physicists I have ever known, forever eager to learn of new developments from younger people...(and) to be prepared for a surprise. (In these respects he was entirely different from Einstein.)

Bohr...had an unparalleled talent for discerning...how progress could be made by judicious use of experimental data. That is what Heisenberg had in mind when he said; "Bohr was not a mathematically minded man. He was, I would say, Faraday, but not Maxwell."

He (Bohr) explained how he had to approach every new question from a starting point of total ignorance. It is perhaps better to say that Bohr's strength lay in his formidable intuition and insight, rather than in erudition.

And Einstein wrote: "That this insecure and contradictory foundation of (quantum physics) was sufficient to enable a man of Bohr's unique instinct and tact to discover the major laws... appeared to me like a miracle."

"He (Bohr) utters his opinions like one perpetually groping and never like one who believes he is in possession of definite truth"

Pais tells about Bohr and Einstein being together:

My first direct experience of the impact of Einstein on Bohr happened a few weeks later (1948)...when Bohr came into my office at the Institute (IAS)... He was in a state of angry despair ... He told me he had just been downstairs to see Einstein. As always they got into an argument about the meaning of quantum mechanics. And, as remained true to the end, Bohr had been unable to convince Einstein of his views. There is no doubt that Einstein's lack of assent was a very deep frustration to Bohr. It is our good fortune that this led Bohr to keep striving at clarification... And not only that: it was Bohr's good fortune too.

Einstein appeared forever as his leading spiritual sparring partner; even after Einstein's death he would argue with him as if he were alive.

Part 2 Niels Bohr's Leadership, at The Institute for Theoretical Physics Institute and the Copenhagen Interpretation

Bohr and his associates led the development of Quantum Theory on, what might be called a 'European-wide scale', with certain universities in the United States also participating. The worldwide central location for this work was The *Institute for Theoretical Physics* at the University of Copenhagen.

The Institute was awarded to Bohr by the Danish government and the Carlsberg Foundation in 1921. In 1965 the name of the institute was renamed the Niels Bohr Institute. In addition, it was expanded to encompass the study of astronomy, electromagnetism and geophysical sciences.

The analysis and the thoughts about the new Quantum Theory (or Quantum Mechanics, the name it transitioned to during this period) emanating from this institute were referred to as the *Copenhagen Interpretation*. (And indeed, it was an interpretation—and not necessarily always a purely scientific-based one. (The essentials of the Interpretation are described in the next part, Part 3.)

Note: Whether Bohr and his team realized it or not, I consider the use of word 'interpretation' to be a masterful marketing ploy. It gave Quantum a certain flair, or panache. End of note.

The name itself, *Copenhagen Interpretation*, set up Bohr's Copenhagen Institute as the center of quantum's worldwide activity. Physicists from all over the world would come to Copenhagen, to visit, study and work at this center of quantum activity—and meet with his eminence, Niels Bohr. Bohr was the Director—on top of this organization and held that position for four decades

I can picture Bohr as the maestro of a large symphonic orchestra, leading and directing the various players. But 'output of his orchestra was the documentation and any auxiliary systems that presented the *Copenhagen Interpretation*:

The Copenhagen Interpretation (According to Pais)

"Heisenberg and Schrödinger invented mathematical formalisms, which provided correct answers to all the various problem of (non-relativistic) quantum theory. It was Bohr who uncovered the underlying significance of the theory; in particular he showed how the profound conceptual difficulties encountered as the developed—the so-called wave-particle duality, the totally neoclassical nature of the uncertainty principle—could be neutralized by a revision, or more accurately a generalization, of our use of physical concepts."

...He (Bohr) saw Quantum Theory as demanding fairly drastic reformulation of our usual physical concepts; this was required, not to provide a physical picture of any sort of 'explanation' of quantum theory, just to avoid logical confusion and give us a rigorous basis for discussion of experimental results.

Briefly stated, Bohr's main points are these: Quantum mechanics renders meaningless the question: Does light or matter consist of particles or waves? Rather, one should ask: Does light or matter *behave* like particles or waves?

Complementarity is the realization that particle and wave behavior are mutually exclusive, yet that both are *necessary* for a *complete* description of all phenomena. (Italics added.)

Bohr's formulation of the complementarity concept, which he kept refining from 1927 on, makes him one of the most important philosophers of the twentieth century. As such, he must be considered the successor to Kant, who had considered causality as a 'synthetic judgment *a priori*', not derivable from experience ...this rule must now be considered passé. Since Bohr, the very definition of what constitutes a phenomenon has wrought changes that, unfortunately, have not yet sunk in sufficiently among professional philosophers.

Again, according to Kant, *constructive concepts* are intrinsic attributes of the *Ding an sich* (things actually), a viewpoint desperately maintained by Einstein, but abandoned by quantum physicists. In Bohr's words: 'Our task is not to penetrate into the essence of things, the meaning of which we don't know anyway, but rather to develop concepts which allow us to talk in a productive way about phenomena in nature.'

Probably no other physicist, or even any scientist in the history of science, ever had such a monumental role! Or spoke in such terms!

To consider his eminent role in physics, we should compare the early findings of Planck and the theories of Einstein and their impact on physics in general. Planck and Einstein considered their findings in quantum to be inconsistent with classical physics—the physics of

Newton and later physicists. *But Bohr's Copenhagen Interpretation model—made a clean break with classical physics!*

Bohr, however, frequently showed that even he was not fully convinced of what Quantum meant. Starting in 1913, from the period just after he presented his ideas of the quantum model—and for many years after that, it is said that he would ask physicists whom he met at various meetings, "Do you believe it?" "

Bohr's Personal-Personality (with lesser mortals)

The author has worked with many senior executives in industry and university professors. I have worked with directors of major industrial laboratories. I have worked with senior members of the military who had one or two stars on their uniform. During all these relationships, no matter what my question or comment was, I was never bullied by anyone.

But, I admit, I have never worked for long, arduous periods with persons while they were performing the roles described in this book, such as Planck, Einstein, Bohr, Heisenberg and Schrödinger. You may ask, "Why am I making such an introduction?" Because in my readings about Bohr and his personal dealing with some fellow-scientists, I gained the strong feeling that he was an intellectual bully toward lesser mortals. (That is, with the exception of people such as Einstein.)

In the category of lesser mortals, I include the Dutch physicist, Hendrik Antony Kramers (1894-1952) who was hired by Bohr in 1916, and left the Institute in 1926 to become Professor at the University of Utrecht. Kramers, originally, may have thought he could be the successor to Bohr at the Institute. But when Kramers had once expressed his disagreement with Bohr, Kramers was 'bullied down'. "Kramers became Bohr's yes-man," is the comment of the American physicist, John Clarke Slater (1900 – 1976), who was at Bohr's Institute in the 1920s and became MIT's first Institute Professor in 1951.

There was another situation that can be chalked up to Bohr's bullishness. This concerns a person who is also known in the physics world, the American physicist, Hugh Everett III (1930-1982). (I had met with Hugh Everett one or two times in the early 1960s when he was a director within

226

the Weapon Systems Evaluation Group (WSEG) a strategic organization in the Department of Defense, based in the Pentagon.)

In 1954 Everett entered Princeton University and gained an MS in mathematics, studying under the guidance of his advisor, the Nobel Prize recipient, Eugene Wigner. The next year Everett switched to physics, worked towards a PhD, where his thesis advisor was the highly respected physicist, John Wheeler (1911–2008). Everett was awarded his PhD in 1957 and continued working at Princeton in Quantum Theory. Under Wheeler's guidance, Everett developed the concept of multiple universes, which is usually referred to as 'Multiverse.'

Wheeler was so impressed with Everett's work that he arranged for Everett to meet with Bohr in Copenhagen, hopefully, to get Bohr's blessing concerning Everett's new concept—or, at least to learn Bohr's thoughts on Multiverse. Bohr, however, was negative about such a concept and he did not get that blessing. As an article in Wikipedia about the visit, says, "The visit was a complete disaster."

The source http://space.mit.edu/home/tegmark/everett/everett.html#e11

says, "... the 75-year-old patriarch (Bohr) was not inclined to discuss seriously "any new (strange) upstart theory" and, it seems, did not give Everett a chance to express himself. Everett has only the most gloomy memories of this meeting, and was rather reluctant to recollect it at all."

This action by Bohr greatly affected Everett. He left the field of physics research, but performed original systems-research involving physics and mathematical modeling for the Department of Defense in the Pentagon. Multiverse is still in the air in Quantum and more alive than ever—and many articles about multiverse frequently include a reference to Hugh Everett.

Part 3 Einstein's Challenges to Bohr,

We will be examining this *Battle of Titans*, which consisted of three separate confrontations between Bohr and Einstein. Bohr won the first two, and Einstein won the third. The third event, referred to as *The EPR Challenge,* may be considered to be a seminal event in the history of quantum physics since it identified, among other 'Quantum Virtues' an important, unique, powerful property of Quantum Mechanics, known as *Entanglements*. You can find that topic on the web by using the two-word phrase, *Quantum Entanglements*.

We now will change to a different guide/writer/source, Professor Andrew Whitaker, source 22, *Einstein, Bohr and the Quantum Dilemma*.

Many of books have been written about those aspects of Quantum Mechanics, (or Quantum Theory) which deal on one side with Niels Bohr— his deep thoughts about Quantum and his Copenhagen Interpretation; and on the other side, Albert Einstein. Einstein had certain concerns, one of them was, "Is it complete.?" The issue was never questioning whether it was, 'not correct'; rather, the issue was "Was it complete?"

Note: The website, http://www.benbest.com/science/quantum.html lists the following persons, who were leaders in their respective fields, as also being initially opposed, or hesitant, about the *Copenhagen Interpretation* at that time: Erwin Schrödinger, Louis de Broglie, Max Planck, David Bohm, Alfred Landé and Karl Popper. This is probably a small sample from a much larger group of similar-thinking scientists, mathematicians and philosophers. End of note.

The first two confrontations between Bohr and Einstein took place at the Solvay Conferences in 1911 and 1930. The Solvay Conference was traditionally held in Brussels, Belgium, the location of the conference's founder, Ernest Solvay, a Belgium industrialist. The guiding idea behind the Solvay Conference was to provide a major, international venue for chemists and physicists to present and discuss their ideas. It was originally planned to hold the conference every three years, but bigger gaps interceded.

The third confrontation, which was an informal one (not listed on the Solvay web site), was in 1935. Europe was then starting to be in turmoil due to the rise of Hitler and Nazism; the Solvay, in Belgium, was no longer, necessarily the best place for this third encounter, but it was held there, anyway.

Notes: 1. It appears that the on-coming Nazi-generated turmoil did not affect the 1935 conference. 2. A complete list of the twenty-six Solvay

Conferences and their subjects can be found on the web at https://en.wikipedia.org/wiki/Solvay_Conference. The last one was in 2014; its subject was 'Astrophysics and Cosmology'. The previous subject, in 2011, was 'The Theory of the Quantum World.' End of notes.

Now, back to EPR. Einstein had been driven out of Nazi Germany in 1932 and had settled at the Institute for Advanced Study (IAS) in Princeton, New Jersey. To prepare for the 1935 Solvay Conference, he worked with two colleague physicists, Boris Podolsky and Nathan Rosen, and together they prepared a technical paper that attempted to show that Bohr's *Copenhagen Interpretation* describing Quantum Mechanics *did not completely describe Quantum Theory*. That is, it was *missing* a necessary part of the theory, or an important 'ingredient'. Hence, the famous question (challenge) is named after the initials of the three university physicists: The EPR Challenge.

They were attempting to prove the *incompleteness* of Quantum Theory—not that it is wrong. It was Albert Einstein who would present the verbal-challenge to Niels Bohr. Their challenge of 1935 had been prepared in the paper, "Can Quantum-Mechanical Description of Physical Reality Be Considered Complete?" The paper was published in the May 1935 issue of the *Physical Review*, and a pdf copy of that paper is on the web.

(If that paper is no longer on the web, or, if you have a problem accessing that paper, there are others, such as the Stanford University site, titled, "The Einstein-Podolsky-Rosen Argument in Quantum Theory.")

Varied Issues To Help Understand The Three Confrontations:

First, a statement about the friendship between Bohr and Einstein: They got on well, and they greatly respected one another—(in spite of the fundamental disagreements over the role of probability in Quantum Theory). While the Solvay Conferences provided a Grand Stage to display their differences, there were many meetings between the two position-leaders that were outside of formal conference meetings.

One of Einstein's concerns about Quantum involved how measurements are made in Quantum, and was based on Bohr's own concern about 'discontinuity' and it impact on measurements. The following is a quote from source 22, with the author's parenthetic insert added for additional clarity.

"For Bohr, all the problems of the quantum world began with discontinuity. Bohr has written that 'every notion, or rather every word, we use is based on the idea of continuity, and becomes ambiguous as soon as this idea (of continuity) fails'.

... Bohr always commenced his discussion of discontinuity by using his own 1913 model of the atom...where there were what he called stationary states, in which the atomic system remained unchanged, (that is, the orbits of electrons circling the nucleus) and between which, ... and very important for a clear understanding of Quantum—transitions were caused by discontinuous emission or absorption of a photon. .....discontinuity must inevitably imply a loss of determinism."

Bohr, therefore, had to develop theories, concepts and explanations to explain what happened at *discontinuities*, and those explanations are an important part of the Copenhagen Interpretation.

Let us consider what Bohr might have been considering when a measuring device is measuring a particle: Bohr postulated that prior to the measurement of the atom, all is fine and classical mechanics rules. When, however, at the moment of measurement, there is a change in the system (energy, momentum...) and thus there is a discontinuity.

Bohr considered that while the first item being measured may provide a correct (classic) answer, any subsequent measurements could be incorrect due to the disturbance (discontinuity) in the measuring—regardless of the relative sizes of the measured particle and the measuring system.

We can say that once a measurement is made, there is a major change as follows:

- The deterministic equations are no longer valid.
- The measurements have been replaced by probabilities of the momentum and position of the particles.

This limitation of not being able to get a second, accurate measurement, was one of the many explanations of Quantum Mechanics that caused Einstein to believe that Quantum Mechanics did not contain a *complete set of explanations*—and needed additional theories or explanations in order to make it a *Complete* Theory.

230

We now briefly review Bohr and Einstein's first two, earlier confrontations, and then describe the third confrontation, universally referred to here, as EPR. First, the issue being considered in all three: *Was Quantum Theory, or Quantum Mechanics, as articulated by the Copenhagen Interpretation, a complete theory, or did it have any gaps, holes or lapses*? (Or, it had an equivalent problem, referred to as *hidden variables*, but this problem was ruled out as being a problem, early in the post-conference discussions on EPR.)

Now, to the conferences themselves: The first confrontation occurred at the fifth Solvay Conference in 1927. This conference is perhaps the most famous of the twenty-eight that have been held since, with the last Solvay Conference being in 2013. The historical significance of the 1927 conference was primarily due to the presence of the many distinguished, physicists of the time, who had come there to discuss Quantum Theory. Admittance was only by invitation and this was the first time for the meeting of all these august, scientific persons. It can be said that the conference's significance was not necessarily due to its content, but to its attendees.

The specific issue at the 1927 conference was whether a measurement could be made more accurately than Heisenberg's *Uncertainty Principle* would allow. Einstein set out to prove that such measurements *could* be made, and he provided a 'thought (*gedanken*) problem' to Bohr. Thus, it was not that Bohr would have to prove that the Uncertainty Principle was a valid and complete theory factor; rather, it was Einstein who was trying to disprove it —by demonstrating a an incompleteness within the Quantum Theory.

Rather than providing a blow-by-blow description of that confrontation, a summary would conclude that Bohr *did* provide a *valid* explanation that the Heisenberg's Uncertainty Principle *does provide the limiting accuracy*. Hence Quantum Theory was intact and shown as being complete. In addition, Bohr's brilliance was shown at that first confrontation, and his personal and technical stock among his audience rose.

The second confrontation was held at the sixth Solvay conference in 1930. The problem that Einstein presented to Bohr was a much deeper one than the one presented at the first confrontation. Initially, Bohr had a difficulty with the problem, but he finally solved it. And the physics world recognized once again the utter brilliance of Bohr. As a matter fact, Bohr utilized an idea from Einstein's *General Theory of Relativity* to solve the problem posed by Einstein.

Just to repeat once again, the *raison d'etre* for these 'mind-battles', was critical to the physics world's *unquestioning* acceptance of quantum mechanic as articulated by the Copenhagen Interpretation. The question was whether the theory was complete (as maintained successfully by Bohr, up to then)—or not complete. The latter position was Einstein's position at *all* times—*it was not complete.* Einstein may be considered to have been a 'leader' among those who did not *fully* accept Bohr's Copenhagen Interpretation.

You should not think that Einstein did not believe in Quantum Theory; that would be incorrect. Einstein had been an early and star contributor to it, indirectly, by his defining light as being either a particle (a quanta) *or* a wave. Also, when Einstein learned of Bohr's achievements in the early days of Quantum Theory, before 1920, he was very complimentary toward Bohr concerning his achievements in Quantum Theory.

Now, the Solvay conference of 1927: The EPR Challenge. As indicated above, Einstein had essentially lost in the two previous rounds of the Bohr-Einstein debate, but in both those earlier meeting Einstein had proposed *multiple* measurements that interfered with one another. Now, in their third meeting he proposed *only one* measurement. But Bohr, his team—and the Copenhagen Interpretation *were not prepared for that one-measurement situation.* Thus, the Copenhagen Interpretation *was missing* a way to handle EPR. It *was* incomplete, as predicted by Einstein. This *incompleteness* was fundamental to the success of the EPR challenge.

The basic challenge of EPR is as follows, (from source 22): "One particle at rest decays into two particles. From conservation of momentum, these two particles must move off in opposite directions, the magnitudes of their momenta being equal, both p, let us say. This implies that at any moment their positions are related: if you measure the distance particle 1 has traveled, x_1, it is clear that the distance traveled by particle 2_x will be equal and opposite to x_1

232

Allan Karson

This is true in Classical Physics.. The new point in quantum theory is that, until a measurement is made, according to orthodox Quantum views p, x_1 and x_2 have no values. And *emphatically* one of the particles *cannot have precise values for both momentum and position.* Thus, there were *unexplained* differences (or incompleteness) on the part of Quantum Theory that could not account for these measurements in the above-defined challenge.

But, to this author, that was not the important result of the conference. Yes, Einstein showed the Copenhagen Interpretation was *incomplete. But* the physic model, or physics system, that Einstein used to prove his point of incompleteness, was a very rudimentary, simple, small communication system made up of two distantly located particles in space, in communication with each other. And they did show that that *a Quantum system consisting of* (a basic) *communication could act in an unprecedented way—it also suggested that such a system can maintain communication contact under many types of conditions.*

Niels Bohr's reaction to the EPR challenge was quite dramatic, according to the American physicist, Leon Rosenfeld, (1904-1974) who was working with Bohr in Copenhagen at that time. According to Rosenfeld, "as soon as Bohr had heard of Einstein's argument, everything else was abandoned: we had to clear up a misunderstanding at once. Its effect on Bohr was remarkably clear that the EPR paper had, as Einstein hoped it would, challenge many of Bohr's arguments with great effect."

The very significant *real* output of that EPR meeting was to establish that, within a Quantum-based, there was *could be* a simple 'communication system', which was based on *unpredicted, unseen, unanticipated entanglements* between the two particles, that enabled the system to continue operating properly. But Bohr's documentation on quantum systems, (i.e. The Copenhagen Interpretation did not mention or recognize or know of this important, hidden-till-then feature of Quantum technology, *Entanglements.*

Yes, as shown in the EPR meeting, Quantum Mechanics does include some non-intuitive (spooky) features. An important feature, that was shown to exist in EPR, is that "spooky" actions *do occur* between 'distant' particles. They are made possible by that special, unique quantum feature known as "Entanglements."

233

The thought experiment, known as the EPR experiment, is no longer an issue within Quantum Mechanics. The former enigma in Quantum Theory, entanglements, that is associated with EPR, is now accepted as a realistic feature of the 'world of quantum'—and, in the future, products and systems that use or incorporate the "EPR" features of quantum technology will, become a part of our daily lives. Bohr's Copenhagen Interpretation, today, is alive and well. (For those interested in the EPR challenge, the (complicated) details can be found on the web at site: EPR Paradox.)

Entanglements are now seen as the key to the next 'Technical Age", with the present *age* being *The Electronic Age* and—the next one expected to be *The Quantum Age*. We will now trace some of the steps, or experiments, leading to the acceptance of EPR, or, more correctly, of entanglements. But first, a 'look' at the work of the mathematician, John Stewart Bell, who was called upon to analyze the EPR situation. He provided a 'justification/explanation' of the power of quantum systems. (I believe, however, that Bell's reasoning would convince no one if it were not backed up with later, actual experiments that *prove* EPR functions—with its entanglement feature, as proposed by Einstein and his team—and accepted by Niels Bohr.)

Andrew Whitaker, who has been quoted above, says "…in 1964 John Stewart Bell (1928 – 1990), a mathematician and native of Northern Ireland and a staff member of CERN (European Organization for Nuclear Research), picked up the quest for resolving the EPR issue. Bell provided the first mathematical proof to the physics world that they should have more confidence in Quantum Mechanics and they should disallow (the negative) points of EPR."

Bell's test does, actually, attempt to show the deficiencies in Copenhagen's Quantum Theory. But what Bell actually developed was a theorem that said that *there might be hidden, or unseen variables, in Quantum theory—but that is within the nature or characteristics of the way Quantum Mechanics works. Thus,* as Einstein had said, *Quantum Theory is (outwardly) incomplete.*

According to the Wikipedia's Bell's theorem web site, the theorem, loosely stated, is: "No physical theory of local hidden variables can reproduce all of the predictions of quantum mechanics." That is, one might push the issue and say, "Quantum mechanics can produce so many non-intuitive,

234

unforeseen, results that can never be pre-attributed to hidden, or unknown, theories of Quantum Mechanics."

The Bell-accepted EPR test has been performed many times by numerous, different organizations. And the result has been always the same. Thus, Quantum Theory continues on, unblemished. And Bell's theorem has provided physicists to new, remarkable experiments that have led to remarkable achievements. Professor of Mathematics and Physics, Brian Greene, of Columbia University cites such an achievement in his book, *The Fabric of the Cosmos, Space Time and the Texture of Reality* ((Source 41).

> The (Bell) test shows that in some way two particles can communicate with one another under conditions where it is assumed to be impossible. An example of this is where the distance between two particles is such that they are communicating with the use of messages that would appear travel at speeds that are greater than the speed of light.

Professor Greene is referring to a French team of physicists that performed a series of tests in the early 1980s. Professor Alain Aspect of the Ecole Polytechnique, Paris, France, directed the team. Parts of Professor Aspect's tests are described later in this section.

Entanglements

As previously stated, the "tying together" of particles over long (or even very short) distances is called "entanglements." Remember that word well. It will play an important role in future quantum computers and in quantum-based communication systems. The following are descriptions of three colloquia that the author attended during the period 2008 to 2011 that discussed the entanglement feature in quantum systems. The information presented in those colloquia descriptions showed how entanglements will give rise—in the future, to quantum-based products and systems.

1. From Einstein's Intuitions to Quantum Bits: A New Quantum Age

French Professor Alain Aspect of France's *Ecole Polytechnique*, who is also Director of Research at the French National Center for Scientific Research (CNRS) presented the results of his testing EPR in the Alpine mountains of Switzerland at a physics colloquium at Columbia University, September 2008.

Professor Aspect has performed many experiments testing, what he now calls, 'The Bell's Inequality'—which he considers "to be one of the profound scientific discoveries of the century". The abstract of his presentation is presented in order to get a perspective of his presentation and what his findings indicate for the future. (Note: Qubits, discussed below, are a unit of quantum information—the quantum analogue of the classical bit in binary computers. It is a two-state quantum-mechanical system.)

"Based on that concept, a new field of research has emerged, *quantum information*, where one uses *entanglements* between qubits...Large scale practical implementation of such concepts might revolutionize our society, as did the laser, the transistor and integrated circuits, most striking applications of the first quantum revolution initiated at the eve of the 20th century."

In one of his charts, Professor Aspect presented an overview of the sequence-of-tests toward proving that entanglements works. Key statements within the charts are presented :

Three Generations of Tests of the Bell Inequality:1st Generation: The pioneers (1972-76): Tests performed by John Clause at UC Berkeley, F. M Pitkin at Harvard and Edward Fry at Texas A&M,

First results: Contradictory but clear trend in favor of Quantum mechanics.

Experiments significantly different from the ideal Bell scheme.

2nd Generation: Institute d'Optique, University of Orsay, France (1975-82)

A source of entangled photons of unprecedented efficiency.

Schemes closer and closer to the ideal *Gedanken* Experiment.

Test of quantum non-locality (involved relativistic separation).

3rd Generation experiments (1988-): Performed in Maryland, Rochester, Malvern, Genève, Innsbruck, Los Alamos, Paris, Boulder and Urbana-Champaign.

New sources of entangled pairs

Closure of the last loopholes

Allan Karson

Entanglement at very large distance

Achieved entanglement on demand

The overall significance of Professor Aspect's presentation is that in the Geneva experiments, performed in 1998, he used the optical fibers of a commercial telecommunications network and proved entanglements at a distance slightly greater than 10 kilometers (6.2 miles).

2, Quantum Computing

Presented by Seth Lloyd, Professor of Quantum Mechanical Engineering at MIT

(Presented at a physics colloquium at the City College of New York, October 2008.)

The entanglement feature of Quantum is fundamental to the development of Quantum Computers. In the present system of electronic digital computers, instructions are performed sequentially. In the Quantum Computer, most instructions are performed simultaneously, a feature made possible by entanglements—which leads to a significant speedup in performance when compared with today's silicon-based computers.

In Professor Lloyd's book, *Programming the Universe*, source 56, he presents interesting comparisons between the time it takes present computers to solve standard problems and the time it would take a Quantum Computer to do a similar task.

Factoring: Determining all the possible combinations of two numbers that, when multiplied, equal the number being factored. For a 400 digit number:

Classic Computer

No known way (takes too long)

Quantum Computer

With a few 1000-qubit machine: "with ease" (Simulated at ATT Laboratories, 1994)

An update on quantum computers from Physicsworld.com April 7, 2011: In 2005, a group of researchers led by Rainer Blatt at the University of

237

Innsbruck in Austria set a new record by entangling eight qubits formed of calcium ions in an electromagnetic trap. That alone represents 128 or 256 dimensions—and was performed in a few seconds. It would take a week on a classical computer.

3. Description of numerous experiments set up and operated by Professor Anton Zeilinger and his teams—using Quantum technology

(Presented by Professor Zeilinger, a member of the Faculty of Physics, University of Vienna, at a physics colloquium at The City College of New York, April 2011)

The use of entanglements is fundamental to these experiments. Perhaps the most impressive experiment is the establishment of a communication network that is 144 km long (~90 Miles) and, at that time, was the longest free-space link on earth using quantum technology. The link is from La Palma to Tenerife on the Canary Islands archipelago.

The network uses photons to represent bits. It polarizes (or does not polarize) the photon to encode the message to represent the standard '1' and '0'. This was a new use of this technology within communication systems. An important and significant result of the test held on that network showed that quantum communication with satellites is feasible. An important feature of Quantum networks is the incorporation of quantum cryptography, using Quantum Key Distribution (QKD). QKD enables the a shared encryption key to be shared between two parties *without a third party learning anything about that key*—even if the third party eavesdrops on the communication.

A caveat: Researchers at least two universities have shown that undetected quantum hacking *might* be possible in a variety of implementations of quantum key distribution systems. Thus, the overall design of the system is still critical and more studies of this subject are still needed.

Note: In 1989 a team of three distinguished physics professors, consisting of Professors Anton Zeilinger, Daniel M. Greenberger (The City College of New York) and Michael A. Horne (Stonehill College, Easton Massachusetts) wrote the first paper ever on entanglement that involves at least *three* subsystems (particles), i.e., beyond two particles. End of note.

Allan Karson

Thus, the EPR debate—as a debate, it is really no longer alive.

In the long run, I would say that Bohr won. It was Bohr—and Bohr alone, who steadfastly was the world leader in the development of Quantum—from its initial inception, till the time of his death in 1962. His desire to learn, to approach new subjects, to conceive of new concepts—may be unparalleled in the history of Science.

The EPR discussion is one of the most written-about issues in the recent history of physics. Its demise, and (wary?) acceptance of quantum, might be considered a major historical event, comparable to the discovery and first harnessing of the strong nuclear force and the United States' first orbiting space craft. In the early 1930s, Bohr changed the direction of his Institute from Quantum Physics to Nuclear Physics. This change in direction provides an introduction and reason for Part 5 of this chapter.

Part 5 The Bohr and Heisenberg Meetings During World War II, and the Atomic Bomb

You may have seen the theatre-play titled, *Copenhagen*, which first played in England in 1998 and then played in the United States in 2000. Or you may have seen the BBC film based on that play, which has been shown on the PBS network. *Copenhagen* is about the 1941 meeting, during World War II, between Bohr, a Dane, and Heisenberg, a German who is high up in the German war machine. Bohr, whose mother was Jewish, is a citizen of Denmark, a country defeated and occupied by Germany. Heisenberg was a former close and favorite disciple of Bohr, but Heisenberg is now the head physics-scientist for the Germany's attempt to build an atomic bomb.

The meeting took place in Bohr's home, a lovely mansion with expansive green grounds and a lake. Only Bohr, his wife and their 'friend', Heisenberg, are present in the scenes. A person who saw the film may infer from the size of the house and grounds that Danish physicists are very well paid. This is not necessarily so, since it was the Carlsberg Foundation, that derived its funds from the Carlsberg Beverage Company, that helped support Bohr's Institute for Theoretical Physics. But it was not clear how it affected Bohr's personal life.

(To answer that last question about how the support of the Carlsberg Foundation many have affected Bohr's life, the author sent an email to

239

Carlsberg's headquarters in Copenhagen to clarify the issue, and the response explicitly described the background and setting and location of the film' rendition of Bohr's home. It said, "Up until 1997 the founder [of Carlsberg] J. C. Jacobsen's residence house was given as a free residence for life to a man or woman deserving of esteem from the community. Niels Bohr lived in the house from 1931-1962."

In Pais' book on Bohr, it is noted that Bohr opened his home to royalty (Danish and British) and to young physics students, as well. An aside: It is also the author's understanding that the Tuborg Beverage Company supports playgrounds and similar facilities in Denmark. U.S. corporations could learn from the Tuborg/Danish example.

Now, back to Copenhagen. The meeting, as portrayed in both the play and film, gives the impression that Heisenberg 'dropped in casually' to see Bohr. According to source book 26, *Heisenberg and the Nazi Atomic Bomb Project, A Study in German Culture*, neither the play's nor the film's rendition tells the complete truth. The whole visit was a carefully staged visit by the Nazi German military and Gestapo establishment.

A quote from source 27 to show the ambience and emotion in that meeting: (This quote also shows a completely different emotional setting from that was shown in both the stage play and film.)

> This "unforgivable" visit was arranged in the framework of German attempts to inveigle Bohr and his institute into collaboration with the German army of occupation.... Bohr tried with all his powers to resist the efforts of Heisenberg and Weizsacker to involve him publicly in this scarcely disguised collaborationist venture...with Heisenberg blithely justifying the German attack on Russia and the domination of Europe and smoothly rationalizing the Nazi regime. The abysmal climax to the visit came when ...Heisenberg raised the possibilities of constructing nuclear weapons. Bohr was deeply shocked (about this, since) it appeared that (Heisenberg) was seeking to recruit Bohr to the project...

What *chutzpah*!

And continuing a quote from that source: ... some years later at a reception attended by Heisenberg and Weizsacker, Mrs. Bohr burst out from across the room, "I don't care what they say—their visit in 1941 was a hostile act!"

I refer you to Appendix 9 for more on the Atomic Bomb and its early theoretical development prior to, and during, WWII. Most of the early development was performed in Europe. That is, Appendix 9 describes how Lisa Meitner, a Jewish scientist exiled to Sweden, directed an experiment to be performed in her former laboratory in Berlin—that created a fission process. It also describes how a German science writer, who was also a spy, but working for the British, made sure the world knew that fission had occurred in a laboratory. This eventually led to President Roosevelt being informed of that event—and its possible consequences.

This chapter closes with a quote from an article about Niels Bohr in the New York Times in 1962, the year of Neil Bohr's death, which appears in *Inward Bound*, source 11.

'The tentative character of all scientific advance was always in his mind, from the day he proposed his hydrogen atom, stressing that it was merely a model beyond his grasp. He was sure that every advance must be bought by sacrificing some previous 'certainty' and he was forever prepared for the next sacrifice.'

Chapter 11. Three Physicists of the Early Quantum Era: Max Planck, Albert Einstein and Erwin Schrödinger

This chapter discusses the professional and personal lives of three important contributors to 20th Century physics, Max Planck, Albert Einstein and Erwin Schrödinger. But first, once again, a quote by Pais in *Inward Bound*, "…. the rapidity of advances in the decade 1895 to 1905, being unparalleled in the history of science, remains true to this day, especially since one must include the contributions of Planck and Einstein."

Planck was the first person to 'open the door' to Quantum physics. Albert Einstein followed Planck's contributions to quantum, as described in his Special Theory of Relativity. And about ten years later, Einstein replaced the concepts of Isaac Newton of gravity with his General Theory of Relativity,

You probably know far less about Erwin Schrödinger, but you will learn of his versatility and his unorthodox personal life with its many twists and turns. The author considers Erwin Schrödinger to be a 'special person'—and physicist—for many reasons, soon to be described.

11.1   MAX PLANCK (1858–1947) **— The Reluctant Revolutionary**: The first person who stumbled into the World of Quantum.

A quote from Planck's biographer, J. L. Heilbron, in *The Dilemmas of an Upright Man*, source 13.

> "For several years around the turn of the century, Max Planck, was the dean and definition of theoretical physics in Germany. In 1900 his long and relentless battle with an obscure technical problem culminated in a revolution that he neither wanted nor welcomed: He initiated the quantum theory, which conflicts with ordinary physical ideas—but forms the foundation of modern physics."

Allan Karson

In most discussions about Quantum Theory, or even about the cosmos, the name of Max Planck, (Max Karl Ernst Ludwig Planck) and the "Planck Constant" and "Planck Length" and other Planck numbers appear frequently. We will now expand on the discussion presented in Chapter 9 concerning Max Planck's discovery of, what I will refer to in this section as, *Quantum Behavior.* That is, at the time of Planck's actual discover, I consider it would have been too premature to call it *Quantum Theory.* (This is the author's own assessment of the situation, and I have not seen similar discussions elsewhere about this issue of *Behavior* verses *Theory.*)

Planck made his entrance into the world of Quantum through, what we might call, a 'back door'. That door was opened to Planck by the classical physics discipline, *Thermodynamics.* Planck sought to solve a problem in Thermodynamics (heat transfer) and he solved that problem. In addition, in the course of refining his solution to that first problem, he made a new, momentous observation in physics, namely, Quantum Behavior.

1 Background Leading to Planck's Investigation: We start with the investigation of the relationship between electromagnetic radiation and heat energy. It begins with a brief history of Gustav Kirchhoff (1824-1887), who was a highly gifted professor at the University of Heidelberg, Germany, from 1854 to 1875. In 1875 Kirchhoff later accepted the first 'Chair' position dedicated to theoretical physics at the University of Berlin. (The next person to occupy that chair position was Max Planck, beginning in 1889.)

Note: The reader may find that the following text concerning Max Planck similar to that which is described in Chapter 9, a brief history of the development of Quantum Theory. The author considers that Planck's role in the birth of Quantum Theory may not, possibly, be fully appreciated in the *overall* context of Quantum Science—but the introduction of that *momentous new concept, Quanta,* deserves some repetition. End of note

An aside about Kirchhoff: Electrical engineers are well acquainted with his name since Kirchhoff formulated the basic equations that are used for the analysis of electrical direct current circuits. Kirchhoff also produced fundamental equations in the physics discipline of Fluid Mechanics. Many of those equations are still known as the 'Kirchhoff equations.'

243

It is 1859, while teaching at Heidelberg, that Kirchhoff raised the following *key* question: "What is the relationship between the *frequency distribution* of the heat radiated (say, from the Sun, or from a hot kettle) and the *energy* of that heat?" *And Planck provided the answer to that question in 1900, forty-one years later.* Pais says in source 8, "If Planck, Einstein and Bohr are the fathers of the Quantum Theory, then Gustav Robert Kirchhoff is its grandfather."

Heat, *per se*, is measured as a characteristic of matter. Radiation was not yet described as a wave in space, but the radiation, under investigation by Kirchhoff, was known to have frequencies. Hence, what was being sought was the relationship between matter and a form of heat that had a frequency. (Aside: Clerk Maxwell, in 1859, was the first to say that Heat Radiation should be considered to be made up of heat *waves*, the same year that Kirchhoff raised his 'classic' question.)

Now, back to Planck. First, the Data: It was in 1894 that Max Planck developed an interest in attempting to provide an answer to Kirchhoff's question. Fortuitously, Planck had access to a large amount of data that was gathered in what are called 'blackbody' experiments—including that formulated by the "Dean of the Analysis of Heat", Prussian-born William Wien (1864-1928).

Note: The following is a slightly expanded version of the description of a blackbody that was provided previously. A 'blackbody' is a universal term used to describe a closed system (an insulated box, for example) whose six inside walls are both perfect absorbers and perfect emitters of electromagnetic radiation. All the walls are at 90 degrees to one another. The 'black body' box also has a small hole from which radiated heat (in frequency form) could be introduced into the box—and samples of the frequencies could come out of the box and be measured.

Planck's data had been developed by his colleagues at the University of Berlin's physics laboratory and at Germany's future Bureau of Standards, the *Physickalisch-Technische Reichsanstalt*, which was probably the world's best-equipped physics laboratory. That experimental data, along with Planck's ingenuity, determination and perseverance, enabled him to make the following *monumental* physical finding (or, hypothesis): *At fundamental (small) levels, the Universe operates according to Quantum Theory.*

Allan Karson

As mentioned in the earlier discussion about Planck in Chapter 9, the year 1900 was an important year at the *Reichsanstalt*. During early that year, new experiments were performed that provided new data in the lower frequency, infrared range—a critical frequency range in Planck's analysis. This range had not previously been analyzed in detail by any organization.

In the summer of 1900, when Planck learned about the new data that had been gathered at the *Reichsanstalt*, he intensified his investigation—and completed it by December 1900. Obviously, this new data provided Planck with the insights that were necessary to develop the complete answer to Kirchhoff's question—i.e. energy and frequency—over all frequencies.

We can also assume that from when Planck started this investigation in 1894, he probably developed or learned many of the mathematical procedures and 'tricks' that might be needed to modify equations that he had been developed in seeking the answer to Kirchhoff's question.

For example, there was already an historical-technical base to Planck's attempt to answer Kirckoff's question: The physicist, William Wien (1864-1928) had earlier performed measurements similar in purpose to Planck's later work, but at the time, there were some disagreements over some of Wien's findings. It may be said that Planck sought to rectify and produce 'the correct' theory. Note: Wien was awarded the 1911 Nobel Prize for this class of work, and his work is still discussed and highly respected.

Also, during this period, Planck developed ideas about 'which were the important varied constants' in thermodynamics and physics, and Planck's *own*, universal constants that might be involved in his formulae. (We shall discuss those important constants again, after we discuss Plank's answer to Kirchhoff's question.)

Planck developed the answer to Kirchhoff's question (repeated here):

What is the relationship between the *frequency distribution* of the radiation and the *Energy* of the heat that is generated from that electromagnetic radiation—over a wide range of frequencies?

Planck presented his results at two meetings of Physical Society of Berlin, as follows:

245

On October 19, 1900, he presented his 'first' draft of the answer to the question. On December 14, 1900, he presented the final energy equation, known from then on as the Planck Equation. He also presented his observation of quantum behavior within the black-box experiment.

Concurrent with those presentations, Planck prepared two articles for the *Annalen der Physik* that form the basis for his observations of quantum behavior:

- On the Theory of the Law of Energy Distribution in the Normal Spectrum, 1900, and
- On the Law of Distribution of Energy in the Normal Spectrum, 1901

The following presents what Planck did in those two steps. The description is from what Planck wrote in his *Scientific Autobiography and other papers*, source 1, and in the two cited articles in the *Annalen der Physic*.

2. Step 1, October 19, 1900 Presentation: A *Premature* Answer to Kirchhoff's Question:

Planck presented an equation that consisted of two terms, one for low energy and one for high energy. (Planck, however, did not maintain an interest in the idea of separate low and high energies; he later dropped the idea.) Also, while he had actually developed a significant portion of his subsequent thoughts about quantum behavior *before* his October 1900 meeting, he did not mention anything at that meeting about quantum behavior being present-and-measured in his experiments.

Note: The author makes this distinction between the differing accomplishments of the two distinct steps, October and December, since some source books and web sites indicate that Planck accomplished much more than he actually accomplished in Step 1. Or, they imply more than he actually found or said during Steps 1 and 2. And, as in most of the cases, they do not indicate that there were, actually, two distinct steps in his presentation. End of note.

3. Step 2, December 14, 1900 Presentation:

1. Provides his *Final* Answer to Kirchhoff's Question
2. Presented His Observation of Quantum Behavior

Allan Karson

Planck presented the revised, final version of his equation:

$$\text{Radiated Power} = (2\,h\,v^3)/[c^2\,(e^{hv/kT} - 1)]$$

where

h is his newly derived Planck constant, called, the elementary quantum of action,

k is the value of Boltzmann constant that was determined by Planck. (Boltzmann had accorded k the value of '1'; Planck determined the appropriate number.)

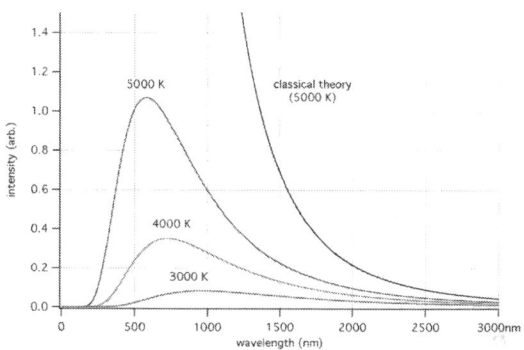

Graph: 27 Planck Amplitude vs. Wavelength, according to Planck Equation

4. The Boltzmann Entropy Equation and Planck's Observation of Quantum Behavior

In his autobiography, Planck specifically says he wanted to expresses his sought-after equation, i.e., the answer to Kirchhoff's question, by using 'first principles'. That is, Planck's goal was to develop an equation that was *not to be based on any assumptions or experimental results*—as the Planck equation of *Radiated Power*, which is shown above, is based.

To do this, Planck considered the *Entropy* of the system. At this time-period in physics, most physicists did not understand or appreciate the study of Entropy and its significance—but Planck was an expert in the understanding and calculations of Entropy. This is mainly due to the fact that the study of Entropy was the basis for his doctorate thesis. Thus, in doing this work, Planck also became a leader in drawing attention to concept of Entropy within the world of physics. There will be more, later, about Planck's relationship with Entropy—and with non-believer fellow-physicists.

To attain his non-assumption or non-experimental equation, Planck used Boltzmann's Entropy equation, which was described in Chapter 2 and is repeated below. Up until then, the use of the entropy equation was limited by Boltzmann and his colleagues to examining *only the entropy of a gas*, by their looking at the state of the gas's molecules.

Planck, however, adopted it to look at states of *individual frequencies*—and he would define the actual value of *k*—not being '1', as Boltzmann had accorded it that value. Since Planck continued to use the name of '*k*', and continued to call it the 'Boltzmann constant', it was sometimes not clear who is the 'deriver' of k in Boltzmann equation. But Planck did derive it for The Boltzmann Entropy equation,

$$S = k \log W,$$

where S stands for Entropy, and *k*, Planck's new constant—whose value was determined by Planck, and W, that stands for Probability.

Note: *k* is *still* known as the Boltzmann constant—and is considered "a bridge between macroscopic and microscopic physics". The present numeric value of the constant 'k' is

$1.38064852 \times 10^{-23}$ m^2 kg/ s^2 K, where 'm' represents meters; 'kg', kilograms; 's' is seconds, and K, temperature in Kelvin.

The development and use of these mathematics by Planck was intense, involved a *great deal* of formula, and was somewhat complicated in reasoning, and, therefore, is not described in this text. (For those persons who would like to know, or learn, the actual, step-by-step mathematical approach that Planck used to make his determination of Quanta, I refer you to Source 8, *Subtle is the Lord, The Science and Life of Albert Einstein*, by

Abraham Pais. It starts in chapter 19, on page 364, and the detailed presentation of what Planck did, continues to page 372.)

Let us now consider the significance of Planck's findings, using the Boltzmann equation. Through his development of the Boltzmann equation of Entropy, or $S = k \log W$, Planck recognized that radiant energy *can*, or *may* occur in steps, which he called *quantum* steps!!

And, as Planck wrote in his *Annalen der Physik* 1900 paper, (italics added):

> "If an (object's) E (energy) is considered to be a continuously divisible quantity, this distribution is possible in infinitely many ways. We consider, however—this is the most essential point of the whole calculation—E (energy) to be composed of a *well-defined number of equal parts* and use thereto the constant of nature $h = 6.55 \times 10^{-27}$ erg seconds."

In the above the phrase, well-defined number of equal parts, describes what Planck later refers to as, 'Quantum Behavior'. That is, there are 'parts' (or steps)—and not a smooth, continuous curve. Thus, *The Birth of Quantum Theory*. And h is known as Planck's constant.

(An example of the *Energy* of a Quanta: If the value (or amplitude) of a quanta is A, then A x h is the amount of energy in that quanta. Thus, for a quanta having a value of 20, its energy is

$$20 \times 6.55 \times 10^{-27} \text{ erg sec, or } 131 \times 10^{-27} \text{ erg seconds.}$$

(An 'erg' represents a unit of work or energy, equal to the work done by a force of one dyne when its point of application moves one centimeter in the direction of action of the force. Thus an 'erg' represents a very small amount of work.)

Digression 1: The author would like to make a 'respectful' comment concerning the generally *significant—but unattributed* contributions implicitly made by Ludwig Boltzmann (1884–1906) to the development of Quantum Theory. Planck incorporated Boltzmann's Entropy Equation, as just described. And, then, there is also the use by Einstein of Boltzmann's statistical formulas when Einstein determined that the light photon could be considered to be either a particle or a wave, cited in Chapter 2. End of digression.

Planck's presentation on December 14, 1900 to the Berlin Physical Society was the introduction of a new and important era in physics—and, ultimately, to the world at large. As Max von Laue, one of the attendees to that historic December meeting, later said, "That was the birthday of Quantum Theory."

Reviewing what Planck effectively did, *'this was the first time anyone declared* that the world of Physics/Nature acts with a different set of physics rules from what was considered 'normal' or 'classical' or 'smooth, continuous' physics.'

Today, we consider that the range of 'small distances', as applied to Quantum Theory, to extend from 10^{-7} (or .0000007) centimeters to 10^{-33} (or . 000,000,000,000,000,000,000,000,000,000,001) centimeters. That is, in that smaller region, the world exhibits 'Quantum Behavior.' We can assume that Planck never had the occasion to explore that complete range of distances due to the limitation of instrumentation-measuring equipment that was available during his time.

The author has not attempted to describe the mathematical and physical rules that Planck used—and sometimes broke, in deriving the Boltzmann equation (and its constants), the Planck equation—and in his hypothesizing Quantum Behavior. I will, however, present two quotes, the first by Professor Pais in source 8, written *circa* 1982, and the second by Niels Bohr, written in 1938, that describes the overall importance of Planck's findings.

> "Thus the only justification for Planck's two desperate acts was that they gave him what he wanted. His reasoning was mad, but his madness has that divine quality that only the greatest transitional figures can bring to science. It cast Planck, conservative by nature, into the role of the reluctant revolutionary... he made the first conceptual break that has made twentieth century physics look so discontinuously different from that of the preceding era. Although there have been other major innovations in physics since December 1900, the world has not seen since a figure like Planck." (By Pais.)

"Scarcely any other discovery in the history of science has produced such extraordinary results within the short span of our generation as those which have directly arisen from Max Planck's discovery of the elementary quantum of action. This discovery has been prolific, to a constantly increasing degree of progression, in furnishing means for the interpretation and harmonizing of results obtained from the study of atomic phenomena, which is a study that has made marvelous progress within the past thirty years." (By Bohr.)

This discussion of Planck's masterful contributions closes with a review of the general method Planck used: Within both steps above, Planck attacked the problem by *considering the Entropy* of the system. Planck was an expert in the understanding and calculations of Entropy. From Planck's autobiography, source 1, are Planck's comments about his and his associates' (poor) knowledge of Entropy (bullet-format added for clarity):

- In an irreversible process the terminal state is… more important that the initial state, Nature preferred it to the latter (the starting state). I saw a measure of this 'preference' in Claudius' entropy.
- I found the meaning of (Entropy) in every natural process the sum of the entropies of all bodies involved in the process increases.
- I worked out these ideas in my doctoral dissertation at the University in Munich, which I completed in 1879.
- The effect of my dissertation on the physicists of those days was nil. None of my professors at the University had any understanding for its contents.
- But I found no interest, … even among the very physicists who were closely concerned with the topic…. Kirchhoff disapproved of its contents… I did not succeed in reaching Clausius (A German professor in the field of thermodynamics, as per Chapter 2.)
- However, deeply impressed as I was with… this task, such experiences could not deter me from continuing my studies of entropy, which I regarded as next to energy the most important property of physical systems.

Following his producing an answer to Kirchhoff's question and his discovery of Quantum Behavior, Planck spent the following eleven years trying to derive his formula for blackbody radiation using a 'classical physics' explanation. He did not succeed, however, and he finally gave up on his quest to unite Classical Physics with Quantum Physics.

An example of Planck remaining a classical physicist occurred in 1913 when Planck introduced Einstein to his colleagues. This was on the occasion of Einstein's recent appointment to the Kaiser-Wilhelm Institute for Physics. Planck essentially said that Einstein had one of the most fertile minds of the 20th Century as evidenced by his *Theory of Special Relativity,* so he could be forgiven for his ridiculous theory of the photoelectric effect.

As we have seen and will see, many physicists did not accept did not accept Planck's idea of Quantum, just as many, later did not accept Einstein's idea, published in 1905, of the duality of what light is made up of—a wave, or a particle, now called the *photon.* Such disagreements became a testy issue when the award of the Nobel Prize was being considered during the beginning of the 20th Century. Thus, Planck essentially 'had to wait' from 1900 until when he was awarded the Nobel Prize in 1918 for his discovery of Quantum. (Einstein, similarly, waited a relatively long time, from 1905 until 1921, to be awarded his Nobel Prize for his theory of the nature, or characteristics, of Light.)

Today we would agree Planck was brilliant, dynamic and creative, but a died-in-the-wool, classical physicist.

4. Planck's Persona

An overall description of Max Planck: "An unblemished character with an international reputation." There are many stories and personal and professional accolades that can be found out about Max Planck in various books and on the web. They abound with stories of Max Planck's life, his warm family environment, what he contributed to the world of physics following his initial definition of 'his world' of measurement—and the various institutes and awards that were named in recognition of his work.

Planck's held the position Professor of Physics, University of Berlin from 1902 to 1926. As noted previously, Kirchhoff had held that same position from 1875 to 1887, was later followed by Planck in 1902, and later followed by Schrödinger—the last two being major contributors to Quantum Theory.

252

Allan Karson

Planck was considered to be an excellent lecturer. From source 13, *The Dilemma of an Upright Man, Max Planck as a Spokesman for German Science,* by U. L. Heilbron, "People used to the disjointed and often bumbling of the English style of instruction were particularly impressed. ... The chemist, J. R. Partington, judged Planck to be the best lecturer he had ever heard. The Indian physicist, D.M. Bose (of the Bose-Einstein theory), who had studied at Cambridge...called it a "revelation" in Berlin.

In addition to all those plaudits, Planck was the first accomplished physicist to recognize the outstanding work of Albert Einstein. This was during the period when Einstein was unknown to the world, working quietly in the patent office in Bern, Switzerland. And this was *before* Einstein had his theories published in the German journal, *Annalen der Physik* in 1905, Einstein's *Miraculous Year.*

The following describes a few examples of Planck, as being a wonderful human being. The first example is based on his relation to Albert Einstein and how he helped launch Einstein in the world of physics.

According to source 13, U. L. Heilbron, Planck's biographer:

In 1913 Planck, and his colleague, Walther Nerst, put together a package to entice the mercurial and unconventional Einstein into the staid circle of Berlin physicists: an ordinary membership at high salary in the Berlin Academy, a professorship without teaching obligations, at the University (of Berlin), the directorship of the nonexistent (to be built) Kaiser-Wilhelm Institute for Physics.

Planck had a large house and spacious garden in a select superb of Berlin, countrified Grunewald, favored by university professors. He was also a gifted musician. Planck's house was full of music. He also had many guests. He played piano, organ and cello, and composed songs and operas. He entertained Einstein at his home, where Einstein played his violin."

I also cite the warmth that Dr. Lise Meitner, a physicist, expressed concerning Planck, described in source 21. Dr. Meitner was a young physicist, working at the time, under the overall guidance of Planck. Meitner's biographer, Professor Ruth Sime, writes, "Of all the people in

253

Lisa's circle in Berlin, the one she admired with near-reverence was Max Planck."

And additional, important, excerpts by U. L. Heilbron in source 13—going from great stature to unhappy years, and ending in utter decline:

> Max Planck also was a staunch believer that woman should not be denied the right to study at a university. His support for women's rights may have contributed to the fact that by 1913-1914 there were 770 women studying in Berlin and throughout Germany there were almost 3,500, about 6% of the university population.

> Following the defeat of Germany in World War I in 1918, Planck was essentially given the role of leader of the German scientific community, in its attempt to get back into the world of the European scientific community. The Europeans could not forgive the German scientific community for the following reason: When World War I started, ninety-three German intellectuals wrote an open letter, titled, "Appeal to the Cultured Peoples of the World" ... It declared the leaders of German art and science to be at one with the German army and refuted the charges that the German army had committed atrocities in Belgium.

> The non-German European scientists were not pleased with that letter and they remembered it well in 1918. Thus, it was up to Planck to made amends and he devoted a considerable amount time effort to this cause of the German scientific community. ....After the rise to power of Hitler, becoming Chancellor or Germany in 1933, Planck, being the leader of German science, had to work with him. This was a distasteful experience and way of life for Planck.

Continuing J. L. Heilbron's book, source 13, after Hitler had shown clearly his anti-Semitic hate, "Planck's most notable move behind the scenes was an interview with Hitler, probably... in May 1933. Planck hoped to convince the Führer that the forced emigration of Jews would kill German science and the Jews could be good Germans. ... Planck's recollection, Hitler replied that he had nothing against the Jews, only against communists, and then flew into a rage."

With all the honors he was given, Max Planck had a difficult and sad personal life. He lost his first wife of twenty-three years of marriage, but remarried. He had five children from his first wife, but two of them died under various circumstances. During World War I, his son, Karl, was killed at Verdun. During World War II, his home was destroyed in an air raid. Not a thing could be saved—his library, diaries, correspondence...

In late 1944 Erwin, Planck's son, was judged guilty of complicity in the famous failed attempt on Hitler's life on July 20, 1944, which was led by Count von Stauffenberg. The Nazis executed Erwin; it is said that the command to execute him was given by Hitler. This brought the number of deaths to four, all children from his first wife.

Erwin had not been in the conference room with Hitler. There were actually many Germans involved in planning the conspiracy, and only a few of them were in the conference with Hitler.

I end this section on Max Planck with another edited quote from Heilbron:

After Erwin's death, Planck took no pleasure in life... Planck's vertebrae fused, literally doubling him over in pain. Scarcely able to walk, the old man had to leave his sanctuary in Rogatz, (Germany). He and his wife hid in the woods and slept in haystacks... At last came the rescue by American officers alerted to the Planck's probable plight by Robert Pohl, professor of experimental physics at Göttingen. Utterly destitute, the Plancks found refuge with a niece in Göttingen, which had not been damaged by the war. ...After five weeks in a hospital, Planck could shuffle about...

(After the war ended) ...another long stay in the hospital, he accepted ... the invitation of the Royal Society of London to participate in July 1946 as the only German, in a belated commemoration of the three hundredth birthday of Isaac Newton.

There is much more about this humane, venerated man in source 13, *The Dilemmas of an Upright Man, Max Planck as Spokesman for German Science*, J. L. Heilbron, University of California Press, 1986.

Part 2. Albert Einstein (1879-1955) Theories That Changed Our Understanding of the Universe

Hundred of books and articles have been written about Albert Einstein by the world's best biographers, scientists and scholars. He is mentioned in almost any book that deals with physics. In the year 2000, *Time Magazine* named him the *Person of the Century.*

I hope to give you an appreciation of what he revealed about the Universe. I also hope that you will go on to read about this phenomenal person, whose complete life is described in many biographies—some of which are identified in the source list at the end of this book. In that list, I identify those that may be considered non-technical biographies.

First, there certain things about Einstein from which you can say that he was an ordinary guy. He liked to play jokes and he liked only certain subjects in school. He was not a great student, probably due to the way subjects were presented—but he was very bright. He did not suffer fools. He liked to sail, loved music and played the violin and he liked girls. And later, when he became a professor, he liked to be with his students, to talk about physics and learn what they were thinking. A regular guy. But the thing he liked to do the most was to question himself—and to think.

There are some biographers of Einstein who imply that he was 'just an ordinary student.' While there were certain subjects he was not interested in and only received average grades, that should not lead us to believe that his peers and teachers did not recognize his intelligence.

To help set the record straight, from Einstein's teen-years on, he was considered to be *very* intelligent—probably at genius level. In Ronald W. Clark's 1971 biography on Einstein, (source 4) there is a quote of a young medical student who was guiding Einstein in his reading material during Einstein's youth:

> "After a short time, a few months, he (Einstein) worked through the whole book of Spieker, (a popular textbook on math.) He thereupon devoted himself to higher mathematics, studying all by himself Lubsen's excellent work on the subject.... Soon the flight of his mathematical genius was so high that I could no longer follow. Thereafter, philosophy was often a subject of our conversations.

256

I recommended to him the reading of Kant. At that time he was still a child, only thirteen years old, yet Kant's works, incomprehensible to ordinary mortals, seemed clear to him. Kant became Albert's favorite philosopher after he read through his (Kant's) *Critique of Pure Reason* and the works of other philosophers."

Einstein's Early Life-History

Most Einstein biographies describe that he got his first professional job at the quiet Patent Office in Zurich (1902-1909). He actually had problems finding that job, or any job, since his scholastic records were not the greatest, and some of his professors would not give him a good reference. In addition, he was an independent person—not the best characteristic to show to potential employers.

But consider that he may have sought a job that would give him time to allow him to think. In that case, the Patent Office was an ideal for him, since he could be reviewing and evaluating the *creative side* of the (local) world. He was assigned to evaluating patents submissions, mostly for electromagnetic devices, which was ideal for him. In his free time he developed some of his most famous theories, and he did this without much contact with other professionals in the world of physics, nor with references to the available physics texts.

Before delving into those three articles, it is necessary to remedy a misconception. It appears many people believe, and also some science writers imply, that Planck's first inquiry to Einstein 'came out of the blue.'. This is not so. The refutation of that false impression can be found in three sourcebooks: source 4, *Einstein, the Life and Times*; source 8, *Subtle is the Lord*; and source 52, *Einstein's Miraculous Year.*

According to source 4, (*Einstein, the Life and Times,* by Ronald W. Clark a great, wonderful biography of Einstein) "The two men (Einstein and Planck) had been in correspondence since 1900. At first, it was desultory, but Planck was increasingly impressed by the young man who had boldly taken his Quantum Theory into fresh fields."

Note. According to source 8, the scientific biography of Einstein by Abraham Pais, Einstein's *first* published article was in 1900 edition of the *Annalen der Physik*, and it discussed the topic of intermolecular forces.

Since Planck had a working relationship with the *Annalen* at that time, as an adviser, Planck may have known Einstein since the publication of that article. End of note.

1. Einstein's Miraculous Year—1905

In 1905 Albert Einstein, age twenty-six, wrote three monumental papers,— each in three different domains of physics. Each, by itself, could easily be a candidate for a Nobel Prize. All three were published in *one* issue of the highly respected German physics journal, *Annalen der Physik*, 1905. (The list of papers published in that *one* 1905 edition of the *Annalen der Physik* is also presented in Appendix 10.

Einstein actually wrote and published five physics papers that year. The book, *Einstein's Miraculous Year*, source 52, is a relatively small book of 195 pages, and contains all five of those articles. The first of those papers is his doctoral paper, which was not published in the *Annalen der Physik*. Its title is *A New determination of Molecular Dimensions*. The other four were published in the *Annalen der Physik*, with two of the original papers being lumped together into one paper. (See below for their arrangement in the *Annalen*.)

Einstein's Miraculous Year was edited by John Stachel, an 'Einstein scholar'. Stachel notes that in 1894 Max Planck had acquired an official responsibility for all of theoretical physics for all of Germany. Planck, therefore, would, presumably, be an advisor on such matters to the *Annalen der Physik*.

Digression: I was interested to know the subject of a pre-1905 paper, and I did eventually learn about it. This brings me to my brother, Marvin Karson. My brother, Dr. Marvin Karson, Professor Emeritus of Statistics, University of New Hampshire, had (fortuitously) sent me a modern review of a pre-1905 paper by Einstein, and that review provided specific information concerning the possibility of a pre-1905 Planck-Einstein connection.

The review of Einstein's pre-1905 paper appeared in the American Statistical Association's journal, The *American Statistician*, November 2007, and was written by Dr. Boris Iglewicz, (1940 - 2015). Dr. Iglewicz was Director of the Biostatistics Research Center and Professor of Statistics at Temple University.

Allan Karson

Einstein's pre-1905 paper was written in 1900 and deals with molecular forces and uses statistical techniques—a subject that might also have been of interest to Planck. Dr. Iglewicz writes that Einstein's paper "was submitted to deal with molecular forces." There were suggestions that this paper was reviewed by Max Planck and, thus, it may have opened the door for Einstein's other, later, 1905 publications in the *Annalen der Physik*.

To continue the discussion of Einstein's life-path, according to source 4, the first time Einstein and Planck met was at a conference in Salzburg, Austria, in 1909. In the years following that meeting, Einstein's professional life changed significantly. For example, he developed relationships with other physicists such as Madame Curie, Arnold Sommerfeld and Max Born. And he became a Professor at the University of Berlin in 1914, a position proposed by Max Planck—and he also had a brief meeting with the German Kaiser, Wilhelm II.

Now, back to the three seminal articles (or, equivalently, papers) that appeared in the 1905 volume 17, of the *Annalen der Physik*. A list of the articles is presented below. The italicized words used in this list are those that Einstein used to describe each article, translated into English, as found in source 52, *Einstein's Miraculous Year: Five Papers That Changed the Face of Physics, Albert Einstein*. The purpose of each article is described in the parenthetic phrases.

- *1. On the Motion of Small Particles Suspended in Liquids at Rest* (To show the existence of atoms.)
- 2A. On the Electrodynamics of Moving Bodies (To all observers, the speed of light is the same, and the size and mass of a body changes as the body approaches the speed of light.), and

  2B. Does the Inertia of a Body Depend on Its Energy content? (This article introduces the equivalency of matter and energy through the well-known equation, $E = mc^2$.)

- On a Heuristic Point of View Concerning the Production and Transformation of Light (Einstein's hypothesis that light exists as particles *and* as waves.)

Before these articles are considered, I would like to clarify the word 'articles'. It actually represents *subjects* or *theories* that Einstein wrote

259

about. That is, there were four subjects and these were described in the series of three articles. Since authors use the word 'articles' and the word 'papers' for the same purpose, the two, equivalent, terms are used here.

The modest-size book starts with his PhD. Doctoral dissertation, *On the Determination of Molecular Dimensions.* This is followed by **Article 1**, which appeared in volume 17 of the *Annalen Der Physik,* wherein Einstein defined an experimental solution to the problem of predicting 'the effect on molecules of pollen grains, when suspended in a liquid solution.' This paper provided the first quantitative treatment of the cause of what is known as *Brownian motion.*

Note: An experiment was later performed in the laboratory by the French experimentalist, Jean Baptiste Perrin (1870-1942), and confirmed Einstein's predictions. (Einstein visited Perrin's laboratory during his first visit to Paris, *circa* 1914.)

From *Encarta Encyclopedia*: Perrin observed the irregular wiggling of pollen grains suspended in a liquid (a phenomenon called Brownian motion) and correctly explained that the wiggling was the result of molecules of the fluid colliding with the pollen grains. This experiment showed that the idea that materials were composed of real molecules in thermal motion was in fact correct. (Note: The Encarta article neglects to cite an additional result of Perrin's work: Perrin *actually counted the number of molecules* in a drop of water.)

The Second Article was titled *On the Electrodynamics of Moving Bodies*. It later came to be known as The *Special Theory of Relativity*. This article actually consisted of two different, but related, parts. In the following, the first part of the original article is denoted as Article 2A, and the second part as Article 2B.

In Article 2A in the *Annalen*, Einstein introduced the concept that the speed of light is *always* the same as seen by any observer. That is, any viewer, measurer or traveler, will observe that ONE speed—no matter how fast he/she is moving. This was a truly revolutionary theory and it changed the world of physical science forever.

First, a possible question that you might ask: What provided Einstein with the idea that the speed of light is constant? Did he, or someone else, ever observe this phenomenon? Or did he, or someone else, ever measure this phenomenon?

The answer is a definite No! The answer is that Einstein saw that in Maxwell's system of four equations dealing with electromagnetism, in the fourth equation the speed of light is constant in a defined medium such as in the space in our Universe. (Air is a medium and it is different from water, from glass. Thus light does move at different, but constant speeds, according to the medium.) There are no changes or modification or accommodation that could 'adjust' that speed, no matter what is the speed of the 'frame of reference of the system.'

A (long) digression: Wolfgang Pauli (1900 – 1958), winner of the Nobel Prize in physics in 1945 for his *Exclusion Principle*, was clearly identified as being at the physics-genius level while he was still at University. At the age of twenty-one, he was chosen by the eminent German Professor, Arnold Sommerfeld, then editor of the *Mathematical Encyclopedia*, to "give a complete review of the whole literature on relativity theory existing at that time (1921)" for that encyclopedia.

This included both Special Relativity and General Relativity. Pauli' work was later published in 1958 as a freestanding book, titled *Theory of Relativity*, source 2. I recommend it for its *complete* presentation on Relativity, but, at the same time, I recommend it only for the advanced-in–physics reader.

Also, according to Sommerfeld, the particular volume in the *Mathematical Encyclopedia* in which Pauli's article appeared, was devoted to physics and physics' applications. Pauli's article starts with the works of Fitzgerald, Lorentz, Larmor, and Michelson.

I quote from a note written by Pauli in his introduction to his 1958 book, source 2. The note clearly relates the work of Einstein to *the flow of physic's history*. (Italics added in previous sentence and parentheses added in following.)

Pauli wrote,

"By its epistemological analysis of the consequences of the finiteness of the velocity of light... the theory of special relativity was the first step away from naïve visualization.

Without this general critical attitude, (in) which (Einstein) abandoned naïve visualizations in favor of a conceptual analysis of the correspondence between observational data and the mathematical quantities in a theoretical formalism, the establishment of the modern form of quantum theory would not have been possible.

Einstein also showed that nothing could move faster than (or even near) the speed of light. He also showed that if two groups of people (one going a normal speeds, and the other in a rocket ship, for example, going near the speed of light), their respective clocks would go at different rates. They would show different times! Each person has her/his own clock; each travels in one's own frame of reference.

And, as Pauli says, so succinctly in his book, "The velocity of light is independent of the motion of the light source." End of Pauli digression.

At the time, 1905, one could say that this new theory of Special Relativity came into the physics world as a bombshell.

Special Relativity is now proven and accepted. From the theory's inception, however, it was considered so revolutionary in concept that there is a whole literature, folklore, background stories and intrigues associated with it. Hence, a little more space is devoted to this theory than to the others.

To help you understand that there are different moving systems, consider what is called a 'Frame of Reference.' A Frame of Reference is the space around you in which you are all standing still or moving at a constant speed —a car, a railroad train, an airplane or a space ship. It is what you and your colleagues and your friends are all moving in.

No matter what (constant) speed your fame of reference moves at, you—the 'local observer', will not observe or detect any change in size or mass. I refer you to Chart 27 for a list of 'the rules' of Einstein's theory concerning the speed of light. I also refer the reader to Appendix 11 for an explanation of what is a Galilean 'Frame of Reference'—and its history.

262

Chart 27 Some of the 'Rules' in Einstein's Theory of Special Relativity
(Note: To insure that all points are clear, there are certain redundancies in this chart.)

The speed of light is the same in all constantly moving, non-accelerating (Galilean) Frames of Reference. While moving in a Frame of Reference, you can never detect the speed of light at any other speed than 186,000 miles per hour. Thus,

- You will *always* see light going away from you at the speed of 186,000 miles per hour.
- That is, even if you travel at a speed that you believe is close to the speed of light, and if you measure how fast light is going away from you, it will be measured at 186,000 miles per hour, going away from you.
- An observer who is standing still and is watching you travel near the speed of light will see you have shrunk along the direction of travel and you will look heavier to that observer.
- Your wristwatch, or any other clock that is in your vehicle (Frame of Reference), is going slower than the observer's. You will not know that it is going slower.
- Your body's timing rhythms (heart rate, breathing…) slow down as compared to a non-moving observer. You, however, do no observe any slow down in your body's timing.
- Even if you are traveling at speeds close to the speed of light, you will see yourself as looking normal.
- When you are traveling near the speed of light, the mass of the vehicle you are traveling in will increase. You will not detect this change.
- When you will return from a trip traveling at speeds close to the speed of light, your body slows down, (your wrist watch goes slower than the person standing still on Earth), you will not have aged as much as the observer who did not go with you in your (fast-moving) frame of reference.

The following chart, (still referred to as part of chart 27) shows how much a fast moving 100 foot spaceship (Column 1) would contract in the direction of travel, (Columns 2 and 3) and how much heavier it would become (Column 4).

For example, the one hundred foot space ship traveling a speed of only 10% of the speed of light (18,600 mph) would shorten by .5% and would gain mass of .5%.

But the one hundred foot space ship traveling a speed of only 40% of the speed of light (74,400 mph) would shorten by more than 8% and would gain mass of 9%.

Column 1	Column 2	Column 3	Column 4
100 foot spaceship is moving at speed of	Spaceship is Percent of Start-Length	Spaceship's Length in feet	Spaceship's Mass is increased:
Standing still	100%	100.00	0%
10% of c	99.5%	99.50	100.5%
20% of c	98%	97.98	102%
40% of c	92%	91.65	109%
60% of c	80%	80.00	125%
80% of c	60%	60.00	167%
90% of c	44%	43.59	229%
98% of c	20%	19.90	503%

Note: As explained below, the equations that form the basis for the calculations of the numbers in the above chart are frequently referred to as the "Lorentz Equations", even thought both Lorentz and Einstein developed them independently from one another—and during the same general time period.

Appendix 13 presents those equations in detail, along with a methodology to enable you to make the calculations in a non-complicated way. End of note.

Allan Karson

A frequently asked question concerning The Lorentz Transformation Equations and the Einstein Transformation Equations: *Are They the Exact Same?* The answer: Yes. But the reason for the development of the equations is different. Einstein developed his system of equations to determine the transformations that occur when a frame of reference approaches the speed of light, c. This was different from Lorentz's purpose.

It is a twist of fate that another group, consisting of two physicists, had developed, independently, what is the *same system* of equations—but for a different purpose. These are:

Hendrick Lorentz (1853 – 1928), a Dutch physicist who held the chair of theoretical physics as the University of Leiden (and a dear friend and sometime-mentor to Einstein), and

Joseph Larmor (1857-1942), born in Ireland, Lucasian Professor of Mathematics at Cambridge University and later was Member of the English Parliament from 1911 to 1922, representing Cambridge University. Their project is discussed in Appendix 13.

Even though they did their development of the equations in a time frame that is close to and just before that of Einstein's development period, there has always been some confusion over who knew what and whether any team benefited from work done by the other.

At the time of this work, *circa* 1905, Einstein did not know Lorentz, personally, but they later developed a warm friendship, mutual professional respect and mutual admiration.

We can resolve this 'mix-up' by referring to a book written by Max Born, source 5. Born cites the fact that Einstein developed his set of transformation equations before 1905—and they were eventually published in the *Annalen der Physik* in 1905. And Lorentz published his system of equations in the same year, 1905. Hence, we assume Einstein did not have knowledge of Lorentz's transformation equations. Einstein's 'prime' biographer, Pais, also says that Einstein was unaware of the other group's similar equations

But to complicate matters even further, the physics community continues to refer to Einstein's transformation equations as Lorentz transformations—

even though the community knows of the two different sources for the same system of equations. This is understandable however, since the two systems of equations are identical.

Refer to Appendix 12 for a more complete and interesting discussion of purpose of the Lorentz and Larmor experiments. In their case, the ether was being investigated by the both of them, and there also previous histories of their trying to prove the existence of the ether. That appendix will also tell you about the well-known Michelson-Morley experiment that was trying to (1) justify the presence of the ether in space and (2) show its effect on spatial measurements, while Einstein was showing what happens to measurements while the speed of light is held constant under *ALL* conditions.

Digression. At the time, *circa* 1905, the esteemed French physicist, Jules Henri Poincaré (1854 – 1912) had conjectured that there was no such thing as *absolute time*. Source 8 states that Einstein definitely knew of Poincaré's theory and this may have provided Einstein some (moral? but not technical) support to the introduction of multiple time-clock systems and the unequivocal declaration that time is another dimension. End of Digression

Article 2B, Part of Einstein' Formulation of $E = mc^2$

This article also considers the speed of light, which is usually referred to as 'c', but it considers light in a very different way from article 2A. In this part of article 2, Einstein established the analytic relationship between the energy (E) of a body and the mass (m) of that body. He developed the universally known 'explosive' equation: $E = mc^2$

Aside: For ALL its power, Article 2B is incredibly short in length; it could possibly fit into one 8" by 10" page!

In 1905 Einstein derived the formula that was based on both his detailed analysis and his feeling about the relationship and equivalency of matter and energy. But he was provided with the prior insights and speculations afforded by a host of physicists.

For example, even in Newton's time, the 1600s, physicists had discussed the relationship between energy and mass. But the formula or proof of the relationship showing the specific relationship had never been made—until, by Einstein.

If we jump immediately to 1905, we have just seen how Einstein was also publishing a paper about the speed of light. In that theory, he showed that there was a definite, but undefined, relationship between energy and mass.

Remember that if a person is going tries to go fast, near the speed of light, that act of moving very fast, explicitly requires more energy to go that fast. And that additional energy causes the person to become heavier. More mass. Thus, *some* relationship exists between mass and energy. But what was that definite or specific relationship?

A conjecture: Another way he might have looked at this equation is to consider the kinetic energy (moving) of a moving mass. In a high school first-year physics course we learn that: E (kinetic)$=Mv^2/2$. Einstein knew, of course, that classic equation relating the kinetic energy, 'E', of a mass 'M' moving at a velocity 'v'. This equation calculates the energy the body would have, for example, if it hit a wall. Or the braking energy needed to stop the moving body.

Intuition may have also allowed Einstein to think of substituting the velocity 'v' with 'c', the speed of light. He was dealing with fundamental parameters (or universal relationships) in the way that recalls Planck's derivations, using fundamental parameters—and '*c*' could have been 'appropriate' for him. But in that case, Einstein would not have had a proof. He would have a conjecture. The idea of using a kinetic equation to describe what was later used to describe a nuclear reaction would have been a little (a lot?) off the beaten path.

I realized he would have to use the same transformation equations you read and observed in the preceding section: the equations that Einstein and Lorentz both developed independent of one another, that are known as the Lorentz transformation. But I asked: how could he transform mass into energy? There is no 'energy' term in the Lorentz-Einstein equations.

I did find the answer to my original question—how Einstein, himself, defined that famous equation. I found it when I referred to *Einstein's Miraculous Year*, source 52, Page 161. It is paper 4 in the *Annalen der*

Physik, consisting a *mathematical development of only four small pages* in the source book's small (9" x 5.5") format.

The mathematics involves the same type of equations he used to determine the size or mass of an object traveling a speed different from light—the same used in the previous charts. And it was only toward the end of his very short article, in which Einstein stated the amazing significance of this paper.

"The mass of a body is a measure of its energy content; if the *energy* changes by L," the *mass* changes in the same sense by $L/(9 \times 10^{20})$ if the energy content is measured in ergs and mass in grams."

(In the above formula, *'c', the speed of light, equals 3×10^{10} meters per second and $c^2 = 9 \times 10^{20}$ meters per second squared)*

If we are so audacious as to modify Einstein's above statement to show the equation if we *change its mass* to determine its impact of the *change on the energy,* it is as follows:

The energy content of a body is a measure of its mass; if the mass changes by M," the energy changes in the same sense by $M \times 9 \times 10^{20}$ if the energy content is measured in ergs and mass in grams."

The relationship stated in the equation, $E = (mc^2)$ was verified experimentally at the particle, or laboratory-level, at the Cavendish Laboratories in April-May 1932. The proof involved splitting the lithium atom by bombarding it with high-energy electrons, protons and neutrons. Two future Nobel Prize winners, the physicists John Cockcroft (1897 – 1967) and Ernest Walton (1903 – 1995) performed the experiments. According to author-physicist Graham Farmelo in source 64, Einstein happened to be lecturing in Cambridge during that time, and he had the opportunity to observe the experimental proof of his equation.

That equation, or relationship, means that a mass of gram has the energy of 25 million kilowatt-hours, or 21.5 kilotons of TNT energy. There are many sites on the web that show the relationship between mass and energy. The author adds that that equation was not fully experimentally proven for large masses until the first atomic bomb test by the U.S. Army, which was performed by an actual detonation of a bomb on the sands of Alamogordo, New Mexico, July 16, 1945.

An important note about Einstein's derivation: The way that Einstein derived the equation did not consider 'mass' directly, as may be commonly believed. Rather, Einstein considered what happens when a body *emits plain, every day, or ordinary, waves of light*, where those 'plain waves of light' have a defined energy. He was speaking of "how a body gives off energy in the form of electromagnetic radiation." By his working along this (amazing) line of thought, he came up with the equation $E = mc^2$. End of note.

Einstein must have been amazed or fascinated by this result—and its implications. In that amazing year, 1905, he wrote to his friend, Conrad Habich, "The line of thought is amusing and fascinating, but I cannot know whether the dear Lord doesn't laugh about this and has has played a trick on me." (Source 8, bottom p148)

In the previous section of this chapter in Pais's book, Einstein said that light always goes at *one* speed. A constant speed. A Universal constant. And we just read about the universality of $E = mc^2$. I now quote the well-know author and Professor of Theoretical Physics, Michio Kaku, at the City College of New York. Professor of Kaku wrote in his book, *Parallel Worlds*, source 49:

> Thus the energy source of the stars themselves was revealed to be the conversion of matter into energy via this equation, which lights up the universe. (Also) the secrets of the stars could be derived from the simple statement that the speed of light is the same in all inertial frames.

Now, on to Einstein's third article of that year, 1905, the Quantum Theory of light

First, an important comment about Einstein's innate understanding of the Universe. Einstein had an intuitive understanding or feeling of the Universe worked, and he promulgated that relationship in a clear, unequivocal way. We will now see this in the most astonishing way, as he develops the Quantum Theory of light.

To introduce the topic, recall that in Chapter 2 that Boltzmann had a great influence on Einstein. Boltzmann's thoughts and equations were all in the domain of classical physics. But here Einstein uses those equations, derived by Boltzmann—classic statistical equations, in a completely different field from thermodynamics, and arrives at the conclusion that light can cause the (quantum) photoelectric effect—plus the concept that light can be a particle or a wave!!!

The Third Article deals with the theory of light, and the photoelectric effect. This was an effect that was first observed in by the German physicist, Heinrich Hertz, in 1887. Hertz had greatly expanded on electromagnetic theory that, originally, had been developed by the Scottish physicist, Clerk Maxwell.

Up to the appearance of Einstein' article in 1905, physicists considered that light was a wave—and only a wave. Einstein's paper strongly questioned this 'no interaction' concept. (Some physicist, especially Pais, consider that *the idea of how light, interacting with matter, was Einstein's major and revolutionary contribution to physics*. It may be that the Nobel Prize committee also felt this way,)

First, a note about the photoelectric effect: We frequently meet examples of the photoelectric effect in our daily lives. Photoelectric systems consist of a beam of light that shines on a metal detector, and when the beam of light is interrupted by an intrusion, the metal detector causes something to happen. In a bank or a museum, the something is the setting off of alarms; in office and apartment building elevators, it opens the door. And there are myriads of other examples. End of note.

This work also provided another set of ideas that added to the idea that the world of physics operates 'in quanta'. This work showed that light operated as a wave and had 'smooth flow'. But it also showed that light could also be considered to be made up of particles. Hence, it might not *always* be a 'smooth flow' as a light-wave is.

Einstein used 'his own way of handling things' in coming to the photoelectric effect. As previously mentioned, he started with equations that came directly from the physics discipline, Thermodynamics. These equations considered gas molecules, entropy and other issues of thermodynamics. After two and one-half pages of equations, he wrote, (as above, from source 8):

270

Light-quantum hypothesis: Monochromatic radiation of low density (…) behaves in the thermodynamic respect as if it consists of mutually independent energy quanta of magnitude!

He then formalized this thought, based on his heuristic principle:

The heuristic principle: If the above hypothesis is true, then this suggests an inquiry as to whether the laws of the generation and conversion of light are also constituted as if light were to consist of energy quanta of this kind.

(This all reminds the author of the similar words of Planck, when he first identified the quanta in heat radiation.)

On the basis of the heuristic principle, Einstein proposed a simple picture for this effect:

A light-quantum gives all its energy to a single electron, and the energy transfer by one light quantum is independent of the presence of other light-quanta.

Light interacts with matter in the photoelectric effect! Before Einstein, this was never hypothesized. (Note: In his 1905 article, Einstein did not hypothesize that 'one light quantum' was a particle.). That occurred later, in 1916.)

Following the publication of the Third Article about light, this concept of interaction between light and matter did not receive strong support within the world of physics. Thus, even without that peer support, Einstein joined Planck in defining a key feature of the new, not necessarily accepted domain of physics—Quantum Theory.

Einstein continued his studies of light as quanta, and in 1916 he postulated that light was made up of particles, later known as photons. This may be considered to be the completion of the work that was originally published in his third article in the 1905 *Annalen der Physik*. The following is what Pais thought of this revolutionary hypothesis:

The genius of the light-quantum hypothesis lies in the intuition for choosing the right piece of experimental input and the right, utterly simple, theoretical ingredients. One may wonder what on earth moved

Einstein to think of the volume dependence of the entropy (thermodynamics) as a tool for his derivation.

In 1921 Einstein was awarded the Nobel Prize "for his services to Theoretical Physics, and especially for his discovery of the photoelectric effect". The photoelectric effect, however, was but one part of his many studies of light. Even with his being awarded the Nobel Prize, Einstein did not immediately gain 'universal peer-belief' with his truly revolutionary postulations about light. The situation changed, however, when his theories about the fundamental nature of light were proven in a series of experiments performed later, in 1923.

The results of the 1923 experiments became known as the *Compton Effect*, or *Compton Scattering*. They are named after Arthur Holly Compton, Professor of Physics at the University of Chicago, who performed the experiments. In 1927 Compton was awarded the Nobel Prize for this work, which consisted of experimental proof of Einstein's theories on the duality of light.

Einstein had started his study of light in 1905. Compton proved his hypotheses twenty-two years later. (Another example of 'Physics time.') Actually, there was a set of experiments that were performed before Compton's. The results of these earlier experiments firmly corroborated Einstein's theory; but the 'words of the experimentalist', Robert Millikan, made even his experimental findings questionable, as we shall now see.

That 'earlier' experimentalist was Robert Millikan (1868-1953), who was president of the prestigious California Institute of Technology (Caltech) from 1921 to 1945. He had previously also been a professor at the University of Chicago. During 1909 he worked on an oil-drop experiment in which he measured the charge on a single electron. This was one of the bases for Millikan receiving the Nobel Prize in 1923. The other was for his work on Einstein's theory of light, performed in the time frame, *circa* 1916.

When Millikan started his experiments on light—as defined by Einstein, Millikan, was convinced Einstein was wrong. But Millikan's experimental results confirmed Einstein predictions! (From Wikipedia) "and as late as 1916 he (Millikan) wrote, "Einstein's photoelectric equation... cannot in my judgment be looked upon at present as resting upon any sort of a satisfactory theoretical foundation," even though "it actually represents very accurately the behavior" of the photoelectric effect."

What a round about way for Einstein to be proven correct!

Einstein's third article (on light and, essentially electromagnetism) puts the electromagnetic force in the same 'class' as the two nuclear forces, (the strong and weak nuclear forces): The three electromagnetic forces—now including light, electricity and magnetism, could be considered to be made up of particles (such as of protons, neutrons, electrons for the first two, and photons for light (later, in 1916)). They all were somewhat similar and all could fit within the 'quanta community.'

Note that Einstein did not say that light could not also (simultaneously) act as a wave: it retains this duality—particle *or* wave. As mentioned previously, only a few years later, in 1923, the French physicist, de Broglie, hypothesized in his PhD thesis, how a moving particle of matter might have an associated wave. de Broglie wrote, "After long reflection ... I suddenly had the idea, during the year 1923 that the discovery made by Einstein in 1905 should be generalized by extending it to all material particles and notably to electrons.' A scientific revolution that Einstein started in the world of science!

A note on de Broglie's hypothesis: As previously noted, photons are different from the other particles since they are weightless, travel at the speed of light and appear to have an infinite life—if undisturbed. Normal matter, such as human, animals, coffee cups, is quantitatively different— heavier and its related quantum and energy equations would make the actual possibility impossible—much less likely to represent ordinary matter by waves. End of note.

Einstein considered that Article 3 to be his most radical among the set of three articles in that one 1905 volume of the *Annalen der Physik.*

A thought by the author about Einstein's selection of equations used in thermodynamics for his derivation of the 'light quanta': Einstein had to start *somewhere* to analyze light. The study and findings of thermodynamics, which Einstein knew very well at that time, appears to be the only discipline in physics that contained a multitude of subjects/topics that related—in some way, to light, such as being described by energy and temperature.

Also, thermodynamics was a relatively new field of study—starting with the analysis of heat and steam in the steam engine, *circa* 19th Century. But, importantly, *there was no other comparable, integrated* field of study in physics at that time (1905)—except for Thermodynamics! He might have thought that it could help provide him his answers to his study of light— which it did.

And it brings up to a final point: When we look at the thousands of mathematical equations that are used in physics, chemistry, cosmology, biology.... we must realized that they are related to one another—even if, *only distantly.* End of notes.

Digression. I interrupt the flow of fact and information about Einstein to present a thought that is personal to the author. This has to do with 'which previous physicists' have provided information that assisted Einstein—in some or any way, in his work in the development of Quantum Theory. Pais, in source 8, cites many people who are in such a line, such as Kirchhoff, Wien and Planck. The author considers that there should have been another —Ludwig Boltzmann.

It was Boltzmann who generated a system of statistical mechanics that gave Einstein the ability to review it, analyze it and to select those parts that were applicable to his needs. If this 'portfolio of statistical mechanics" did not exist for Einstein, he might have had to invent such a system.

Another relationship between Boltzman and Einstein is described in Abraham Pais's scientific biography on Albert Einstein, source 8. First, an explanation about 'what is a 'scientific biography'. A scientific biography is different from a standard' biography; it is written primarily for the scientific community. A scientific biography usually contains descriptions of scientific facts, events and theories and equations, along with the description of the life of the scientist.

Pais's scientific biography on Einstein, Chapter 4, *Entropy and Probability*, contains a Section 4b, titled *Maxwell and Boltzmann* and a Section 4d, titled *Einstein and Boltzmann's Principle.* No other physicist's work is treated in this book in such a concentrated way as Boltzmann is, in this book. The section describes Einstein's use of Boltzmann's equations and parameters, and also shows how Einstein modified Boltzmann's entropy equation to make it more suitable for his special needs.

And, finally on the importance of Boltzmann to Einstein, as stated in the previous chapter, "Einstein had a profound knowledge of Statistical Mechanics and wrote as many as forty papers using Statistical Mechanics— mostly for investigations into subjects related, in some way, to Thermodynamics.

After the publication of those three articles in the *Annalen der Physik*, Einstein was recognized worldwide as a creative genius and was offered professorships at leading European Universities.

Finally, a note about the publishing world before World War II. The journal, *Annalen der Physik*, was published in the German language—as to be expected. Before the World War II, most major articles in physics were first published in German. And this was the situation not only for physics, but also for other scientific fields. Thus, before WWII, university students, who were planning for a science degree, were advised to learn German. It was only after the war that English became the leading language for technical publications. End of Note.

3. Einstein and the U.S. Atomic Bomb Program

The formula $E = Mc^2$ first became famous on August 6, 1945—and generally known to the general public. That day, August 6, 1945 is the date that the first atomic bomb was dropped on Hiroshima, Japan. The formula was frequently mentioned in the press following the use of the first and second atomic bombs, the first against the Japanese city of Hiroshima, and the second, August 9, 1945, on Nagasaki in World War II.

We then also learned that Einstein had previously written a series of letters to President Roosevelt in which he stated that he believed that the United States should undertake the building of atomic weaponry. There were three such letters, written before the U.S. became directly involved in the war, August 2, 1939, March 7, 1940, and April 25, 1940.

(Actually, these letters were written by the Hungarian physicist Leó Szilárd, who was living in the United States at the time. He had learned of the experiments of Lisa Meitner and her team from the February 11, 1939 issue of *Nature* in which she had indicated she had proven the possibility of the fission process.)

While none of Einstein's letters to Roosevelt contained in writing that famous equation, $E = Mc^2$, the press referred to that equation as being the basis for the power of the bomb. Only the two bombs, both used in August, 1945—and the secret, previous test, blown up as in Alamogordo, New Mexico, July 16, 1945, named *Trinity*, came close to validating this powerful formula- relationship, $E = Mc^2$.

I repeat here that Appendix 9 describes the events leading up to explosion of that first atomic bomb, along with the activities of the European and American scientists who performed the research that enabled the building of that bomb.

4. Einstein's General Theory of Relativity

In the period 1907-1912, Einstein started developing what he later called, the *General Theory of Relativity*. In this work, he sought to identify what truly is the *Force of Gravity*—and, and as importantly, *what are some of its distinctive effects on the Universe at large*. Einstein achieved his mastery of his development and analysis during a later phase, 1913-1916. He continued working on it even after that, but it may be said he 'stopped' when his two most-significant predictions were proven. These predictions, that were dependent on Gravity, were:

1. The prediction of the perturbation over a century, or observed change, of the perihelion of the planet, Mercury. The perihelion is the point in Mercury's orbit around the Sun where it is closest to the Sun.

2. The prediction of the degrees of arc that light would be bent by the Sun during the eclipse of the Sun. This was to be observed and measured by the British astrophysicist, Sir Arthur Stanley Eddington, on 29 May 1919. (A brief description of Eddington's experimental observation is

The two predictions were compared extensively to the data and proven correct—providing unequivocal proof of the General Theory of Relativity. As the physicist and science editor for the *New York Times,* Dennis

Overbye, says of the period 1915 to 1917 in his very readable scientific book, *Einstein in Love*, source 66:

> ... the stretch from 1915 to 1917 in Albert Einstein's life represented arguably the most prodigious effort of sustained brilliance on the part of one man in the history of physics.

Two Notes—Note 1: In 1916 when Einstein first announced his General Theory of Relativity, it was said "there were only ten (or only a few) persons in the world who fully understood the theory." Even today, I would guess that the quantity of zeros that could be added on to that original '10' is no greater than four or five zeroes.

Note 2. In the development of this theory, Einstein changed his working style to some extent. During the preparation of the articles for 1905 volume of the *Annalen der Physik,* he was basically an unknown physicist (except for Max Planck), and he worked primarily alone. Following the general acceptance of his articles in the *Annalen*, he became a "venerated" physicist. Thus, in the development of this later theory, describing Gravity, he was able to consult with such established physicists as David Hilbert, Hendrick Lorentz, Willen de Sitter and Arnold Sommerfeld. End of notes.

Gravity is *the* subject in his General Theory. Gravity is not so different in *basic* function *in the world of physics* from the other three basic forces. It too, is a force. *But it is significantly different in what it does to the Universe, or in its "implementation."* Gravity's actions can cause *a change* in the space-time "coordinate system" (referred to as the Metric) due to changes in the (total) Matter (its location and individual mass) and the Energy in the Universe.

In the figure below, we can conjecture that some, unseen, body (or energy source) caused the depression in the space-time metric (shown by the distorted horizontal and vertical coordinates), thus *causing* the observed, round body to sink.

If we use terms that we are comfortable with, Gravity, (in an overall sense) involves the pushing or pulling of bodies (or particles) on one another, as the other three forces also do. But as described in Chapter 3, Gravity's action is effected over far greater distances than the other three forces and what I will refer to as—only here, the 'unit gravity force', is miniscule in comparison to the strength of the other three forces.

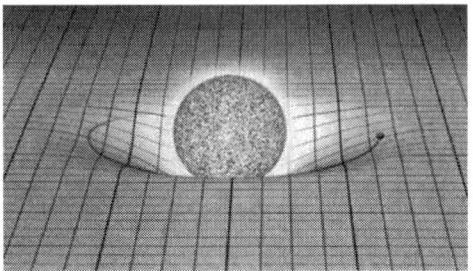

Figure 29: A Round body 'Stuck' in a Distortion/Depression in Time-Space

It is here that I will provide some description of what constitutes the General Theory of Relativity—or, as popularly known, 'The Theory of Gravity.' I have 'broken down' the overall description of the system to small packets, in order to assist the reader to understand and appreciate this system.

We start with a historical description of a "system involving gravity in space"; this is followed by a question whose answer is provided by two historical, conflicting views—what Isaac Newton would have answered, and what Ernst Mach would have answered. (After reading these two views, try identifying the correct view.)

The question: A non-spinning bucket is suspended far out in space. There is water in the bucket and it is at rest and has a flat surface. The bucket is set to spin; it is started to spin. As the water begins to spin in the bucket, the surface of the water becomes concave. When the bucket's spinning is stopped, the water will continue to spin. The water's surface remains concave. Question: The water's concave action is due to a gravitational relationship *with what*?

Isaac Newton (*circa* mid-1600s) 'might' have said it was due to the water's relation with absolute space; Ernst Mach (*circa* late 19th Century), said "It was due to the water's relation with the stars and *all* the matter that is out there." Mach proved to be correct; Newton might have had something to learn.

A note about the above statement of what Newton *actually* believed: In Appendix 11 there is an explanation of what Newton actually believed. He did, however, believe the same way that Mach did, but Newton—a scientist of the 17th Century, *may* have considered it would be too difficult to present that the 'relative' part of the new idea of gravity to the scholastic and scientific community of that era. That is, he probably could not envision quantifying—or even considering, the *whole* (outside) Universe. End of note.

Before we enter in the specific items in Einstein's Theory of General Relativity, it is necessary, first, to describe the ideas of the Czech-Austrian physicist, Ernst Mach (1838 – 1916), who had an appreciable impact on Einstein's thoughts and concepts about the action in Space—and which made it so different from his 'predecessor's (i.e. Newton) on the *concept* and *action* of Gravity. Thus, it is considered necessary, at this point in the text, to first have, a brief digression about Mach:

Digression: We mainly know Mach through his association with the Speed of Sound (The Mach Number), but his far greater contribution was his views of how the complete Universe acts and is presented to us. His position was that "We know the world through elements." The elements include what our sensory perceptions tell us—plus '*all* the other things in the Universe that have (appreciable) Mass.'

Mach applied this line of thought to the way the items in the Universe, or Cosmos, such as stars, planets, comets, etc., were analyzed or considered. Mach was firm in his belief that the TOTAL cosmos must be taken into account when anyone makes an analysis of items in space. (This includes *all* stars, *all* celestial bodies, etc.) Mach ideas, therefore, made a profound impact on Einstein when Einstein developed his mathematically-based *General Theory of Relativity.* In summary, Mach *implicitly* said, "The Geometry of Space (or presumed *Force* of Space) comes from the relations with the *all* the stars and *all* the matter that is out there in space." End of Digression.

I will now go through the overall process that Einstein used to answer this fundamental question. I will concentrate on issues and will stay away from being explicit about the mathematical equations he used. (But I will describe the *type* of mathematics he used.)

As Einstein Saw the Problem of General Relativity

In *circa* 1907, Einstein realized that he had a problem in incorporating the force of Gravity into his-then Theory of Relativity, which dealt with Light. The problem was that force of Gravity is thought to travel *instantaneously*, between two different objects, which conflicted with his new theory of light. Also, his theory of light dealt only with constant–velocity motion, and did not account for acceleration—which the Gravitation force produces.

In addition, he thought about including Ernst Mach's ideas that require a sort of 'interaction' between *all* items with Mass and located *anywhere* in space. He also realized that a 'new' mathematical system representing gravity (to be identified, later in the text, as Riemann geometry) would take, at least, a minimum of five year to complete. He completed task, from his first-thoughts on the matter to publication, in eight years, in 1915.

According to Einstein, he gravity force has similar counterparts to those in the other three forces. It, too, is believed to have a "particle and/or version", called *Gravitons*, and waves, called *Gravity Waves*—which he anticipated to be very small in amplitude. There are presently two huge land-based gravity detectors in the United State, seeking to detect gravity particles and waves. They are known as LIGO Detectors (Laser Interferometer Gravitational-Wave Observatory) are located in Hanford, Washington, and 1,900 miles away in Livingston, Louisiana.

The first time that Gravity Waves were detected was in September 14, 2015. It is believed that these waves were (indirectly) caused by the multiple, successive collisions of a pair of Black Holes circling each other. The larger Black Hole was 36 times the mass of our sun, and the smaller one, 29 times the mass of our sun, and this occurred about *1.3 billion light-years* from Earth.

The intense gravity between the two Black Holes accelerated the Black Holes to half the speed of light, pulling them closer and carving distortions in space and time. In a fraction of a second, the pair collided and merged into an irregular (blob) shape. The unstable blob smoothed into a sphere, a process called a 'ring down'. Three solar masses worth of energy were vaporized into the energy necessary to create such a storm of gravitational waves, distorting space and time—and leaving a new Black Hole, 62 times the mass of the sun.

Note: The author has seen on the web that Gravity Waves are *directly* caused by the collision of the two Black Holes in space. The author, however, provides an alternate, more-fundamental theoretical basis for the source of Gravity Waves, as follows: Gravity Waves are *directly caused* by *changes that are made by changing the geometry of space-itself*—and in this case, the *change in geometry of space* was caused by the collision of the two Black holes. End of Note.

More About Gravity and its Mathematics.

According to Ernst Mach, in very basic and simple terms, "Any body in Space acts as a force on any other planet, a star, or any other body in space." Thus:

1. The Overall Gravitation Force of the Universe is built-up from the *total system* of masses in space, and their locations in space, which manifest themselves by affecting and *dynamically* changing the *geometry of space-itself*, (according to the rules and methods and equations of Einstein's *Theory of General Relativity)*,
2. Space has different, varying *curvatures (or undulations)* in it structure, which *change* the location and path and speed-of-movement of the numerous, varied bodies that are in Space.

Thus, he built a system of what Mach *implicitly* said, "The Geometry of Space" comes from the relations with the *all* the stars and *all* the matter that is out there in space."

A simple method that is sometimes used to describe, or visualize, this phenomenon is to consider a ball rolling on a presumed-flat surface. If the surface is near other large bodies, there will be undulations in the surface, and the ball will follow those undulations. Figure 29 is one such example of an *unseen* heavy weight-force causing a major downward-bump in the surface.

If we consider a 'blind ball', the ball does not know whether it is being 'pushed' by a force or whether there are undulations in the surface. Planets, human bodies, objects in space, similarly act like a 'blind ball'. That is, we do not know whether there is a force pulling us, or whether there are undulations in our path. We cannot tell the difference.

Einstein showed that when something is attracted to another item, it is because *the space between them is curved*—and it is the space-curvature that brings them together.

The next step following this realization was to express the 'undulations' or space curvatures mathematically. This was a problem, even for Einstein. Einstein was acquainted with the same mathematical geometry you and I learned in school, known as Euclidian geometry. But Einstein realized that Euclidian geometry did not provide him with what he wanted to express for gravity—curvature of space. Einstein asked the assistance of a friend and former fellow classman and mathematician, Marcel Grossman, and through Grossman, Einstein learned of a mathematical geometry invented by the German Georg Bernhard Riemann (1826 - 1866).

When you see pictures of Riemann space geometry, you see surfaces that curve and bend. It is a geometric system where parallel lines *can* meet. Riemann geometry, for example, can describe the curved surface of a horse's saddle. Euclidean geometry cannot do that; it can only express a two-dimensional world. Riemann geometry can express an N-dimensional world—where N can be any number.

To help you visualize what the geometry of space can be, I suggest you go to http://en.wikipedia.org/wiki/Hyperbolic_geometry. (Or go to a similar web site found by your search system, using words such as 'Riemann space' or 'hyperbolic paraboloid'.) You will see a picture, a triangle immersed in a saddle-shape plane. This shows how 'flat space' can be warped. In space— all around us, *the force of gravity causes this warping*.

Thus, the theory of gravity appears to be very different from the other three forces (the two nuclear-based forces (the strong and the weak) and the electromagnetic). Those three have particles and waves. Relativity has geometry—and *may* have particles and waves.

Allan Karson

Two Fundamental Assumptions Made by Einstein Concerning Gravity —and What He Did

These assumptions provided a valuable tools (or methodologies) that enabled him to perform the many and varied calculations that later performed—and to correlate the different sets of data he was bound to accumulate.

1. Prior to Einstein's detailed study of Gravity, he perceived/felt physically that the action of gravity provides the equivalent, or same sensation, as a person who is in an elevator that is accelerating downward. (Einstein used the elevator example in many articles and books for layman to describe his ideas about light, gravity, weight and other phenomena.)

2. He predicted that physical laws should retain their characteristics in all forms of Reference Systems. This would include those that are stationary reference systems, those moving at constant velocity—as well as those whose velocity is changing (accelerating or decelerating). This assumption, #2, is fundamental to the General Theory, and he later developed it into what is known as the *Principle of Equivalence*.

A note, continuing with Pauli, source 2:

The implication of the Principle of Equivalence was crucial to Einstein's work: If he could—as he did, develop calculations in one system, in, say, a homogenous gravitational field, it should be able to a translated to a uniformly accelerating field.

It is this feature, which renders the Principle of Equivalence *very* powerful. In this way, Einstein derived the result that the rate of clocks at points of lower (weaker) gravitational potential is slower that that for higher (stronger) gravitational potential.

Note: In the above paragraph, this is what occurs when clocks go slower in a lower (weaker) gravitational field—such as in a light space ship in space or on the planet Venus, and those same clocks would go faster on the surface of the earth (and even faster in the center of the Earth), or even faster on the heavier planet, Jupiter. End of note.

Background of Georg Bernhard Riemann

Georg Bernhard Riemann (1826-1866)—an innovative German mathematician, was brilliant in high school and was always outstripping the rest of his fellow-students.

The story is told that the principal of the school gave him a huge book of 859 pages, *Theory of Numbers*. (Remember that the Greeks of the Golden Age started this field of study.) Riemann finished it in six days. After he finished his education, he worked for Karl Friedrich Gauss (1777-1855), who is called 'The Prince of Mathematicians'. (Gauss was not the man to suffer fools, and he must have realized Riemann's innate, intellectual abilities.)

Gauss gave Riemann the assignment to develop a new geometry to replace or be an alternate to Euclidian geometry. Euclidian geometry, founded by Euclid, had been the one and only geometry since the time of the Greeks. (What an assignment!!)

At the time, the mid-1800s, mathematicians believed that Euclidian geometry was based on first principles, i.e., established based on *fundamentals of nature*, called axioms in the Euclidian geometry. These, so-called *fundamentals of nature* have since been shown not to be necessarily true, even for Euclidian geometry—nor, especially for Riemann geometry.

Do not worry about this, since in the period of the end of the 19th and beginning of the 20th centuries, many other disciplines in mathematics were found not-to-be-based on first principles. For those who may be interesting in such fundamental of mathematics, there is much more about the general problem of proofs-in-mathematics in Source 9, *Mathematics, The Loss of Certainty*.

Gauss, too, may have had doubts about some of those axioms, since Gauss had strong doubts about whether Euclidean geometry could express all the features of Nature *as they truly are*. Gauss might have developed the new geometry himself. But showing that nature could 'support' another geometry would be against a basic belief of his colleagues, whom Gauss considered to be 'narrow-minded'. And Gauss did not to be the 'first bearer' of such unorthodox ideas.

(I have read at some web sites that Gauss did not like Riemann's first presentation of Riemann's 'new math', but I attribute that to a 'cover-up' of Gauss's true evaluation of Riemann's work.)

Source 61 presents a different story of why Riemann developed his new geometry; the initiative came from Riemann himself. "Riemann conceived of the idea that matter would warp geometry.... Riemann also concluded that geometry can cause force—and was a limitation to math at the time. At that time this was, was a momentous insight." Authors' note: Both ideas are fundamental to Einstein's Theory of Relativity.

The development of this new geometry took a heavy toll on Riemann, involving physical and mental stress. Articles and books describing this process all speak of its stressfulness on Riemann. After many months of study and development, Riemann successfully delivered an 'Earth-shaking lecture'. He described a geometry that contains multiple dimensions and was capable of expressing the surface undulations—eventually required by Einstein.

The *General Theory of Relativity* is considered to be Einstein's greatest contribution to physics—and probably Riemann's too.

Websites on Riemann Geometry, or The Mathematics of The Universe

The 'mathematically inclined person can easily perceive the many, many different features that are included in this geometry at: http://en.wikipedia.org/wiki/Riemannian_geometry,

And about the Gauss-Riemann professional relationship: http://www.usna.edu/Users/math/meh/riemann.html.

Varied Notes.

Note 1: Verification of Einstein's Theory of Relativity: The theory was first experimentally verified in 1919 by the English astrophysicist, Arthur Stanley Eddington (1882-1944). Eddington had organized an expedition to observe an eclipse of the Sun by our Moon and chose to observe the eclipse on the island of Principe, which is located close to the equator at 1.37

degrees north. The island is located off the coast of Africa, in the Gulf of Guinea, where Nigeria joins Cameroon.

Eddington observed that light from stars behind the Sun was *bent in* toward the Sun. This was caused by the huge mass of the Sun—a display of gravity-in-action. But, according Einstein's theory, it was not light that was bent; the *space-time coordinates were bent*. The results of this expedition—strongly suggesting that Einstein's Theory of General Relativity was correct —and Eddington's verification made immediate headlines in newspaper *throughout* the world.

Note 2. The difference between inertial mass and gravitational mass: This note is intended for those of you who may have come across these two forms of mass and have been confused about them—as so many persons have been confused. A body's inertial mass is a *constant feature* of the body. It is the same mass wherever the body is—in space, on Earth, on Jupiter... It is the mass that resists change-in-position. It is the mass of a body located on a lake of ice (or in space) that that resist the force attempting to push it on the smooth ice (or move it in space). It is a constant value of the body—no matter where it is.

A body's gravitational mass is a *body's weight*. It is variable; it changes according to where the body is—on Earth, on Jupiter, in space—and so on. It is calculated by multiplying the body's inertial mass by the Gravitational constant of the planet/spaceship/star that the body is located on.

Note 3. When Einstein completed his General Theory of Relativity (or Theory of Gravitation), he was disappointed that *nowhere* in the mathematical model he constructed, did any of the *concepts* of Ernst Mach appear 'outright' in the final version of his mathematical model—even though he followed Mach's ideas and incorporated them in the mathematics of the model. That 'absence continues today.'

Note 4. When Einstein wrote his seminal article about the theory of light in *Annalen der Physik*, in which the photoelectric effect is discussed, he became one of the founders of the Quantum Theory. (Max Planck is considered the first founder; Einstein, may be considered the 'second founder.' At least, that is the way many science writers describe him.)

Note 5 The Amazing Pair: Max Planck and Albert Einstein. If you investigate further the work of these two men, you will learn that both made their initial contribution to Quantum Theory by developing theories that were based mainly on a blend of intuition, correct guesses and the correct selection of experimental information and equations. Amazing individuals!

Note 6 The Cosmological Constant—(A more difficult note): In your reading of texts concerning Einstein and Gravity, the topic called the 'Cosmological Constant' is almost certain to be mentioned. Even after you learn about it, you may consider it to be a strange concept—at least an ambiguous one. The following briefly attempts to describe what it is about. But first, before you know anything about the Cosmological Constant, let it be known that it is a very small number—but could have an immense effect. It is '1' x 10^{-52} (meters)^2.

When Einstein developed the General Theory of Relativity in the early 1900s, it was a time when physicists and cosmologists believed that the Universe was a stable, non-expanding Universe. This was before Edwin Hubble's surprising discovery in 1920 that the Universe is expanding. From source 40, *The Fabric of the Cosmos* by Brian Greene, what Einstein felt *before* 1920, the appearance on the scene of Edwin Hubble: (The parenthetic phrase are added for clarity.)

> "Einstein... .. realized that the equations of general relativity showed that the universe could not be static; the fabric of space would stretch or it could shrink, but it could not maintain a fixed size. ... Einstein balked at this consequence of general relativity, because he and everyone else 'knew that the universe was eternal and... unchanging." Einstein (made a) modification to his .... that conformed to the prevailing prejudice... the inclusion of the (presence of the) Cosmological Constant (in his equations explaining Gravity)."

Note: Some information sources say that Einstein's original model of the Universe indicated that the Universe could eventually contract due to the enormously powerful force of gravity, so that the Cosmological Constant would be an (expected-to-be-actually-present) anti-gravity force—and, which Einstein 'conjectured' *was needed 'in real life' to prevent such a collapse*. End of note.

Following the discoveries of the astronomer, Edwin Hubble, in the 1920s that said that the Universe is expanding, Einstein considered he made an error by injecting the Cosmological Constant into the system of cosmic equations. It appeared to him, given that it *was* expanding, there was no need for it. Einstein called the inclusion of the Cosmological Constant in his equations 'the biggest blunder' he ever made.

If you assume Einstein's purpose for the Cosmological Constant was to combat the possibility of the Universe contracting, it would also give 'body' throughout the Universe, as if there were actually 'mass' distributed throughout the Universe. If such a mass were distributed in space, its total gravitational pull would pull all the other elements of the Universe together —to keep it in a non-expanding state.

It turns out, however, that there is already in space such a strong matter distributed in space (usually referred to as Dark Matter,)—and there was no need for such a 'fudge-factor constant' to be put in by Einstein. (Einstein died before Dark Matter in space was identified as being present.)

The 'substitute' for the cosmological constant comes from the *Dark Matter* that was first mentioned in Chapter 4, in the section describing The Composition of the Universe. Satellite probes have indicated there is both dark matter and dark energy in outer space. Neither of these two entities has ever been seen—only inferred. These dark matter and energies, therefore, should be taken into account to understand the expansion rate (or the possible contraction rate) of the Universe.

Dark matter and dark energy are now an integral part of what might be still referred to as the Cosmological Constant No.2—which now it lives on 'in its own right.' (So you can now relax, knowing that Einstein was basically correct in his realization that *some sort* of Cosmological Constant *is* required—and was *already* there, physically.)

I will alert you to the fact that discussion of the composition of the Cosmological Constant still goes in the world of cosmologists. Such discussions frequently lead to heady, deep, philosophical discussions that are beyond the scope of this book. End of Notes.

A brief note to end, temporarily, the discussion of Einstein: Before the advent of Quantum Theory, physicists worked only in 'classical physics', which was basically started by Sir Isaac Newton. Physicists consider that Einstein was one of the last, great, classical physicists—and among the *First* in Quantum Theory.

Part 3 Erwin Schrödinger (1887–1961) The Person Who Identified the Quantum Mist and Created a Revolution in Quantum Physics

We will first consider the background and experience of Erwin Schrödinger at the University of Vienna. His scientific mentors at the university were Frederick Hasenohrl, a theoretical physicist, and Franz S. Exner, an experimental physicist. In 1907 Hasenohrl was the successor to Boltzman as the head of the Department of Theoretical Physics at the University of Vienna.

At university, Schrödinger exhibited a strong ability in both theoretical and experimental physics, but his strength was more on the theoretical side. He also developed strong abilities in many sectors of physics. All of his papers were highly competent, detailed and contained an inordinate amount of science—both in mathematics and physics. Most of the early papers, however, were 'ordinary' in that they did not introduce new ideas. They might be called "Yeoman papers"—professional, but not exciting.

But a few, however, were highly distinctive and did introduce new ideas. This is especially true of his 1926 paper on *Quantum Wave Theory*, for which he received the Nobel Prize in 1933, sharing it with the English quantum physicist, Paul Dirac.

The following are some examples of papers he produced. You need not feel that you should understand the topics described in each report. The important point to recognize and remember is that the following studies cover a wide range of subjects within physics. And Schrödinger is said to have attained an intellectual and mathematical mastery in each subject-area.

He developed his first set of papers during the period 1914– 1918 while he was serving as an artillery officer in Austrian Army, assigned to the Italian Front during the entire war:

- *On the Dynamics of Elastically Coupled Point System* (Submitted to the *Annalen der Physik*.) A mathematical treatise of atomic trajectories in time and space.
- A course on Meteorology for Austria's principal weather station located outside Vienna that included 1. Standard weather factors—variation of temperature, composition of atmosphere, types of clouds... and 2. A Review of classical theories and derivation of new methods of weather analysis.
- *The Result of New Research on the Atomic and Molecular Heats*, his first paper on Quantum Theory
- Two papers dealing with Einstein's Theory of General Relativity. (In one of the papers he raised the problem relating to the localization of gravitational energy, which continues to be studied today.)
- Two long papers on analysis of random fluctuations in the rate of radioactive decay.

And in the 1918 – 1920 period, following World War I, he produced a series of articles on Color Theory, some of which were published in the *Annalen der Physik*. This work was done in collaboration with his senior professor, Franz S. Exner, who also had published papers on the same topic. (As Schrödinger's biographer, Walter Moore, says, "...modern experts in Color Theory have called one of the *Annalen* papers' 'masterly'." Through this paper Schrödinger became known as a world authority on color.) And in 1925, there was another article on color, *On the Subjective Colors of the Stars and the Quality of Twilight Sensitivity*.

And, in addition to the plaudits concerning color, Walter Moore, Schrödinger's biographer-physicist, says in source 16, *Schrödinger, Life and Thought*, concerning relativity in the list above, "It is remarkable that in his first note on general relativity, he was able to uncover such a deep problem, but it is typical of an approach that became increasingly evident in his work..."

Now, back to Schrödinger's personal life. To do this, we must also consider the situation in Germany and Austria following their joint defeat in World War I. Inflation was running rampant in Germany and this affected life in Austria. As many others at this time, Schrödinger continually sought a better, (life-sustaining) position. As late as 1923, hunger and starvation were still widespread in Germany. Thousands perished, inflation was rampant—all while the Allied blockade stayed in force. His odyssey within the German-speaking universities was:

- Assistant to Max Wien, who was Director of the Institute of Physics, University of Jena, Germany April 1920 to October 1920

- Associate Professor, University of Jena, 1920

- Professor Extraordinaire, University of Stuttgart, Germany, 1920

- Regular Associate Professor, Technische Hochschule, Stuttgart, Germany, winter semester, 1920 - 1921

- Professor for Theoretical Physics, University of Breslau, Germany, summer 1921

- Professor of Theoretical Physics, University of Zurich, Switzerland, 1921, 1922 (A position once been held by Einstein)

- Professor of Theoretical Physics, Planck's Chair, University of Berlin, 1927, where he was successor to Max Planck

During the period 1922 to 1927 he had a seven-month rest cure for tuberculosis, published six major papers in *Annalen der Physik* in 1926, and traveled to the United States from December 1926 to March 1927. These six papers provided significantly new ways to consider Quantum Theory.

His personal life: He was married to one woman all his life, Annemarie Bertel. In 1920 they married when he was thirty-two and she was twenty-three. It may be said, they did not appreciate the 'bourgeois style of marriage'; they each had many affairs all during their lives. As a matter of fact, a second woman frequently was part of their household. At one time, the second woman was the wife of a colleague of Schrödinger. It is understood that Schrödinger fathered three children from these varied relationships. This familial situation, however, may not have been acceptable to some of his potential university employers, so he may have had limited employment opportunities.

Back to the Schrödinger's professional world: In 1926, following the seminal work by Louis de Broglie that is briefly described in Chapter 9, which hypothesized that matter could live a double life, in both particle form and/or in waveform, Schrödinger wrote his most famous article, which consisted of a series of six papers. The papers were titled, subtly, *Quantization as an Eigenvalue Problem.*

These papers defined Quantum Theory in terms of *wave equations*, using differential equations to express the waves. This formulation in waves gave life to the idea that matter can also be expressed as a wave, validating, to some extent, de Broglie's similar concept and providing an alternate way to consider Quantum and Heisenberg's matrix approach. As mentioned previously, it was for this work that Schrödinger was awarded the Nobel Prize in 1933, which he shared with Paul Dirac.

(Schrödinger's formulation of Quantum theory was a parallel development to the work done by Werner Heisenberg. It has been shown that, even with their being a fundamental difference between the two systems, matrix versus differential equations, their results are essentially equivalent.)

Tranquility—such as it was, came to an end in the early 1930s when Hitler came to power. Schrödinger disliked intensely the anti-Semitism of the new regime and decided he could not continue to live and work in Germany. He left Germany in 1933 and briefly held a fellowship at Oxford, England. In 1935 he was invited to lecture at Princeton University and was offered a permanent position there, which he did not accept.

In 1936 he was offered and accepted a position at the University of Graz, Austria, his homeland. He went to Graz and was there through 1937. When the Nazis invaded Austria in 1938, they considered Schrödinger's previous-leaving from Germany in 1933 to be a hostile act directed toward them. He, therefore, was forced to escape to Italy, from where he proceeded to Oxford and then to the University of Ghent, Holland.

In 1939 he moved to the newly created Institute for Advanced Study in Dublin, Ireland, as Director of the School for Theoretical Physics. He stayed in Dublin until his retirement in 1955. While in Dublin, he gave a series of three lectures at Trinity College, Dublin, titled *What is Life.* These lectures *hinted at the concept of a complex molecule with the genetic code for living organisms*—a new subject for him—and for the world-at-large.

He was one of the first to note that it is the *difference in genes* that distinguish one animal species (including humans) from another. He went on to describe the rudimentary genetic process. And in 1944 he wrote the book "What is Life," which deals with genetics. Within that book Schrödinger introduced 'the most fundamental concept in the new science of molecular biology: The *chromosone* is a message written in code.'

I call Erwin Schrödinger: "A Man for All Seasons"—having been deeply involved in both physics and genetic biology.

Francis Crick, one of the discoverers of DNA, credited Schrödinger's book with giving him guidance on how the genetic storage system would work— and also credited Schrödinger for giving him, Crick, inspiration for his work. And, according to James D. Watson's memoir, *DNA, The Secret of Life*, Schrödinger's book gave Watson the inspiration to research the gene, which, along with Crick, led to the discovery of the DNA.

Note: For those who are too old to remember, or too young to know, James Watson, Francis Crick and Maurice Wilkins jointly received the Nobel Prize in Physiology or Medicine for their 1953 determination of the structure of the Gene. End of note

We continue with Schrödinger's direct contributions to physics. In 1947 Schrödinger contributed with a modification to Einstein's *Unified Field Theory*, concerning the laws of gravitational fields. Schrödinger's contribution had to do with our 'cosmic friend', the ubiquitous, *Cosmological Constant*. This modification is known as the *Einstein-Schrödinger Theory*.

I will not be attempting to describe the Einstein-Schrödinger Theory. Rather, I leave the reader with a very brief description of this theory, distilled from the web's Physics Forums site, and written by Russell E. Rierson:

"The Einstein-Schrodinger theory is also known as "Einstein's Unified Field Theory" ... It was developed by Albert Einstein and Erwin Schrodinger, primarily in the 1940s and 1950s. ....It is thought by some to be a unified theory of gravitation and electromagnetism... This was supposedly disproven way back in 1953...."

Before World War II, there was the expression that "The Sun never sets on

293

the British Empire." I would like to adapt that expression to describe Erwin Schrödinger, by saying, *The Sun never set on Erwin Schrödinger.*

Allan Karson

Chapter 12. The Cultural Background of Quantum Theory

Part 1 Quantum Theory's German-speaking Development-Environment

An important factor in the cultural background of early quantum physicists is *where* most of the early successes in the theory occurred. That is, there was a very noticeable scientific advancement for Quantum Theory and Quantum Mechanics that was attained in German-speaking countries and near-Germany, such as Denmark, the home of Niels Bohr.

Also, prior to World War II, the better scientific journals were mostly published in German. At that time, knowledge of the German language was a requirement in getting a technical degree at many U.S. universities. European scientists are said to not have known of, much less read, *any* American technical journals.

This was the situation until 1939, when the leadership in this field was essentially relocated to the United States, where did German-speaking refugees from Germany and Austria continued Quantum's development. Further advancement was aided by the forth-coming development of high quality physics and math education in the U.S. This led to high quality, albeit small, physics communities in certain locations such as California Institute of Technology, University of California at Berkeley, CA, MIT in Boston/Cambridge, MA, Princeton, NJ and Columbia University in New York City.

We should also note that during this transition stage, England also developed a semi-leadership position in this field of physics, and England continues to hold an important position.

Let us now examine the German-speaking countries and their education facilities. First, the countries that made up the German-speaking group: The

German language was used in its original form, or a 'dialect' form, in a wide range of countries.

On that broad geographic-language basis, German-speaking countries include Germany, Austria, Czechoslovakia, and southern parts of Poland and Rumania. In addition, there were parts of Switzerland, Belgium, Holland and Denmark that were German-speaking. In the two latter countries, however, their 'German dialect' was not sufficiently close to German to always enable an easy understanding between the two languages.

In 1900, the two largest of these German-speaking centers were the newly united German nation and the Austro-Hungarian Empire, which included Austria, Hungary, Czechoslovakia, Hungary, Rumania and many smaller, non-German-speaking nations or groups. (At the time, Czechoslovakia was made up of Slovakia, Bohemia and Moravia.)

While many of these nations or entities had their own indigenous language, German was the language of most of the middle-class living in the larger cities, such as Budapest, Hungary.

As will be seen in Part 2, the German-speaking universities essentially operated as a *quasi-unified group*, with instructors and professors moving from university to university, without a concern for national borders.

The following is a list of some of the most influential German-speaking physicists of the first half of the 20th Century (listed in order of their birth: Max Planck, Max Wien, Arnold Summerfield, Albert Einstein, Max Born, Niels Bohr of Denmark, Erwin Schrödinger and Wolfgang Pauli. These persons may have communicated freely their ideas, impressions, criticisms and guidance to one another during this period—until 1939, when WWII started.

Part 2. Two Outstanding European Universities of the Early Quantum Era

A German, German-speaking University, the University of Göttingen, its world-class professors and students and status.

There were two universities located in Germany that were centers for the study of Quantum: University of Munich and University of Göttingen. My choice for *the* leader is Göttingen—for the many world-class persons associated with it, and because it shows how a small town can produce a world-class center of learning.

Göttingen is located sixty miles south of Hanover, one hundred twenty miles north of Frankfort—and fifty miles from the Pied Piper's town of Hamelin. The University of Göttingen was founded in 1734 and is one of the highest-ranked universities in Germany. At least *fifty-five Nobel Prize laureates* have studied, taught or made contributions at the University of Göttingen.

Digression: The following is a direct quote from Wikipedia's web site about Göttingen. It should help clarify an important non-destructive action—on the part of the Allies, at least, that occurred in World War II.

From http://en.wikipedia.org/wiki/Göttingen:

> Nearly untouched by Allied bombing in World War II (the informal (unwritten) understanding during the war was that Germany would not bomb Cambridge and Oxford and the Allies would not bomb Heidelberg and Göttingen.

(The author sought further information concerning Wikipedia's above-item and found it at the *Snopes* web site, that this was not truly adhered to, since Oxford and Cambridge were bombed, but not as heavy as other cities.) End of Digression.

The following is a list of it world-class members of Göttingen's faculty over time: Mathematicians David Hilbert, Richard Courant (after whom the Courant Institute at New York University is named), Hermann Minkowski (who was the first to define the four-dimensions making up one unified system in space, known as 'space-time'), Emmy Noether, the female-mathematician who contributed considerably to the mathematics of symmetry; the physicists Karl Schwarzschild (who was the first to predict Black Holes), Max Born, Werner Heisenberg and Arnold Sommerfeld.

Add to that list earlier faculty members from the 19th Century, such as Carl Frederick Gauss, referred to as The 'Prince of mathematicians', and the mathematician, Bernhard Riemann. Their presence at Göttingen may be one of the reasons Göttingen was referred to as *the Mecca* for mathematical research.

The head professor of the physics department during the years of the development of Quantum Theory (1900-1927) was Max Born, the mathematical theoretical physicist. Born's first assistant in Göttingen was Wolfgang Pauli and his second assistant was Werner Heisenberg.

Born and his department attracted a wide range physicists and mathematicians who were interested in learning about and discussing Quantum Theory. As an example, Bohr came to Göttingen in the early summer of 1914, speaking in his Danish-accented German, to present a three-week long colloquia to present his new ideas on Quantum Theory. (Later, he also presented his new ideas at the University at Munich.)

A short list of visitors and lecturers at Göttingen includes Paul Dirac, Albert Einstein, Kurt Gödel, John von Neumann, Robert J. Oppenheimer, Erwin Schrödinger, Arnold Somerfield, George Gamow, Edward Teller and Martin Schwarzschild (son of Karl Schwarzschild). Göttingen also attracted a considerable number of 'serious' American students, whose names are not listed here.

During the time that Max Born was the head physics professor at Göttingen, the American-based Rockefeller Foundation was offering fellowships to foreign students to study in foreign countries. It stated that Göttingen had become the 'Number 1' physics institution in Europe.

In 1925 Born made a five-month lecture tour in the United States, speaking at numerous universities to tell them about Europe's work in Quantum Theory. According to Born's biographer, Nancy Thorndike Greenspan, in source 50, Quantum Theory was basically unknown in the U.S., and large audiences listened attentively to Born. He crisscrossed the entire U.S. network of major universities to give his lectures.

A similar situation—that quantum was basically unknown in the U.S., was observed by Robert J. Oppenheimer when he returned to the U.S. to continue his teaching career at California Institute of Technology and the University of California, Berkeley. Oppenheimer had done his post-doctoral work in Holland, Switzerland and Germany. During this period, he had worked with some of the 'Masters' in the physics, including Niels Bohr, Paul Ehrenfest and Wolfgang Pauli and Max Born. Oppenheimer's collaboration with Born resulted in the Born-Oppenheimer approximation, which is central to the study of molecular vibrations.

Göttingen is but one example of the key universities that thrived until 1939. Better known of course, were the universities and institutes in Berlin, with such faculty as Planck and Einstein.

2. A Non-German, but German-speaking University—and its World-Class Professors and Students.

The oldest German-speaking University in Central Europe is the University of Prague, founded in 1345. It is located in the capitol of the Czech Republic, which was formally known as Czechoslovakia. It later initiated a Czech-speaking branch in 1920. Its teaching staff and graduates consist of many leaders in their respective fields, including a few Nobel laureates, but most are unknown to people in the United States. Some who are known, however, include the writer, Franz Kafka; the Czech statesman, Edvard Benes; the religious reformer, Jan Hus; and the discoverer of blood types, Jan Jansky.

The second oldest Non-German, but also a German-speaking university, is the University of Vienna, founded in 1365. As mentioned in the last chapter, Erwin Schrödinger started his university education at the University of Vienna in 1906, and it is where he got his first degree. We now review some of its distinguished lineage of the science teaching staff at that university.

The first to mention is Christian C. Doppler, a native of Salzburg, Austria. Doppler was appointed to the Chair of Physics at the University of Vienna in 1850. We know of the 'Doppler effect', in which the sound wave from an approaching vehicle sounds much higher in pitch than when it recedes from the listener. (And we also know of the application of the Doppler effect to cosmology, such as was done by Hubble.)

And on the teaching staff were Ludwig Boltzmann, Joseph Stephan (who, in addition to determining the temperature of the Sun, produced eighty scientific articles, (mostly in the Bulletins of the Vienna Academy of Sciences) and Ernst Mach, whom we met in the earlier chapter on thermodynamics and in the chapter describing Einstein's General Theory of Relativity.

After the second World War, the center of physic switched to the United States, which now has the largest number of Nobel Prize Laureates in the world, and the school that has the highest number is the University of California at Berkeley with To learn more about the Nobel Prize winners and other, prestigious award, go to Appendix 15, that discusses the Nobel Prize, and the Fields and Abel Medals for Mathematics. And you can go the web site, http://nobelprize.org/

to learn more.

From Appendix 15, an excerpt, describing two schools that are worldwide the leaders in the Nobel Prize:

> The University of California at Berkeley has graduated more future Nobel Prize winners than any university in the world, and the Bronx High School of Science, located in the Bronx, a borough of New York City, has graduated more future Nobel Prize winners than any high school in the world.

Chapter 13. The Standard Model—Designed to Incorporate the Four Forces

This chapter describes:

- The Features and History of the Standard Model
- The Beauty (or Ugliness) of the Standard Model
- Unfulfilled Predictions of the Standard Model
- Why the Gravity Force is not a Part of the Standard Model

In Chapter 6 we saw that Quantum Theory deals with equations that describe how subatomic particles act. The first step to apply Quantum Theory broadly, was to organize and apply these equation to each of the four basic forces. This effort was started early in the 1920s and went on through the 1970s.

In the latter part of that period, the 1970s, three of the now newly-quantized forces (The Strong Nuclear Force, the Weak Nuclear Force and the Electromagnetic Force) were incorporated into what is called the *Standard Model*. The understanding of the Standard Model is important in subsequent discussions. The following answers the question to "What is the Standard Model?" Note: Within this chapter we may refer to the Standard Model as SM.

The Standard Model is the incorporation of the following into one system of equations.

- All the known features of sub-atomic particles such as electrons, nuclei, quarks... (Their charge, size, spin...),
- The equations that describe those sub-atomic particles—and the interrelationships of those sub-particles,
- The equations of Quantum Mechanics, such as Schrödinger's equations,

- Full descriptions of three of the basic forces: The Electromagnetic, Strong Nuclear and Weak Nuclear Forces; the SM presently cannot support Gravity,
- Special Relativity (Light).

The Standard Model enables physicists to examine mathematically the interrelationships of particles, and at the same time, to predict new phenomena and not-yet detected particles.

Two reasons why Gravity is 'special' or different:

- Physicists and mathematicians have not been able to develop the special mathematics needed to includes the General Theory of Relativity, to which the Gravity is fundamental,
- Gravity's predicted force carrier particle, the Graviton (a boson), has not yet been observed, so a detailed, mathematical model representing the Graviton cannot be assured.

The leader of the team that developed the Standard Model in 1979 consisted of three men, for which they shared the Nobel Prize for physics for that achievement. They are Sheldon Glashow, Professor of physics at Harvard; Abdus Salam (1926 – 1996), Professor of Theoretical Physics at Imperial College, London; and the Nobel Laureate, Steven Weinberg, Josey Regental Professor of Science at the University of Texas.

We will now describe an overview of the Standard Model's Herculean development and its worldwide and 'heavy industrial' use. We start first, however, in the work begun in the 1920—before any type of Standard Model (SM) was even thought about. That work generally consisted of applying Quantum Theory to the four forces, and that effort evolved into the Standard Model. Quantum Theory was first applied mainly to the particles making up the Strong Nuclear Force—the nucleus, the electrons and other subatomic particles.

The second application of Quantum Theory to another force was in 1929, when it was applied to the Electromagnetic Force. Later, in the 1960s and 1970s, it was successfully applied to the Weak Nuclear Force. During this period, physicists and mathematicians also added certain 'outside of quantum features' to the Strong Nuclear Force.

As noted above, applying Quantum Theory to Gravity has not been successful. Thus, it is not one of the three forces incorporated into the Standard Model. (We come back to that later in this chapter.)

You may ask, "What does it mean to apply Quantum Theory to a force." The answer is that it must incorporate the following:

- The probabilistic and uncertainty principles of Quantum Wave Theory.

- Special Relativity (While Special Relativity was not originally part of Quantum Theory,

   it was applied to make the SM applicable in all relativistic systems.)

As seen by the years associated with each force, this effort was done over many years and performed worldwide by many different groups. The following are the names given to the various "transformations" from freestanding force to incorporation into the SM:

- The Electromagnetic Force is called Quantum Electrodynamics (in its quantum reincarnation) (QED).

- The Strong Nuclear Force: Quantum Chromodynamics (QCD).

- The Weak Nuclear Force: Quantum Electroweak.

The Standard Model has proven to be accurate and correct during a myriad of experiments. Still, realizing that it does not include gravity, it is not seen as the "final theory"—the joining all four forces in one *consistent* model. That "final theory" is sometimes (optimistically) referred to as the 'Theory of Everything,' or TOE.

(Since physicists do not accept it as the final description of the four forces, work continues to achieve a deeper or different understanding of the Universe. Starting in the 1970s, the path that many physicists and mathematician have been following—to achieve a deeper or different understanding, is called *String Theory*. This relatively newer, major enterprise of physics will be described briefly in Chapter 12.

The Standard Model is comprised of three forces. But, and this can be a strong "BUT", even if the Standard Model had incorporated the force of gravity, it still would be lacking something that physicists, such as Paul Dirac would say, "there is a need that the model should possess *beauty*."

The Beauty (or the Ugliness) of the (present version) of The Standard Model

The building of the Standard Model was a magnificent achievement—applying Quantum Theory and Special Relativity to three essentially *different in kind* forces! Physicists, however, accuse the theory of being *Ugly*! So We must ask the question, "Why is it considered Ugly?"

In answering that question, it is interesting to see how physicists evaluate formula and theories—in addition to laboratory tests and mathematical proofs. Some of these are subjective and some are not. They can be summarized in broad terms as Symmetry, Simplicity and Beauty:

- Symmetry: Symmetry requires that the theory is valid for all coordinate systems and dimensions. It appears the Standard Model does meet this requirement.

- Simplicity: The ultimate examples of simplicity are $E_{kinetic}=mc^2$, or $F_{force}=ma$, $E_{kinetic}=mv^2/2$. While the Standard Model is not expected to be as simple, since it is representing a more complicated and diverse set of laws, it still lacks what might be called 'a certain simplicity.'

The Standard Model was not expected to be as simple as these other simple, but powerful, equations cited above, since it also has too many "rough edges".

For example, The Standard Model contains approximately 20+ parameters that describe the various particles and the relative strengths of (only) three of the four forces. (It has yet to incorporate the force of Gravity.) In addition, there are another 10 parameters that are needed to describe neutrinos. The latter group for Neutrinos is a relatively recent addition. For many years it was only 29. (Note: The number of parameters used in this text is 29.) I have seen numbers other than 29 used in other texts—but numbers close to 29.) Many of these parameters were used in relatively recent, successful, search (July, 2012) for the *intensely* sought-and found, Higgs boson cross-section.

These parameters consist of the masses of various particles and various interaction strengths between particles that are caused by the three forces. It is not only the specific numeric value of each of the particles required in an experiment; it is also the ratios of these numbers between one another.

Now, a note about how Physicists use a unit of energy—the electron volt, or eV instead of using kilograms to measure mass. It is the energy that one electron mass has when the electron's rest mass is converted to energy, calculated by the Einstein's equation, $E = mc^2$. One eV, in equivalent weight, is *much* 'lighter' than one kilogram by a factor of 1.8×10^{-38}.) Also, One electron volt (eV) is equivalent to 1.6×10^{-19} joules—a very small amount of energy.

The following shows the various quantity-levels and equivalencies of the Electron volt and the energy of *Leptons* (subatomic particles that respond only to the electromagnetic, weak and gravitational forces) and the *Quarks* (constituents of Proton and Neutrons).

- One MeV is one million, or 10^6 electron volts; and One GeV is one billion, or 10^9 electron volts.
- The Energy of the three Leptons, in eV:

Electron 0.510998910 MeV   Muon 105.658367 MeV   Tau 1776.84 MeV

- The Energy of the six Quarks:

U up   190 Me   C charm   1.40 GeV   T top   172.5 GeV

D down   190 MeV   S strange   190 MeV   B bottom   4.75 GeV

Note: The author considers it strange—and wonders why there is such a large difference in energy among the Quarks, that, presumably, *all* perform similar functions. For example,

$$\text{Top/UP} = 172\text{GeV}/190\text{MeV} = \sim 10^3 \,!!$$

305

Now, *Beauty*: In order to develop further, "what is meant by beauty", we do know what Einstein felt when he compared his General Theory of Relativity with the Quantum Theory—upon which the Standard Model is based. (Einstein died in 1955, many years before the establishment of the SM.) He considered:

- His theory of relativity was akin to smooth marble, since it was clean and beautiful. He considered this even though it is based on the non-Euclidian, complex Riemannian geometry. Quantum, on the other hand, with its numerous, heterogeneous point forces, he considered it to be akin to rough, gnarled wood.

Note: Einstein would, apparently, have also disagreed with more fundamental aspects of the SM, since "he did not believe that *non-relativistic* quantum mechanics (that is the basis for the SM) provided a basis that was *secure enough for relativistic generalizations*." (Source 8, p463.) End of note.

So for now, according to some of these criteria, the Standard Model is 'ugly'.

What about the fourth force, Gravity? Aside from the issues described above, there has been no success in incorporating the force of gravity into the Standard Model. The difficulty has been apparent since the early 1980s.

This unresolved part of the Standard Model is called *Quantum Gravity*. Starting out from when physicists started quantifying the four forces in the first half of the 20th Century, they realized there was an incompatibility between the force of gravity and the other three forces. Some of the reasons for this incompatibility are:

- Gravity is a much weaker force than the other forces, *but* it acts over immense (astronomical) distances, and usually involves huge masses. (Refer to Chart 13 for a comparison of the Four Forces.)
- The Standard Model describes forces by their energy and momentum (velocity, time, mass). This does not mix with the presently known characteristics of gravity.
- Gravity waves, have only been detected in 2016, and presumably interact with each other. That is, a Gravity wave from one mass interacts

with the Gravity wave of another mass. They add up. But within the three other forces, there is no interaction between their waves. For example, an electromagnet wave (EM) emanating from radar does not interact with the EM wave of a TV transmitter. And this holds even if the source and receiver of the EM waves are close to one another.

- Gravity causes a distortion or change in the geometry of the Universe. The other three Forces are more 'direct to their target' and cause no comparable change in the Universe's geometry.

(There is an exception to the above statement about the non-interaction of EM waves: EM waves do interact if they emanate from the *same* source. An example of this is the two-slit experiment described previously, in figure 24.)

This problem cited above, that Gravity changes the geometry of the Universe, is a critical one to physicists. It means that when they do calculations concerning Gravity, they must be aware that the system of coordinates they use, referred to as space-time systems (x, y, z and time, or 'the metric') is different from the space-time systems (x, y, z and time) for the three other forces.

Physicists refer to this problem in the following way:

To do calculations of Quantum Gravity Theory, a Background **De**pendent Space Model must be used in order to observe the changes that Gravity makes to the space-time coordinates (x, y, z and time).

A repeat: Quantum Gravity, or just *plain* gravity, is intimately involved in defining the actual shape of the space-time coordinates (Refer to figure 29 to see the effect that a mass (or gravity) can cause to the coordinate lines of space.)

Note: The 'De' in stressed above in Background **De**pendent in order that it is not confused with the Background-Independent space model—that is used for the other three forces, in which there is *no* interaction by the forces with the space-time system. That is, the Independent space model hides, or would not take into account, actual important events in space caused by gravity. But the Independent space-time model is the model that the Standard Model uses to analyze the three other forces. End of note.

In summary, the gravitation field is very different from the other three fields. It may be said that Gravity is 'at home' with its own space-time system. It lives there peacefully, working with that flexible-to-gravity coordinate system. The other three forces seem to have no relation with their respective space-time system. 'Intruders' or forces without a home?

Within the Standard Model, work continues on Quantum Gravity, but not at the strength or pace of other problems or opportunities in physics. To apply the Standard Model to the gravity force may require a larger segment of the theoretical physics population to work on it—or, possibly, a different approach.

An Interesting Prediction Within the Standard Model—But Not Yet Fulfilled

The following is a brief introduction to an interesting, very large experiment that was started in 1983 and may still ongoing—and being upgraded:

There is also an important prediction made within the framework of the Standard Model that relates to Quantum Theory—and that is still unfulfilled. This non-fulfillment vexes physicists. It has to do with predicting the half-life of a proton. It is predicted that a proton has a half-life of 10^{33} years! (Remember that the Universe is only 14 x 10^9 years old.)

The calculation of the proton's half-life is still considered an important calculation within the Standard Model—but as of this writing, 2016, there is currently no experimental evidence that proton decay occurs.

The experiment consists of having huge vats of ultra-pure water, which implicitly contain billions of protons. These experiments have been going on since 1983 at the Kamioka Observatory, the Institute for Cosmic Ray Research, of the University of Tokyo. The statistics of the planned experiment indicate clearly that at least one proton should have shown decay. So far, not one decay has ever been detected.

Summary of Status of Standard Model

Since the 1970s, the Standard Model has been the centerpiece for studies in physics. Work is still going on to add certain features to it, such as Gravity, called Quantum Gravity. Since then, however, many physicists have been

308

seeking newer models, such as one that might, more easily, incorporate gravity into it overall theory. In addition, there may be other reasons for this 'lightening up' the effort on the Standard Model.

Physicists are waiting for experimental results from the Large Hadron Collider at CERN. They expect to verify many of their particle theories developed in the Standard Model—and to open up new avenues of study.

Many physics believed from the beginning, from when the Standard Model was first defined, that a 'deeper' all-encompassing model was necessary to accommodate gravity. They considered that the incorporations of gravity with the other particles required a new theory. There is now a large group of physicists who are engaged in that new theory, called "String Theory."

Before we consider String Theory, we need first to open up a fascinating topic that is best understood as a "free-standing topic" that is away from other aspects of physics: Dimensions. As will be seen later, the subject of dimensions is an important feature in String Theory. Plus, based on the drawings that are made from 'dimension equations', which we will also see in the next chapter, it can be an interesting topic in its own, independent domain.

Chapter 14. Hidden Dimensions, Introduced by Theodor Kaluza

We learned previously that when Einstein developed his Special Theory of Relativity—the one that says nothing can go as fast as the speed of light, he identified three accepted space dimensions. And he specifically added a fourth— the time dimension. Thus, when we now discuss a position in 'space', we discuss four dimensions: three in space, one in time.

He implied that each and every person has his or her own time dimension. For example, time, as shown on an individuals watch, can differ among 'stationary' people, from 'moving' people who are moving near the speed of light.

In 1919 an event occurred to Einstein that indicated that there might be more than three space dimensions. And since then, physicists have been more open to that possibility. In particular, when physicists work on the Standard Model, they see the possibility or usefulness of considering additional dimensions in setting up the overall mathematics for the model.

And when we discuss String Theory in the next chapter, you will see that the number of space dimensions can become as high as ten—or more. If you try to visualize these additional dimensions, however, you probably will not succeed. No one has ever seen any of them, or, at least, reported seeing them. But most physicists believe that they exist "in some form". Let us see what Einstein (or Kaluza) "saw".

Einstein, Kaluza and Four Space Dimensions

Einstein spent his later years at IAS attempting to unify the force of Gravity with the Electromagnetic Force and with the two Nuclear Forces. He did achieved a limited unification of the gravity force with the electromagnetic force—but in an interesting (unique) way; he did this by expanding the Universe to *five* dimensions, 'four' for space and 'one' for time. So we can say that Einstein did unify them—but conditionally.

310

It was Theodor Kaluza, (1885–1945) who made that original, seminal contribution of expanding the world to five dimensions. At the time of his analysis of dimensions, Kaluza was a young German, unknown mathematician and physicist, working at the University of Konigsberg in Germany. (Konigsberg, Germany, became part of Poland following WWII, and was renamed Chojna.)

In April 1919 Kaluza wrote a letter to Einstein that amazed him. In just a few lines, Kaluza wrote to Einstein that he was able to unite Einstein's theory of the force of gravity with the electromagnetic force by *introducing the fifth dimension in the equations describing the two forces*. After a considerable time, in which Einstein studied the unifying ideaß, he agreed with it and published the results on behalf of Kaluza.

Other aspects of the unified theory, that do not form part of the dimensions concept, did not work out, however. But physicists continued to think about Kaluza's unique contribution—the role of dimensions in uniting the four forces and also in trying to understand the dimensions of our Universe. So, from now on, let us open ourselves to the idea of dimensions—for the sake of Kaluza, Einstein and String Theory.

The Book, *Flatlands*, and Other Dimensions in Space

When physicists describe a space model with more than three dimensions, they usually describe the additional dimensions as being a minute, small un-seeable, 'appendage' dimension—and being an appendages to our three space dimensions.

If this is a new concept to you—living in a Universe that has more than three space dimensions, I have a suggestion. I suggest you introduce yourself into higher dimensions by reading an amusing, insightful, book named *Flatland: A Romance of Many Dimensions*. Edwin A. Abbott wrote it in 1884. I am sure that if you read it, you will learn to think about our three dimensions in a new way. You need not read the entire book. Read a few chapters to get an idea that there can be higher dimensions that we do not see. Just as the people in this book—who live in two dimensions, do not see the third space dimension.

A brief summary of the book: The book is about a two-dimensional world, where characters (people) are such shapes as a line, a square, a circle, a four-sided polygon—and so on. Just walk over to Figure 30 to meet some of the families and friends who live in the *Flatland* world. All live a serene, calm, orderly life—until a higher dimensional caller (a three-dimension person, like us) comes to visit. The people of *Flatland* cannot imagine—much less visualize, what the visitor could look like. (Imagine what would be the case if a four-space dimension person came to visit us.)

Coming back to our world of three-space dimension, our not being able to 'see' the fourth or higher dimensions should not stop us from analyzing them in an abstract way, using mathematics. Einstein learned a lesson from a fellow physicist and recognized this possibility—by considering that additional dimensions can be present in our lives and in the Universe.

What we are leading up to is String Theory. It is believed to be able to unify all four forces (Theory of Everything, TOE). It is presently based on 11-dimensions—10 of space and 1 of time. (At this point we should be thankful that no one has suggested there can be more than one time dimension.)

Figure 30 The Inhabitants of Flatland

Mr. and Mrs. Cranshaw

MR. and Mrs. Tedderfed

The Cranshaw children:
Joe, Henrietta and Alice

The Lone Range and Tonto

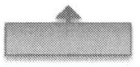 

Animals in the Zoo   **Their Horses, Silver and Scout**

313

Chapter 15. String Theory

String Theory seeks to define the basic building blocks of all the particles we have listed—neutrons, electrons, quarks.... To do this, it is hoped that it will:

- Provide the know-how to enable the four forces to be combined in one model.

- Provide additional insights and theories into the Standard Model.

- Provide a complete time-model of the beginning of the Universe, including the details of what happened during, and possibly before, the first 'moment'.

- Help us understand whether wormholes or other Universes were built in that first 'moment'—or are being continuously built.

- Help us understand the 'unknowns' in the Universe, such as dark matter and dark energy.

- Provide a "Theory of Everything" for particle physics.

A History of String Theory

Previous chapters have described the basic particles that physicists must contend with in their various models of the Universe. That list of particle and their peculiarities are repeated below:

- Electrons, protons, neutrons, photons, quarks [and quarks come in three 'colors' and six 'flavors'], anti-quarks, bosons, leptons, gluons, neutrinos, pions, muons, and so on.... and their antiparticles
- And some of these spin to the right, some to the left, and at different rates of spin. Some have no spin.

314

- They all have different mass from one another. One, the photon, has no mass at all. Also, some act as force carriers of information or forces between two other particles.
- Some have no electrical charge and some can have positive charge or negative charge.
- And so on...

Physicists should be driven crazy by having to keeping track of all the particles, their individual unique characteristics, their spins, their sizes, their relative energy levels, and their distinct interrelations within this family of particles. There must be a better solution, a more 'homogenous' solution.

The world communities of physicists, who work in the domain of String Theory for the past 40 to fifty years, consider that String Theory is that better solution. When I first read about String Theory, I came across write-ups that spoke of strings that acted as rubber bands. I viewed them in the same way as I did in arithmetic classes in elementary school. where we learned about fractions and about a 'least common denominator' or LCD. Aha, the string is that LCD for the Universe, I thought.

Frankly, I was very disappointed when I later learned that the term, *String Theory,* includes a host of many different objects—that are basically similar. It includes:

- Strings that move freely in space,

- Strings that are tied to a post and strings that are tied at one end to a surface,

- Membranes (or flexible sheets of matter) that are called branes, and

- Branes can occupy up to nine space dimensions!

Given the complexity of items above, I gave up the hope to have a truly LCD.

In addition, while this LCD may be simpler in concept than the plethora of parts found in the list of subatomic particles listed in a previous chapter, it comes at a cost—greater complexity in both String Theory, itself, and the mathematics that is involved to describe the theory.

Even though String Theory has been around since 1970, it may be referred to as 'The theory of the 21st Century'. Since those beginnings, it has gone through a number of transitions, to where it now seems to have settled down to where it is now a 'unified' String Theory'—so it may actually be a theory of the 21st Century.

As we did with Quantum Theory, we will first describe its history and then some of the elements of the world of strings. Unlike Quantum Theory, however, where its history was relatively clear while its theory was not, in String Theory, both are not too clear as yet, to the outsider, at least.

Unless otherwise noted, the descriptions into String Theory in this chapter are mainly from two books, *The Elegant Universe*, by Brian Green, 2003, and The Cosmic Landscape: *String Theory and the Illusion of Intelligent Design*, by Leonard Susskind, 2006.

First, I would like to cite briefly two significant events in String Theory's early history.

(The following is from

http://physics.stackexchange.com/questions/28211/gabriele-veneziano-strong-nuclear-force-and-beta-function:

> Gabriele Veneziano, a research fellow at CERN (The European particle accelerator lab) in 1968, observed a strange coincidence—many properties of the strong nuclear force are perfectly described by the Euler beta-function, an obscure formula devised for purely mathematical reasons, two hundred years earlier by the mathematician, Leonhard Euler.
>
> And Veneziano was using general principles to deduce what form a tree-level self-consistent scattering amplitude in a theory of infinitely many particles on straight-line Regge trajectories...

that, presumably, looked-like, or resembled a String. Thus, String Theory. Veneziano must have tipped his hat to Euler when he learned this. And that is but one of the reasons why, later in Chapter 16, I refer to Leonhard Euler as the 'Master Violinist of the Strings.'

And to go to the second major event in String Theory's history, a quote from a web site that is, no longer on the web:

> Yoichiro Nambu, Leonard Susskind, and Holger Nielsen independently discover that the dual resonance model devised by Veneziano is based on the quantum mechanics of relativistic vibrating strings, and String Theory's life begins.

We will now continue the history, which describes some of the features of String Theory itself. Seen from out present vantage point in the middle of the second decade of the 21st Century, it appears complicated. The interplay of String Theory's history and its theory causes some of those complications.

(If String Theory continues to be the dominant path among theoretical physicists, I assume that people who will read about the history of String Theory in, say, 2040, and will perceive it all as a smooth progression of events, rather than a progression of retreats and leaps and starts.)

A brief explanation of what Strings are is helpful (or necessary) to understand its complicated history. Strings first came in various forms of two-dimensional bands. Strings are on order of the size of a Planck Length, 10^{-33} centimeters. Strings can fuse together, making a longer string. And in its early history, the equations told the investigators that String Theory requires nine space dimensions and one dimension for time!

Now, a bit of String Theory's overall early history: By 1985 there were five different versions of String Theory. By then, it was called Superstring Theory. (But many in the physics community continue to refer to it as String Theory.) The new version had both open strings and closed strings and some with 'rigid' ends (i.e., a closed string), as opposed to free form strings. (Note: The force Gravity, represented by the (as-yet-undetected) graviton particle, in String Theory, is considered a closed string.)

Some of those five versions of String Theory appeared to be leading to different laws of physics. Horrors! Thankfully, these runaway versions were eventually realized to be within the bounds of 'our' world of physics. In 1995, at a major international conference, Dr. Edward Witten, a mathematical physicist at IAS at Princeton, New Jersey, showed that all five versions all belonged to one family. He called the family "M-theory."

An important output of the meeting in 1995 was to consider that the strings could also be membranes (called by the physicists, branes) that resemble elastic sheets of rubber. These branes are in addition to the one-dimensional rubber bands. Also, the branes could be of dimensions greater than the standard 2—and the number of space dimensions expanded to 10.

And literally, many physicists, worldwide, continue to expand the bounds of String Theory. And now, what is String Theory?

String Theory—What It Is

As previously noted, String Theory is more difficult to explain in its details since it involves considerable ideas of deep theoretical physics and mathematics. Thus, we will describe it only in an overall sense.

The basic components of String Theory are the strings or membranes themselves. There is no consideration of what may make up a string. It is it. It is the 'true' atom, the indivisible unit, as predicted by the pre-Socratic Greek philosopher, Democritus, (470-380 B.C).

We shall use the word 'string', but it could be a membrane as well. A membrane looks like a string when it is a very narrow membrane.

Chart 31 Characteristics of Strings:

• The size of a string is on the order of a Planck Length, 10^{-33} centimeters.

• The strings/membranes (or branes) vibrate.

• String Theory accounts for all four forces.

• Strings have different vibrating patterns, and it these pattern that determine the particular features of the particle it represents. (Charge, spin…)

• Strings can vibrate in 10 dimensions.

• The energy of a string is proportional to its tension and is a function of the string's vibrating pattern.

- Strings can combine.

- There is a world of 'string' shapes and descriptive names that are part of the string lexicon, such as, branes, membranes, anti-strings and so on.

- String Theory obeys quantum law in that energies exist only in discrete levels.

- String Theory incorporates Heisenberg's Uncertainty Principle and therefore uses probability to define virtual strings, string jitter and paths after collisions.

- String Theory is inflexible and is very precise about what can comprise the description of a string. This is in direct contrast to The Standard Model—which is flexible and allows a wide range of descriptions to be incorporated into it.

- String Theory provides a different depiction of the beginning of the first moment of the Big Bang as described by the Standard Model. (That depiction provides an account of the reason there are 11 dimensions and accommodates the four forces. A description of that depiction is outside the scope of this book.)

- No experimental results have ever been obtained for any part of String Theory! (One of the reasons is presently due to the fact that 'strings' operate at Planck length, which is smaller that the size that present-day colliders can observe.)

- Questions still arise about whether String Theory is a true description of the particles and forces in the Universe.

And finally: At present, String Theory exists only in a mathematical world. There is, as yet, no 'Physical World of String Theory'. The following is a quote from Wikipedia at https://en.wikipedia.org/wiki/Superstring_theory:

According to the theory, the fundamental constituents of reality are strings of the Planck length (about 10^{-33} cm) that vibrate at resonant frequencies. Every string, in theory, has a unique resonance, or harmonic. Different harmonics determine different fundamental particles. The tension in a string is on the order of the Planck force (10^{44}

newtons). The graviton (the proposed messenger particle of the gravitational force), for example, is predicted by the theory to be a string with wave amplitude zero.

Superstring theory is an attempt to explain all of the particles and fundamental forces of nature in one theory by modeling them as vibrations of tiny supersymmetric strings.

Since the second superstring revolution (that generated five versions of superstring theory), the five superstring theories are regarded as different limits of a single theory tentatively called M-theory, or simply String Theory.

In addition to various other questions concerning String Theory, we can ask, 'When will there be colliders or equivalent systems that can verify String Theory? The new CERN Collider is not considered adequate, however, to perform that verification task, so questions remain.

These issues confound many physicists. And there are many theories and heated discussions about what all this means. Debates concerning this topic also involve the Cosmological Constant—and the idea of an Anthropic Universe (a Universe whose universal constants are *designed* (by someone) to support life, continues and continues.

A note about String Theory and 'Physics Time': As noted previously—more than a few times, Time moves very slowly in the world of physics; it crawls. We saw that the initial development of Quantum Theory took thirty years. The next phase, the work that will be required to put all of Quantum Theory in Standard Model, may take about twenty to twenty years. String Theory has been around since only 1970—and is still has not attained the level of possessing a *stable system concept*. Hence, no forecasting for when it is expected to be 'accepted' and part of the Standard Model. Just patience.

Part II

A Reprise: The Language of Physics— Mathematics; and the Tools that Physicists Use to Observe, Detect, Measure and Analyze Elements and Particles

Chapter 16. Mathematics, the Language of Physics

This chapter discusses:

- The Role of Mathematics within Physics

- Three Great Mathematicians: Leonhard Euler, Srinivasa Ramanujan and John Von Neumann.

The Role of Mathematics in Physics

First, two comments:

1. Roland Omnès, an esteemed Professor of Theoretical Physics in France and graduate of the élite *Ecole Normale Supérieur* in Paris, states in the beginning of his 2002 book "Quantum Philosophy":

> If I had to name the greatest thinker of all times, I would say, without hesitation, Pythagoras, who lived on the Greek island of Samos, 6th Century B.C.).... He said that numbers rule the world.

2. To the aspiring Scientist or Mathematician: Follow the numbers, no matter where they lead you. They hold the truth to the Universe.

This chapter describes the important working relationship between the world of physics and the world of mathematics. This relationship is both synergistic (work together or enhance one another) and symbiotic (two different systems live together). This relationship is a key operating feature in the advancement of science.

First, note that there are essentially two branches of mathematics: Theoretical Mathematics and Applied Mathematics. The discussion in this chapter will be about Applied Mathematics. But, in deference and respect to the Theoretical Mathematician, I would first like to explain what a theoretical mathematician does. The following paragraph is from the U.S. Department of Labor's web site:

"Theoretical mathematicians advance mathematical knowledge by developing new principles and recognizing previously unknown relationships between existing principles of mathematics. Although these mathematicians seek to increase basic knowledge without necessarily considering its practical use, such pure and abstract knowledge may eventually apply to new-found scientific or mathematical principles."

Examples of Where Applied Mathematics Relates to The World

There are many people in the world who are trying to understand what happens in the physical world. For example, there are:

- Weather and climate analysts, who seek to understand what makes and drives the various weathers and climates.
- Oceanographers who seek to understand currents and tides in the water.
- Engineers who build buildings and calculate what forces of weather can or will knock down a structure.
- Aeronautical engineers determining the strength of materials.
- NASA engineers designing space-flights trajectory to the planet Mars.
- An automotive engineer designing a new type of contour (curvature) for the hood of a new sports car.
- Physicists analyzing a substance when the temperature of the substance approaches zero degrees Kelvin, or analyzing how electricity flows within the substance, also near zero degrees Kelvin.
- Computer architects trying to reduce the number of components (molecular or smaller) on a memory chip or a logic board.
- Physicists trying to understand the inner features of Black Holes.
- Physicists developing String Theory.
- Physicists attempting to incorporate Quantum Gravity into the Standard Model.
- Physicists trying to harness and apply Quantum Physics to a wide range of objects, including computers, communication systems, medical detection, monitoring systems…
- And thousands more.

These are all people who rely on mathematics to describe the process they are analyzing. Thus, that is why this sector of mathematics is called Applied Mathematics.

They first try to understand the process. When they achieve that, they seek to describe it by mathematical formulae, called equations. Sometimes the formulae are known and available; sometimes there is no formula available. In the latter case, they try to derive it themselves—or they wait for a separate group of mathematicians to develop the formula. Frequently, the development of the mathematic equation provides to the developer, additional insights in the physical process they are investigating.

Mathematics is a usually a stand-alone profession performed by mathematicians; they produce mathematical formulae. Since not many people 'buy' mathematical formula, mathematicians work in, or with, other professions such as being instructors and professors in colleges and universities, and in government and not-for-profit organizations (where mathematical studies are performed) and in industrial and research organizations.

A modern application is the development of a spectacular range of mathematical formulae to describe—and predict, the events and activities of the stock market, the bond market, the futures market, the real estate market, the credit market, the mutual fund market, and so on.

Many of us learned about Pythagoras, a Greek mathematician (*circa* 500 B.C.), who looked at a right triangle and applied an equation to it that we call the *Pythagorean theorem*. The theorem is that the sum of the square of the sides is equal to the square of the hypotenuse. That equation enables people to calculate the sides of a right triangle for all different size right triangles.

I would to dispel some of the miss-conceptions we may have of Pythagoras. He lived on the island of Samos in the 6th Century B.C., and there, he founded, organized and directed *a group* of mathematicians and philosophers. And all their work of analysis and development, work that is now known as being the work of Pythagoras (alone), was actually done by the members of the group, including Pythagoras.

324

In the case of Pythagoras—or a fellow member, he would see a physical object and realized the object could be described by a mathematical formula. (Just as Kepler did, but Kepler did it for the planets.)

There are many different, well-known equations, and they frequently are referred to by the name of the founder. For example, there is the Bessel Function, named after **Frederick Bessel** (1784-1846). Fortuitously, The Bessel Function can be used—and is used to describe the FM radio wave that we listen to on the FM band of our car, iPods and home radios.

And then there is the example that always intrigued me. It has to do with a favorite mathematician of mine, **Leonhard Euler** (1707-1783), considered the greatest mathematician of the 18th Century. Engineers in the 19th and 20th Century were building tall buildings. They would put a steel beam vertically into the ground. It was held in the ground by concrete or some other type of 'building glue'. They also attached steel beams to other beams, extending their height. Or they would place a steel beam horizontally, supported by other beams.

In these cases they wanted to know how far the steel beam would bend if various forces pushed on it. The deflecting forces could come from a variety of outside forces such as the weight of the structure itself on top of a beam, as well as wind, rain, an earthquake or a flood. They sought an equation that can *calculate the load-carrying and deflection characteristics of the beams*.

Leonhard Euler had developed such a formula in about 1755. This was long before steel was even thought about. Thus, Euler had no idea what a steel beam was. But his formula was used one hundred fifty years later to predict the bending of the beam if or when a force pushed acted on the beam. And civil engineers used the Euler equation as they created new buildings and new cities.

The first major test of the Euler equation (now called the Bernoulli-Euler formula) was it application to the building of the Eiffel Tower in the late 19th Century. Euler developed the equation because it was mathematically interesting. Pure Math, we might say. We will come back to Euler in a few pages.

The Example of the Application of **Bernhard Riemann's** Math:

Bernhard Riemann developed another set of 'exhilarating' equations in 1854. These equations describe a geometry that was different from Euclid's two-dimensional geometry, which you and I had been taught in school. Riemann's geometry represented 'unflat', undulating surfaces. (The author realizes that this topic was discussed in Chapter 11, but its greater clarification can be important to the aspiring mathematician or scientist.)

In the early 1910s Albert Einstein had developed a theory of space and gravity. But he did not know how to express it mathematically. It was only after he learned (from his mathematician-friend, Marcel Grossman) about Riemann's geometry, that he could describe his thoughts mathematically.

This application of Riemann's geometry became known as a crucial part of Einstein's Theory of General Relativity. Without that mathematical description, Einstein—or any one else, could not have made calculations about gravity, space and Universe for various different situations.

I add the following thought about 'waiting for mathematics': Physicists today work in String Theory. They have many ideas about it, but they may not have—or do not have, all the mathematics to enable them to truly work in it, or be fully productive in their work. Thus, they may be a need for new mathematic system to describe their theories. Or possibly the math may already be published in the literature, but as yet undiscovered by the physicist or scientist who is seeking those certain formulas.

I end this introduction with a quote of Albert Einstein: "The approach to a more profound knowledge of the basic principles of physics is tied up with the most intricate mathematical methods."

1. Three Outstanding Mathematicians: Leonhard Euler, Srinivasa Ramanujan and John Von Neumann

I have long been aware of the extraordinary and interesting lives that many mathematicians have led—and would like to devote many more pages to their personal history and contributions to pure math and it relationship to the world of physics. Since that is not really realistic for a book devoted to physics, I limit my discussion to only three, very different mathematicians, Leonhard Euler, Srinivasa Ramanujan and John von Neumann.

The first is presented for his contributions into many areas of mathematics and physics, the second for his overall creativity and genius, and the third for his creativity and leadership in three very different, modern arenas.

Leonhard Euler, The Most Prolific Mathematician of All And 'Master Violinist of the Strings.'

The first time I learned about Leonhard Euler, the Swiss mathematician (1707-1783), was when I was taking a civil engineering course, examining the strength of steel beams. The professor described an equation that he attributed to Euler.

This was the first time that I identified a mathematical equation in a modern engineering class that was produced by someone of the 18th Century—before steel was even made. For me this was a 'eureka' moment. Later, in many of my engineering courses, I learned of other such 'earlier' math systems that originally had no specific application, such as those of Fourier and Bessel's—but later had have strong roles in modern physics and engineering.

I will not go into Euler's life, since I hope you will read about him on the many biographies that are on the web. They show his varied travels (Switzerland, where he was born, Russia, Prussia...) and the numerous other leading mathematicians and royalty of his time (The Bernoullis, King Frederick II of Prussia, Catherine II of Russia...). Euler prepared twenty-three volumes containing his work. And during the last twelve years of his life he was blind in both eyes—but he continued to formulate and solve mathematical problems!

I list some of his varied contributions below and give a number to each of Euler's 'fields of endeavor'. I believe, however, that there are many more that are not mentioned here.

1. Euler was responsible for "Combinatorial Analysis". An example of combinatorial analysis: How many ways can eight different items be taken from a group of 21 different items? Combinatorial Analysis is fundamental to the development of statistical theories used in Quality Control, a key ingredient of the modern industrial age.

2. Euler is considered the father of "Modern Graph Theory". Euler's graphs were what are called *linear graphs,* which he developed to solve problems. Today we all know of the 'salesperson problem', which these linear graphs solve the problem of where a salesperson must visit a certain number of clients and the routes are analyzed to minimize the trip—and thus, implicitly, to increase sales and profits.

The web abounds with pictures of Euler and the classic "Konigsberg Bridge problem". Euler first considered the problem of a person leaving her home and crossing all seven bridges in the city of Konigsberg (now Kaliningrad). Would the person be able to cross all the bridges just once? The result was 'no'.

3. A branch of the mathematical discipline known as Calculus is "Differential Equations". It appears that there were two important phases in the development of Differential Equations. Differential Equations are the probably most widely used system of equations to express physical situations in which the parameters, time and changing time, appear, such as in liquids and gas flows. In a similar time-varying case, Schrödinger used differential equations to express his time-varying quantum equations.

The first phase contained the contributions of Isaac Newton, Leibniz, and Jacob and Johann Bernoulli. It is said that Leonhard Euler dominated the second phase in the developmental history of differential equations. An example of their use is their use by Erwin Schrödinger as the foundation for his formulation of the mathematics of Quantum Theory, 'bypassing' Heisenberg's matrix formulation.

4. In the sector of physics known as *Fluid Mechanics*, Euler explicitly set forth the concept of the *internal* pressure in a liquid. He also developed the equations for the formulation of a three-dimensional description used for fluid flow. The study of the pressure and flow of liquids is important today in many of our common industrial systems. Examples include petroleum, natural gas and water pipe line networks that crisscross our continent; similar pipelines networks interconnect European and Asian countries enabling them access products across national borders; and there are similar pipeline networks within many of the world's cities, themselves.

Euler, along with Daniel Bernoulli, another famous mathematician and member of the large mathematically-endowed Bernoulli family, contributed greatly to the formulation of *Hydrodynamics*, the flow of fluids. Euler developed the general non-viscous equations of motion of a fluid particle.

Viscosity in a fluid is a measure of the 'thickness' of the fluid. It can go from minimum viscosity such as water and gasoline, to light oils, to honey and to thick oils. Since Euler was setting up a new field of investigation, we have to assume he first took on the easiest problem, the minimum-viscous case, such as water.

Euler also began the development of the modern water turbine. He performed experiments on the mechanics of actual turbine wheels, which were called 'reaction wheels'. He also developed the equations for the analysis of the reaction turbine. It is apparent that Euler had a tremendously strong capacity to analyze and solve numerous, different problems. Thus, let us put some of his other accomplishments in list form, as follows: Euler

- Analyzed problems in Newtonian mechanics and extended his studies to astronomical problems.
- Made the first breakthrough in one of mathematics' "hardest" problem, the solution of "Fermat's Last Problem", or "Fermat's Enigma", for the case where x=3. (Refer to source 26.)
- Developed what is called the *Euler Characteristic*, which is in the field of Algebraic Topology. It deals with spatial objects, shapes and sets.
- Suggested, in 1760, that the ether (the medium that was then thought to occupy space and that enabled the propagation of light) is also responsible for electrical phenomena. This was about fifty-five years before Faraday's experiments on the electrical phenomena. And this was about one hundred years before Maxwell's combining light, electricity and magnetism into one system.

I leave to the end one more item concerning Euler—an important one in modern physics. You recently read about String Theory and Euler's role in providing an explanation of the equations that were established centuries later by Gabriele Veneziano. It is for this contribution, that I call Veneziano, *Master Violinist of the Strings.*

Note: The web site, http://james.fabpedigree.com/mathmen.htm is titled *The Greatest Mathematicians of All Time*, according to the site's author. Leonhard Euler is number 4 on his list, "beaten" by Archimedes of Syracuse, Isaac Newton and Carl F. Gauss. The site also has brief biographies of the mathematicians.

A Creative Mathematician, Srinivasa Ramanujan—and a Genius (1887 – 1920)

I first became aware of the personage and background of Srinivasa Ramanujan in the book, *Hyperspace* by Michio Kaku, source 19. What follows is a unique story.

Srinivasa Ramanujan was a mathematician who had a non-standard personal history, to say the least. He developed many theorems by himself, living in India, and many of these theorems were unknown in Europe and America. Some of his work may help us understand some of the basic problems and unsolved theories that challenge physicists today and in the future. For example, in the part of physics called 'String Theory' people believe that Ramanujan's formula called the *Elliptic Modular Function* will help describe the theory further, mathematically.

For example, from Wikipedia and the math site, usna.edu/Users/math/meh/ramanujan.html, I learned more about him: Ramanujan lived from 1887 to 1920, dying at only age 33. He lived in India. As a college dropout from a poor family, Ramanujan's life was difficult. He lived off the charity of friends, filling notebooks with mathematical discoveries and seeking patrons to support his work.

Finally, he met with modest success when the Indian mathematician Ramachandra Rao provided him with first a modest subsidy, and later, a clerkship at the Madras Port Trust. During this period Ramanujan had his first paper published, a 17-page work on Bernoulli numbers that appeared in the *Journal of the Indian Mathematical Society* in 1911.

No one in the Western world would have known about him, except for the fact that, in 1913, Ramanujan wrote to the English mathematician, G.H. Hardy, at Cambridge University. The following is from that same web site cited above: http://www.usna.edu/Users/math/meh/ramanujan.html

"The ten-page letter contained about 120 statements of theorems on infinite series, improper integrals, continued fractions, and Number Theory. Every prominent mathematician gets letters from cranks, and at first glance Hardy, no doubt, put this letter in that class. But something about the formulas made him take a second look, and show it to his collaborator, J. E. Littlewood. After a few hours, they concluded that the results "must be true because, if they were not true, no one would have had the imagination to invent them."

The two professors realized that Ramanujan had derived, by himself, 100 years of European mathematics!! There were even theorems that had not yet been derived, or known, by the people at Cambridge. They sent for Ramanujan and he stayed at Cambridge from 1914 to 1918.

Ramanujan's arrival at Cambridge was the beginning of a very successful five-year collaboration with Hardy. In some ways the two made an odd pair: Hardy was a great exponent of rigor in analysis, while Ramanujan's results were (as Hardy put it) "arrived at by a process of mingled argument, intuition, and induction, of which he was entirely unable to give any coherent account".

Hardy did his best to fill in the gaps in Ramanujan's education without discouraging him. He was amazed by Ramanujan's uncanny formal intuition in manipulating infinite series, continued fractions, and the like: "I have never met his equal, and can compare him only with Euler or Jacobi."

At Cambridge, Ramanujan developed 4,000 formulas on 400 pages. Much of his work describes formulas that no one to this day has any idea of its use. But, as we know, in the future a physicist or cosmologist may describe something and he will be looking for a mathematical formula to fully describe it.

Perhaps, in the future, in the year 2135, members of a university cosmology department will be discussing a newly observed phenomena in far outer space. They are discussing the phenomena with the department head, attempting to understand what it all means. And at that moment, one of their colleagues enters the room, having just reviewed a section of Ramanujan's equations that had never been identified with any topic in the physical world

—and they all see its relevance to their newly observed phenomena. "Champagne, anyone? For us and for Ramanujan."

John von Neumann (1903 -1957)

While writing this book, I had been considering whom I should select as the third mathematician for this chapter. Names such as Galois (founder of *Group Theory*, who died at age of 21), Gauss (who was called *The Prince of Mathematicians*) and Riemann, come to mind.

During the Bush administration in Washington, (2001-2009) and later, during the Obama administration, there were many confrontations between the Congress and the president. I thought about how this might be handled in Washington on both sides—and realized that this is a classic case to be handled by the mathematics called, *Game Theory*. Thus, I selected the mathematician, John Von Neumann, who developed Game Theory, along with Oskar Morgenstern, an Austrian economist at Princeton University.

John Von Neumann, (1903-1957) a Hungarian mathematician, distinguished himself from his peers—even in childhood, for having superior mental faculties. For example, he had a photographic memory. He was able to memorize and recite back a page out of a phone book in a few minutes. Also, at the age of six he could divide eight-digit numbers in his head. He was notable for having been involved in many aspects of modern physics and engineering.

Von Neumann published his first mathematical paper at the age of eighteen, in collaboration with his tutor, and decided to study mathematics in college. In 1926 he received his Ph.D. in mathematics with minors in physics and chemistry from the Pázmány Péter University in Budapest, Hungary.

First, Quantum Theory

His first entry into the world of mathematics appears to have been in 1927, when Von Neumann applied new mathematical methods to Quantum Theory. His work was instrumental for subsequent interpretations of the theory. From http://plato.stanford.edu/entries/qt-nvd/#1

In the late 1920s, von Neumann developed the separable Space formulation of quantum mechanics, which later became the definitive one (from the standpoint of mathematical rigor, at least).

Part of his expressed motivation for developing these mathematical theories was to develop an appropriate framework for Quantum Field Theory and a better foundation for Quantum Mechanics.

The culmination of this work was the publication of his seminal work, *Mathematical Foundations of Quantum Mechanics*.

According to Princeton University Press, which eventually republished the above-named book in 1995, the work "was a revolutionary book that caused a sea change in theoretical physics. ...(The book) shows that great insights in Quantum Physics can be obtained by exploring the mathematical structure of quantum mechanics."

Second, Game Theory

Game Theory was a new sector of mathematics; it was developed to provide strategies for a wide sector of 'users'—governments and legislatures, the military, the stock market, businesses, etc. Game Theory was invented or started by von Neumann. His article, "Theory of Parlor Games," (1928) was the starting point of that theory. For von Neumann, the inspiration for Game Theory was poker, a game he played occasionally (not terribly well, it is said).

Von Neumann realized that poker was not guided by Probability Theory alone, as any unfortunate player who would use only Probability Theory would find out. Von Neumann wanted to formalize the idea of "bluffing".

At a time of political unrest in central Europe, he was invited to visit Princeton University in 1930. When the Institute for Advanced Study (IAS) was founded at Princeton in 1933, he was appointed to be one of the original members of its faculty. In 1943, during World War II, von Neumann was invited to work on the *Manhattan Project* that called together the quantum scientists to build the first atomic bombs. Von Neumann did crucial calculations on the implosion design of the atomic bomb, enabling a more efficient weapon to be developed.

In 1944 he teamed up with Oskar Morgenstern, an Austrian economist also at Princeton, to develop Game Theory and they published the first definitive book on the subject, *Theory of Games and Economic Behavior.* Although the work itself was intended mainly for economists, its applications to psychology, sociology, politics, warfare, recreational games soon became apparent.

From the very beginning of World War II, von Neumann was confident of the Allies' victory. He sketched out a mathematical model of the conflict from which he deduced that the Allies would win, applying some of the methods of Game Theory to enable him to arrive at his predictions.

By the latter years of World War II von Neumann was playing the part of an executive management consultant, serving on several national committees, and applying his amazing ability to rapidly see through problems to their solutions. He worked in a variety of technical areas. such as hydrodynamics, ballistics, meteorology, game theory and statistics.

Later, he used his methods to model the Cold War interaction between the U.S. and the USSR, viewing them as two players in a game. A final (remorseful) note on Game Theory: Seven Nobel Prizes have been awarded for their contributions to Game Theory. And the 2007 Nobel Prize in economics in "Mechanism Design", shared by three persons, is also related to Game Theory. Albeit, none had been awarded to von Neumann.

(While Game Theory involves a considerable amount of mathematics, there is no Nobel Prize for Mathematics. Hence, the prize is given under the heading of Economic Sciences.)

Allan Karson

Chart 33: Application Areas for Game Theory—Developed by the Mathematician, John von Neumann

	Issues in Game Theory	
Political Country #1 (United States)		Country #2 (Iran, Iraq, North Korea, Russia - Cold War, OPEC, Cuba, Hugo Chavez…)
The Administration		The Congress
Economics Buyer/Leaser of a Corporation, or a Property, or a Natural Resource	Bluffing Coalitions Uncertain Information Auctions Evaluation of Utility Rational/Irrational Strategies Communication Between Players	Seller/Lessee of a Corporation, or a Property, or a Natural Resource
Product Recall Sports Franchises— Steroids, doping… Public Offerings of Stock		Public Fans Stock Market

Third, von Neumann, The Transcendental Computer Architect

During World War II, there was considerable interest in the computation of mathematical tables, such as for bomb, rocket and artillery trajectories. The center for this effort was the U.S. Army's Ballistics Research Laboratories (BRL) at Aberdeen Proving Ground, Aberdeen, Maryland.

Before and during World War II, there was no such thing as a digital computer. People relied on mechanical desk calculators. People who were at BRL at the time have told me that there were groups of people—as many as fifty persons, grinding out trajectory tables using the large mechanical desk calculators.

The first all-electronic machine, called the ENIAC, was built in 1946 for BRL, to replace the desk calculators. Compared to today's machines, ENIAC was primitive. The ENIAC used large vacuum tubes and was fix-wired; no programming was involved. It was a base-10 machine, as opposed to today's base-2 systems. It was capable of performing 5,000 additions, or 357 multiplications, or 38 divisions in one second. At that time, however, its speed was considered fantastic. (Do not laugh. Think of what people will say about our present computers, fifty years from now.)

Neumann was asked to develop tools to improve the situation. Neumann's interest in computers was actually different from most people, since he sought to use computers for the solution of a wide range of applications—rather than the limited application of the development of trajectory tables.

Through von Neumann's work, he was fully aware of the ENIAC at BRL, as well as other developments going on throughout the United States. Such developments included the work at Harvard on the Mark I (ASCC) Calculator and at IBM's Watson Scientific Computing Laboratory located near at Columbia University. All of these were relatively slow. Clunkers, compared to today's machines.

Von Neumann examined the foundations of *all* these machines and came up with one, new, computer system architecture, which became known as *The von Neumann Architecture.* The basic elements of the von Neumann Architecture were introduced to the computer industry in the report titled *First Draft of a Report on the EDVAC*, (1945), which was the *first stored program computer.* John von Neumann was the sole author of that seminal report. The EDVAC was also built and installed at the Ballistics Research Laboratory at Aberdeen Proving Grounds, Maryland.

Even today, his EDVAC design may be the *basic* design of many of the world's computers. It is called "The von Neumann Architecture". For example, EDVAC was the first to introduce the self-modifying "codes", (or self-modifying software) by which a fixed system of wiring could solve a great variety of problems. This is the fundamental concept of present day software programs. And von Neumann also recognized the need for parallelism in computers. Ancient history, but that is the where it all began.

In summary, John von Neumann: A leader in three fields: Quantum, Game Theory and Computer Science. On the web site list of the fifty greatest mathematicians ever, von Neumann is listed as number twenty-four.

Suggested Reading on the Web

If you have attained a degree of interest in mathematics and mathematicians, you may want to read more about the lives and contributions of other mathematicians. An exhaustive list of mathematicians may be found at the web site listed below, where you can look up your favorites. Some of those I would suggest reading about are listed below.

http://en.wikipedia.org/wiki/List_of_mathematicians

- Abel, Niels Henrico (Norway, 1802-1829, source 48, *The Equation that Could Not Be Solved*).
- Galois, Evariste, (French, 1811-1832), also, as above, source 48.
- Gauss, Karl Frederick (German, 1777-1855) Referred to as The Prince of Mathematicians, was supervisor of Riemann.
- Kurt Gödel (Born Austrian, later U.S. citizen, 1906-1978) logician, considered by many to be the greatest mathematician of the 20th Century. A friend and colleague of Einstein.
- Hilbert, David, (German, 1862–1943) mathematician considered "one of the most influential and universal mathematicians of the 19th and early 20th centuries.
- Leibniz, Wilhelm, (German, 1646-1716) a co-founder of calculus, but independently of Newton; an opponent of Newton's concept of absolute

space and a supporter of a relativist world, as later introduced by Einstein.

- Noether, (Amalia) Emmy, (German, 1882-1935) mathematician, considered by many, including as one of, if not *the greatest* woman mathematician. She studied and received PhD under David Hilbert' guidance and taught with him at Göttingen. She developed Noether's Theorem that has been called "one of the most important mathematical theorems ever proved in guiding the development of modern physics." She left Germany when Jews were excluded from teaching there and settled at Bryn Mawr College in Pennsylvania.

- Post, Emile (Born in Russian-controlled Poland, later U.S. citizen, 1897-1954) logician and an inspiring, tireless professor of analytic geometry and calculus at the City College of New York (CCNY). (I was a student of his, and he showed me how to study mathematics: Do *all* the problems at the end of each section in my math book.)
  Professor Post was one of a group of three mathematicians who, in a few months in 1936, independently of one another, created the first abstract designs for universal computers. The other two, who are better known than he, are Alonzo Church and Alan Emil Turing. Turing's work is the one usually referred to in this group of three mathematicians.

- Pythagoras, Greek, 6th Century B.C. Fascinating: source 15 for a brief 'tale' at the beginning of *The Emperor's New Mind*)

- Riemann, Bernard (German, 1826-1866) Inventor of Riemann Geometry, a non-Euclidian geometry that provided Albert Einstein the mathematical tools to describe his General Theory of Relativity

For those who wish to get an overview of the Great Mathematicians, I heartily endorse the web site whose title is: The Greatest Mathematicians of All Time, which is at http://james.fabpedigree.com/mathmen.htm

A final suggestion about a profound book that discusses the relationship between mathematics and physics—and science in general: If such a subject interests you, I suggest reading certain chapters in source 9, *Mathematics, The Loss of Certainty* by Morris Kline. Kline was among the notable mathematicians of the 20th Century.

338

Allan Karson

In Kline's book, chapters 14 and 15 contain a profound and intense discussion concerning the relationship between mathematics and science. Kline specifically addresses the question: Where does mathematics get its most stimuli for growth? Does it get from work in pure math or by working with science? If you have an interest in this question or issue, I recommend Kline's book for further reading.

A Reprise: 2. To the Aspiring Scientist or Mathematician: Follow the numbers, no matter where they lead you. They hold the truth to the Universe. Two brief examples:

1. The following concerns 'imaginary numbers.' Imaginary may not be the proper word for this class of numbers since they are widely used in fields such as electrical engineering, quantum, chaos theory and mathematics in general to solve problems. Mathematicians, physicists and engineers deal with them as if such numbers are real.

The principle feature of the system is the letter, 'i'. 'i' represents the square root of -1. 'i' is considered which is an 'imaginary number.' Obviously, it is inexpressible in 'normal' numbers. So we use 'i'

Who would think of using such an idea as the basis of a system of mathematics? Well, Heron (10 -70 AD) of Alexandria, Greece, is said to have conceived these numbers. But it was Raphael Bombelli (1528 – 1572) who is said to have established the rules and be the first person to put it into use was Girolamo Cardano (1501 – 1576). Both were Italian mathematicians.

2. **Paul Dirac, English theoretical physicist (1902–1984)** followed the numbers—in this case, the plus/minus signs (+/-) that were intimately associated with particles. But in Dirac's mathematical investigations, he was observing the *reverse* from presently accepted norms for such particles. An example is his observing—in his mathematical equations, a positive electron!

Dirac followed up on his equations, numbers and signs, to where he identified a completely 'other sided' world: where electrons have their positive counterpart, now called the positron. (The presence of positrons in in the early Universe has changed some of the prior theories of what occurred in the Universe after the Big Bang.) As noted earlier, Dirac was awarded the Nobel Prize in 1933 for this discovery.

Chapter 17. 1. How Physicists Identify Elements That Can Be Either Close or Billions of Miles Away—And 2. See What is Happening Between Very Small Particles (~10^{-15} centimeters).

In order to learn how physicists accomplish those tasks—finding, seeing and identifying far-off element-particles, we follow a historical road map that leads to Quantum Theory. The path was initiated in the 17th Century by performing spectrum analysis of light that came from elements located on Earth and in the stars.

This was followed, in the late 19th Century and early 20th Century, by shooting particles at particle targets, and then recording and seeing the happenings and the results. These activities, linking experiment to theories, produced much data, insights and theories. Today, this work continues in many sectors of physics with a focus on Quantum Mechanics.

1. Identifying Elements Far Away (Greater than 5,000,000,000,000 miles away)—Using Spectroscopy

Physicist can tell a great deal about an element by looking at the light waves that the element gives off to the observer. And if the element is unknown to the physicist, the physicist can identify an element from its spectrum signature—for which we will describe and show examples.

The observation is done on the light waves light emanating from an object that is close up, or from light that originated far away, such as light from a star. The light waves can consist of both light waves that produce visible colors and light waves that are not visible, such as those in the ultra-violet and infrared ranges—plus rays beyond those 'outer' ranges. To make theses observations, physicists rely on the science of *Spectroscopy*.

Note: It is due to the limitation of the human eye-seeing system that we cannot see outside the 'visible' light range. As far as light and elements are concerned, all light is equal. You can review the contents of the spectrum of natural light at figure 11. End of note.

Physicists use an instrument called the spectroscope to perform "spectrum analysis." We can also say that that physicists study the light and what 'makes up that light'. Let us first review the history of the study of light and its offshoot, spectroscopy. The following is a brief history of spectrum analysis and it finishes with a specific example how spectrum analysis can detect elements in a far-off star.

Discovering the Spectrum of Light

In the second half of the 17th Century, **Isaac Newton** (1643-1727) applied the word "spectrum" to the colored light obtained by refracting sunlight through a prism. You can read about it in the many books that describe Newton and his numerous activities and scientific breakthroughs.

(I suggest source book 17, *The Life of Isaac Newton*, by Richard Westfall. The book is for 'beginners', and is a slimed-down version of Westfall's more-complete, 900+ page, highly respected biography of Newton, *Never at Rest*.)

Newton is considered the discoverer of the decomposition of white light in the spectrum and, also, the founder of the modern science of optics.

Aside: Remember how, when we were children, we were amazed to see all the colors that came out from light going through a simple prism. And we also can see a spectrum when there is a puddle of oil on the ground, which shows the spectrum of the light shining on the oil. And another form of a spectrum is the rainbow in the sky, which presents the spectrum of light in the atmosphere. End of aside

Following Newton on this path, in 1802 **William Hyde Wollaston** (1766-1828) fitted the entrance of his spectroscope (which was probably a homemade device, using the same principles as Newton) with a fine slit to improve resolution. This helped him see, or discover, the presence of fixed black lines within the of light that was given off by our Sun. Aha! Spectrum analysis *from afar*.

Next, the spectrum of the elements in 1826: **Fox Talbot** (1800-1877) and **William Herschel** (1792 – 1871) studied the changing colors of flames when sodium, potassium, lithium and strontium salts were introduced into the flame. Spectrum analysis *from nearby*. Talbot has been quoted to say in 1834, "…optical analysis can distinguish the minutest portions of … substances form each other with as much certainty… than any other known method."

In 1860 **Gustav Kirchhoff** (1824-1887 and **Robert Bunsen** (1811-1899) published the results of their spectrum analysis of the elements sodium, potassium, strontium, calcium and barium in the German scientific journal, *Annalen der Physik und der Chemie*. This provided further impetus to the science of spectroscopy.

In 1859 Kirchhoff found that there was sodium in the Sun. Thus, it is Kirchhoff who is attributed with saying that spectrum analysis can be used for the detection and identification of elements in space. Henceforth, the science of spectroscopy consisted of examining the light given off by an element, to identify that element, located both near and far.

Some notes on the preceding history:

We note that William Hyde Wollaston, earlier than Kirchhoff, was looking at elements in space and saw their signature frequency lines. But it appears he saw them all as being black—not colored lines. (Black lines normally mean that gases absorbed the frequency before reaching the spectrometer.) Thus, Kirchhoff is the person who is considered the founder of spectrum analysis for elements in space.

A minor note: Physicists use spectroscopy to make their determinations for elements that are nearby; cosmologists use it to make their determinations for elements that are far away. There are additional names for specific research areas within those two overall sectors, (such as the one Kirchhoff named, Stellar Spectroscopy, which is for studying stars). End of note.

Finally, a statement of simple fact: Each element has its *unambiguous* set of colors that it gives off when it is heated or when light is incident from that element. Thus, Each element has its own *unique* fingerprint.

To do Spectroscopy:

To do spectrum analysis well and properly, it is necessary to have two system components: 1. A reference-description of the element's color spectrum, and 2. A device to separate and discern the spectrum of frequencies being observed.

Let us first look at the reference-description of each and every element's color spectrum. One source of knowledge was available in the early years of the 20th Century. This was a six-volume handbook on spectroscopy put together by the German physicist, **Heinrich Gustav Johannes Kayser (1853 - 1940),** called, The *Handbook of Spectroscopy*. Its first volume was available in about 1901 and contained 800 pages. According to Pais in source 8, the whole series was more than 5,000 pages!

Kayser was active in various studies related to spectrum analysis. For example, he is known as the first physicist to discover helium in the Earth's atmosphere; it had already been detected in space. In fact, it was the first element found in the Sun, and thus its name was derived from the Greek word for Sun, *Helios*.

We will meet the next contributor to spectroscopy, **Arnold Sommerfeld (1886 - 1951)** Professor at University of Munich. Sommerfeld published *Atombau und Spektrallinien* (1919), (Loosely, *Atomic Structure and Spectral Lines*) which, in the early years of Quantum Theory was *the* "bible" for physicists performing spectral analysis. Sommerfeld is quoted as saying, 'the language of spectra ... a true atomic music of the spheres.' In later years, much more was added to this early data base, but it is instructive to see that much was known early-on, when physics was first starting to spread its broad, powerful wings at the beginning of the 20th Century.

The Problem of Separating the Close Lines in the Spectrum—and Henry Augustus Rowland's Solution.

The other system tool needed in spectrum analysis is the device to see *clearly* and *distinctly* the lines and their color. Given the fact that there was Kayser's handbook in 1901, it might be assumed that that was all that was needed and spectrum analysis would not be too difficult a task. That is, that it might be considered to be more like a typical laboratory assignment. But there was a major problem in all this at the time—the unavailability of *quality* spectrum analysis equipment.

Performing spectrum analysis concerns 'sightings' in space—and, the follow-up, the analysis performed on laboratory equipment in a laboratory. Thus, the examination of the spectrum, itself, using then-standard equipment, did have *major* problems in the early 1900s. The *key problem* to solve was, basically, to *separate clearly* the lines representing the numerous frequencies being looked at in the spectrum. Thus, all spectrum-investigators recognized, early on, that the system-device that was *truly vital* to the process—*but* there was no-such precise equipment that was built or available at that time. It was a device later called a *diffraction grating system*—a device that *clearly* 'spread out the spectrum', so *each* frequency was clearly evident.

One person, **Henry Augustus Rowland, (1848 - 1901)** did solve that problem and developed and produced the technology and product—more than a century ago, and his diffraction grating system is still the basis of present spectral-analysis system that are still in use today. Rowland's first system (*circa* 1882) produced a map of the spectrum of the Sun that was ten times more accurate than any other available at that time period.

Henry Augustus Rowland was the first president of the American Physical Society and the first occupant of the Chair of Physics at the Johns Hopkins University of Baltimore, Maryland. At that time, he was considered a high-quality, world-class physicist, who, albeit, is (still) relatively unknown today, even in his own country, the United States.

Diffraction gratings systems, at the time Rowland started to build his diffraction grating in 1882, were imprecise. Among their problems, they yielded spectra that contained ghostlike, false signals among the true spectral readings. Rowland's gratings, which overcame those problems, consisted of:

- Gratings that are made of metal or glass with a very large number of parallel lines—thousand per inch, with *extreme* accuracy, ruled by means of a diamond point.

345

- A ruling engine for lines—wherein the engine contains a screw of nearly *flawless* pitch.
- A grating that could display *400-800 lines per millimeter* and equidistance accuracy greater than *1/4000th of a millimeter,* and
- Spherically curved plates for the diffraction grating. (Systems built prior to Rowland's grating system used flat plates.)

As mentioned previously, Rowland's first system (*circa* 1882) produced a map of the spectrum of the Sun that was ten times more accurate than any other spectrum analysis system at that time. It is assumed that *some* of the technical features of that system are still incorporated in today's computer-based diffraction grating systems.

Charts 34.1 and 34.2 How Physicists *Can* Identify Elements That Are Billions of Miles Away

34.1 Spectrum of the Sun

The picture of the (visible) spectrum of the Sun and accompanying text is from:
http://www.pbs.org/wgbh/nova/teachers/activities/3113_origins_01.html

Radiation from the Sun is photographed using a spectrometer and is analyzed through the use of a spectrograph. The dark lines in the spectrum are called absorption lines, and are caused by the absorption of radiation by elements in the Sun's atmosphere.

By studying these absorption lines, scientists are able to identify the elements present in the Sun. The prominent line at the red end of the spectrum is one of the hydrogen lines and the lines in the yellow indicate the presence of sodium.

Both of the two charts below show the visible-light absorption spectrum for the Sun. The visible portion is sometimes referred to as the Continuum. The bottom chart shows the amount of light absorbed.

Allan Karson

34.3. Spectrum of a Star

A data processing system analyzed the color spectrum to see where the prominent lines occurred and identified the elements that are present in the star. These elements are listed below.

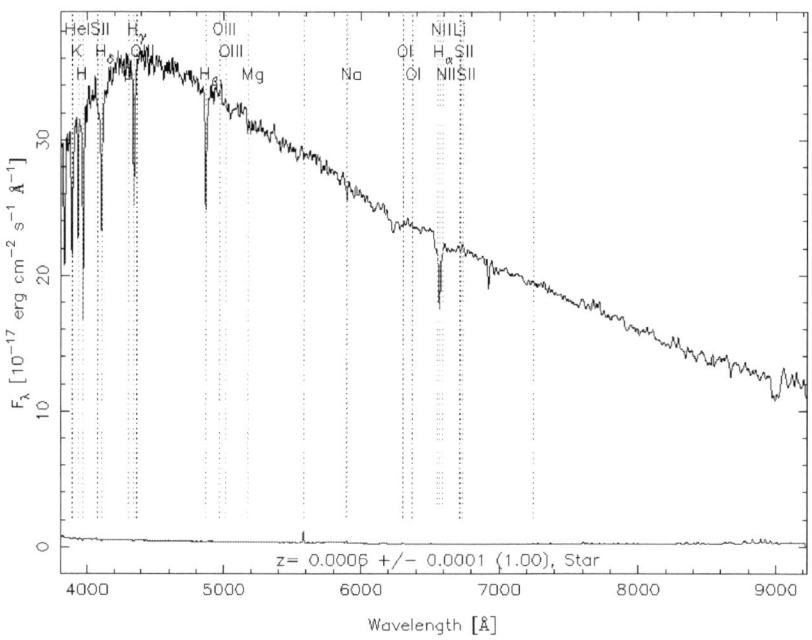

Some of the elements present in the observed star:

Mg	Magnesium	He	Helium
H	Hydrogen	S	Sculpture
O	Oxygen	Ni	Nickel
K	Potassium		

NOTE: This graph was reproduced from the Sloan Digital Sky Survey web site, but is graph is no longer on that site: http://cas.sdss.org/DR6/en/proj/advanced/color/images/full_5820965171516 53917.gif

2. Looking at Tiny Things (Sizes of .000,000,000,000,000,001 meters (or 10^{-18} meters))

First, some guidelines as to why scientists consider that they *must see* the object to in order for them to say they know something about the object— even if the object is very small—as will be the situation we will be examining.

It was the late-19th Century that the Austrian physicist, Ernest Mach, whom you read about in Chapter 2, Thermodynamics, and in Chapter 11's presentation of Einstein's General Theory of Relativity, established the criterion that physicists must *see* things, through their sensory system, in order to say that such a thing exists. Mach called for personally observed accurate measurements. A tough criterion: No guessing and No hypothesizing!

We will now see how the physicists— who have used the same general system concept, from the late-1800s to today—and that continue to be used at CERN's Large Hadron Accelerator, to accomplish this.

First, when physicists observe tiny things, they do not want to have other items to interfere with the operation. They do not want dust, or dirt, or wind or rain to interfere—or unseen gases to interfere. What they want is a 'non-interfered space' in which to observe the tiny item. Thus, in other words, they want a *strong* vacuum.

Now, what is the event that we may be looking at? The event can be simple: hitting (or bombarding) a particle by another particle. That is, a shooting gallery. A small shooting gallery that can be used in a modest-size laboratory room, or a BIG shooting gallery that may be miles long— usually called a particle accelerator.

Vacuum and its glass container

Now, we want to observe the tiny thing, and the way we do this is by first creating a vacuum. Let us first look at how a vacuum—first in glass, came about.

But first, *how we measure vacuums*—and air pressure, in general. When we look at a barometer to find out the current air pressure, we are observing a measurement that is based on the air pressure around us that is pushing up a column of mercury. When normal weather conditions prevail, the air pressure around keeps that column of mercury at about 30 inches high (or 760 millimeters high). Or, we can say, this is equivalent to an air pressure of 14.7 pounds per square inch pressure, or 'one atmosphere'.

A digression about the terminology physicists use to express 'vacuum pressure'. As we know, in physics the methods or names of quantities frequently appear to not be 'straightforward.' In this case the culprit is the way physicists describe air pressure, or a vacuum—for which the term used in physics is 'torr'. The source of name 'torr' is Evangelista Torricelli of Florence, Italy, *circa* 1600, the inventor of the barometer. End of digression

One torr is equivalent to the air pressure of about one millimeter of mercury. Our daily air pressure is, therefore, equal to 760 torrs—the same as the height of mercury during a 'normal' day.

In the Mid-19th Century, **Johann Heinrich Wilhelm Geissler (1814 – 1879)** created a relatively strong vacuum. The vacuum was .1 millimeters of mercury, or $1.3*10^{-4}$ of one atmosphere. A great accomplishment by Geissler. And, adding to that contribution to science, Geissler was the person who invented the glass vacuum tube—which we see in all its manifestations—radio and instrumentation tubes, TV display tubes, X-ray tubes, cathode ray tubes (CRT), and so on.

Note: In the 21st Century, vacuums in the typical vacuum tube-based TV (not flat screen) are built to operate in the vacuum range of about $10*10^{-9}$ of one atmosphere. This is a vacuum 700,000 times greater than that achieved by Geissler. End of note.

We will now see how one of the same tools for observing tiny things, vacuum in glass—starting from the late 1800s, is essentially the same as that used today. But today they are bigger and stronger and have a much stronger vacuum. The next is the shape of the glass. We can visualize something similar to our TV set. A TV tube is very similar to what physicists call a 'Cathode Ray Tube.' In both the TV and shooting gallery, the cathode inside the cathode ray tube is the element that shoots off particles, which are electrons or photons, in most cases. We can call that target area the *anode*, or the plate, or the screen.

The physicist perfects the aim of the electron (or the other-named particles) shot from the cathode, such as photons) by using magnets that surround the vacuum tube. This is very similar to what happens your home TV set; it, too, aims at the target using guiding magnets.

In our TV set, we see the result on the screen in the form of pictures. The physicist does something different: He sets up a target (where the TV screen is) consisting of a particle, or a number of particles. She then shoots particles from the cathode aimed at the target—through the vacuum, and BAM—*something* at the particle level occurs..

The physicist can also has a photographic plate to record what occurred. Or, as we will see with the particle accelerator in CERN's, Large Hadron Collider (LHC), the physicist has *highly accurate data sensing devices* to send the information to a computer, where the data can be analyzed and displayed via a CRT, a motion picture camera, or just simple charts.

Back to the BAM. When the collision occurs, all sorts of particles and sub-particles may result from the bombardment. Thus, such experiments are frequently referred to as 'scattering experiments.' Sub-particles may only exist for only very short times, such as 10^{-22} seconds. Recording the data, or capturing the event on photographic plates, can be very difficult and requires highly specialized equipment.

The data collection system can record or derive all types of data describing the particles generated from the collision at the 'plate', such as: the angle the new particle veered away from the path of the bombarding particle, its charge, its speed and its life-time.

The earlier physicists, such as Becquerel and Roentgen—and all the later experimentalists, needed to move and accelerate the particle from the cathode to the plate. They all set up this similar arrangement—a cathode shooter, particle goes through a vacuum—and hits a plate. Of course, Becquerel and Roentgen and the other experimental physicists of that time, had very simple systems in their room-laboratory, or, sometimes, in the basement of a building.

In contrast with the earlier and even today's small laboratory setup, we now look at the set up at CERN's Large Hadron Collider, or LHC, located in the border area of Switzerland and France. It is the shape of a *big* doughnut—a torus, as a mathematician would call it. It has to guide (very carefully, and with great precision) the protons or electrons or the selected particles that LHC 'projectile gun' sends out on their mission—from its cathode to the target. In this case such 'projectiles' are sent around the complete torus many, many successive times,

In the case of the LHC, CERN went uses a large, metallic vacuum-insulated container. The path for one trip around the torus is 16.7 miles, or 26.88 kilometers. Particles injected in the system make many trips around the torus in order to build up the particle's speed—to produce the anticipated *strong* collisions.

LHC is used to detect tiny, tiny, tiny things or events—mainly by bombardment of a particle by another particle. A BIG shooting gallery, we might say. The LHC requires a vacuum pressure within the metallic container of about 10^{-10} torrs or $1.3*10^{-12}$ of our daily atmosphere, or about thirteen thousand times greater vacuum than our home, semiconductor or vacuum tube-based, TV sets. The system requires such a high vacuum, to avoid any interference with unwanted particles of *any* type, even gas particles.

In the case of LHC, there are many, many humongous magnets along the particles' path to do the guiding—similar in purpose to the guiding that is done in the TV set, but much more so. In LHC's case, the magnets keep the bombarding particles on a circular path around the torus.

Note that CERN's LHC is different from the standard setup in the way it is configures its cathodes. It has *two* cathodes that shoot: They essentially are shooting at one another. The particles from each cathode are pointed in opposite directions; the shooting particles from each cathode go around the torus many times— and they are aimed precisely to hit one another. Refer to Figure 34.

This requires, however, extremely precise magnetic control of the shooting paths of both sets of these microscopic particles—so that they actually hit one another. A very complex targeting task. The advantage of this setup is that the velocity at collision is now much higher than the velocity of the single cathode setup, thereby creating a much larger BAAM. Note: The separate velocities of each group of particles coming at one another is near the speed of light..

About The Shooting Accelerating Voltage

I purposely did not describe above a key, integral part of the whole setup, since I thought it might disrupt the flow of the description of the LHC 'shooting gallery.' That is, in this configuration, there is need to have a voltage difference between the cathode and the point where the bombardment occurs. (The voltage difference can also be referred to as accelerating voltage.)

It is the voltage difference that gives the particles being shot, the speed/energy to go from the cathode to its final destination, the plate. That is, rather than talking about 'voltage', physicists prefer to refer to this as 'the energy (for an electron) an electron gains when being accelerated by one volt. That is, the terminology used is 'one volt gives an electron 1 eV of energy.'

The 'early experimenters' at the forerunners of LHC-type facilities were accustomed to working at about 100 eVs, (which consisted of shooting at *stable* targets) they obtained their voltage from voltaic (liquid) batteries. (Note: To achieve high voltages with voltaic batteries, batteries must be linked together, adding, in series, the voltage of each battery—totally impractical for the LHC application.) In contrast, the LHC—whose mode is a (powerful) *collider*, uses voltages on the order of 10^{12} eVs to attain its collisions.

Comparison of Earlier and Present Shooting Systems

The early world-class physicists—Becquerel, Roentgen, Hahn, Meitner, Bohr, Fermi—and many others, all used some sort of recording device(s) at the plate, to see what occurred at the point of bombardment (or collision, in the case of the LHC).

The recording devices have evolved over the hundred-year span. The recording of the bombardment at the plate started with photographic plates. Now electronic sensing devices are used to capture the data and send it to computers for analysis.

Yes, there are dissimilarities between the past systems of Becquerel and Roentgen's systems and the present systems. This difference is mainly in size and the plethora of equipment and systems that are now used. In that way they are clearly different. But the principles of the operation are just about the same: Shoot from a cathode, or a node, depending on the particle, guide the particle in a vacuum and aim the particle at the target—and record the event for observation.

The physicists of the 21st Century have computers that have sensing systems to feed the computer all the data concerning what happens when the accelerating particle hits its target. She (or He) also has high-speed photography at the receiving end, 'the plate.' They have a plethora of new technologies to record the event. I suggest you refer to Figure 34 and Chart 35 to learn some of the overall features of the CERN LHC. I also suggest that you go to the LHC site on the web and browse around that very interesting, informative, multi-page web site, http://lhc.web.cern.ch/lhc/ whose title is 'The Large Hadron Collider'.

FIGURE 35 CERN'S LHC VACUUM TUBE

1. A 'Typical' Vacuum Tube

Electron Source, Shooting at Target

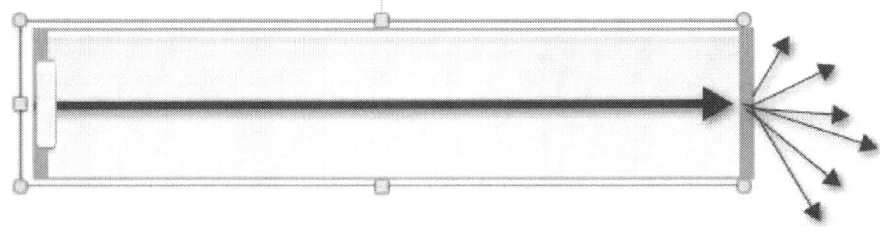

Arrow above represents bombarding particles.

Lines emanating from target at right are the observed results of the bombardment

2. CERN'S LHC "Vacuum Tube"

Two arrows represent two sets of particles introduced into each sets 'own' concentric circle, with each going in opposite directions. The single circle represents TWO concentric circular paths that are eventually guided to meet and collide.

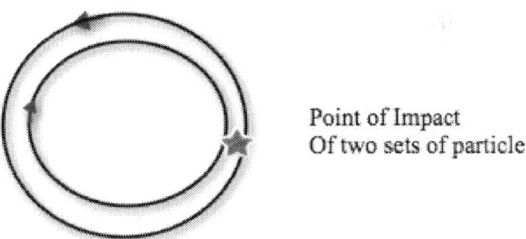

Point of Impact
Of two sets of particle

Note: From the laws of the Special Theory of Relativity:

1. The particles go fast and faster—possibly approaching the speed of light, they get heavier and heavier—and require more and more energy to move them.

2. When they collide, they are both going near the speed of light. Their *relative* speed at impact cannot be greater than c, 186,100 miles per second. The speed of impact is actually in the neighborhood of 12/13 of c, or .923 c, or ~171,000 miles per second.

Chart 36 The Large Hadron Collider

This chart contains information that was copied from CERN's web site on the Large Hadron Collider (LHC), http://lhc.web.cern.ch/lhc/ and from information gathered at a physics colloquium at the City College of New York, December 2008.)

Physical:

• LHC was built in a circular tunnel (a torus) 27 km (16.8 miles) in circumference and is buried around 50 to 175 meters underground. It straddles the Swiss and French borders on the outskirts of Geneva.

• The beams move around the LHC ring inside a continuous vacuum guided by magnets.

• The magnets are superconducting and are cooled by a huge cryogenics system. The cables conduct current without resistance in their superconducting state.

• The beams will be stored at high energy for hours (approximately ten). During this time collisions take place inside the four main areas.

• There are four separate sites along the torus to perform the collision of projectile and target.

Experiments:

• For the first set of experiments, the 'projectiles' will consist of protons (light particles).

- The protons will be in bunches, and a bunch will contain 100 billion protons.

- There will be 3000 bunches to comprise an experiment.

- There are four collision sites on the collider

- The beams at collision point will be focused at two collision sites at 16.7 micrometers and at the other two at 71 micrometers.

- Each collision will produce a large quantity of data—100,000 gigabytes per second.

- *Each experiment will be planned and analyzed by 2000 PhDs.* (The author applied the italics to this line.)

Part III The Future

The 20th Century was the most fruitful century in history for many of the physical sciences—physics, medicine, chemistry, cosmology and biology. Some of those 20th Century discoveries and developments were:

- Recognizing an expanding Universe
- Theory of Big Bang
- Cosmos and its components
- Life Cycle of Stars (and Black Holes)
- Quantum Theory, with all its counter-intuitive ramifications
- Standard Model—uniting three of the basic forces
- String Theory—sometimes described to be the 'Theory of Everything', but as yet far-from-proven or established.
- The Strong Nuclear Force and the Weak Nuclear Force
- Theories of Special Relativity and General Relativity

The Predictable Future

- *Quantum Systems* are near, where 'near' is given in *(*slow*) physics time*. Quantum=based Systems will eventually replace Electronic-based Systems such as used in Computers, Communication and Transportation systems. And there will assuredly be quantum units in hologram-based TVs, cell phones, cameras, telescopes—and so on.
- And *Quantum Systems* will become fundamental, or the *Base-Systems* in Health Care Systems: Cardiograms, MRI, X-ray, systems to test for DNA and body fluids.
- A feature-expanded world-wide Web, The Internet
- Worldwide quantum-based news dissemination and market systems
- Manned space systems and unmanned sensing system sent to 'observation posts' in our galaxy.
- Nuclear systems (Quantum controlled) for power generation, for research and for national defense.

The Unpredictable Future

These are a few personal speculations on where the Global Order *might* go. This consists of products, manufacturing and production—and life-styles in the 21st and future centuries. This part also includes the related economics, world organizations that might accompany such new entities. I repeat: These are pure speculations and are not based on any present theories, findings or discoveries.

Some of these predictions may sound to be 'way out'. I admit that. But think where the world was, in general, and where was the world of physics was at 1900, the turn of the 20th century? Would anyone have suggested or guessed or prophesied anything close to the technology we have today? In all likelihood, the predictions here will probably be too conservative—and will be 'old hat' in the first part of this century.

Expected, Forthcoming Technologies This section includes Nanotechnology (10^{-9} meters) and Millitechnology (10^{-3} meters), the search for life in the Universe, and advanced simulations.

Nanotechnology: During the year 2008 the author first became aware of a new technology, Nanotechnology, which joins together the sciences—physics, biology, chemistry, and of course, mathematics, to produce new products. This may be one of the major technologies of the future, and may reshape or reform the structure of traditional science.

This is a brief list of ideas that were presented at a colloquium at the City College of New York during the fall of 2008, by a professor of the Ecole Polytechnique Fédérale of Lausanne, Switzerland. The web site for the Ecole Polytechnique Fédérale—and its exciting research into the crossing of biotechnology with physics and engineering, is www.epfl.ch/. (Note. This web site is in English, as is the Swiss web site, cited below.)

Nanotechnology involves the design of small, complex structures consisting of combining living organisms and non-living organisms. Nanotechnology deals with structures that are on the order of 10^{-9} meters and are made up of about two to three molecules. Two examples that were described at the colloquium are described below.

Example 1: Nanostructures are injected into the human body and sense what is happening in the interior of the body. They will provide an earlier, quicker and less expensive way to detect the onset of diseases or deterioration to the total person, physically and socially, such as that caused by various diseases. Such structures have been in the research stage since the early 21st century. As we all can appreciate, considerable testing will be required before they could be distributed for use in the general medical community.

How they might work: The structures are introduced through the blood stream and are sent (addressed) to specific parts of the body, the brain, muscles, organs, etc. Once they are in a designated location, they look for and detect changes in the body parts. If there are changes, notification is then transmitted to outside monitoring equipment.

A nanostructure can be designed to *detect* singular or multiple activities, such as change of color, pressure, temperature, or change in chemical composition.

The person who is the *host* to a nanostructure need only go to a local health center to be monitored by an attending technician. The results of the observation are sent to the person's centralized Medical History Site (MHS) and to the applicable local physician who is qualified to resolve the impending health problem.

Example 2: Nanostructures are composed of DNA-derived amino acids and inert chemicals to act as 'photo synthesizers".

The molecular-size items are painted or sprayed on a flat surface to capture sunlight, transform the heat into electricity and thereby perform the function of sun panels. The nanostructures operate at 98% efficiency, are cheap and easy to produce, and have an active life in the 20-year range.

The web site that describes this type of research is http://www.nccr-nano.org/nccr/. That site describes the work at the University of Basel, and the author believes that there are other such research centers working on this emerging technology.

Millimeter Technology

The fundamentals of optical gratings and spectrum analysis systems were discussed in Chapter 7. In that chapter, those systems were described that to enable scientists to perform analysis of matter that is far away. Here we shall consider the application of much-smaller, millimeter systems for matter that is very small, or minute, and presently inaccessible.

A Cancer Detector/Annihilator Cruiser 'Sailing' in Your Body's Highways

This description is based on U.S. Patent 6,240,312 B1 whose lead-author is Professor Robert R. Alfano, University Distinguished Professor of Science at The City College of New York and Director of its Institute for Ultrafast Spectroscopy and Lasers. In 2009 he received the Optical Society's Charles H. Townes Award. (The award was based on his discovery and work on the *supercontinuum*, an *ultrafast* white light source.)

The following are selected parts of Professor Alfano's patent's abstract: Remote-controllable, micro-scale robotic device for use in diagnosing and/or treating abnormalities inside a human body *in vivo*. The (small) device can be introduced into the body either from natural body openings or by injection into the blood stream. .... The device can be guided to different locations …by radio controls and software. End of abstract.

We can all imagine what the electronic system will be that exists outside of the person being treated—a medical team, computer analysts, radio guidance sensors, laser control…. and so on. The following is a list of unit equipment that could be installed inside the robotic cruiser—that is very, very small. There could be a micro-video camera, computer and controller, , battery, propulsion system, internal micro-light filters and optical gratings—and a communication system to maintain contact with the 'outside' world.

The latter items, micro-light filters and optical gratings, would perform spectrum analysis on the body tissues. They would investigate the body tissues and compare the differences in optical signatures of normal and diseased tissues. Thus, optical non-destructive biopsy would replace destructive biopsy.

The patent includes more than one type of unit-item for many of the unit equipment, and they are based on available or foreseeable technology.

2. Personal Speculations: The following are some of the author's speculations about the future. I do not relate these new products or processes to any theory in physics. For now, I suggest that you consider these speculations as a bit of science fiction. A Time Event is provided to each speculation according to when the author considers it *might* occur. We start with a modest device considered to be available by 2025.

362

Language Assistance System Chip—to be used for simultaneous person-to-person translation: The chip is embedded in the ear. It contains a listening device and a low-power speaker to communicate with the wearer. The chip provides simultaneous translation of incoming conversation. The chip also suggests outgoing phrases that are whispered to the wearer and which may be spoken by wearer of the chip.

A High-Density Computer Memory System: Based on what was thought to be a dead-end technique—but may not be, *Chemical Emulsions*, estimated to be available *circa* 2025. The density of new memories will now be very different in size and kind from previous memories: Previously, a unit memory was approximately the size of 10^{-6} microns (10^{-10} centimeters) and contained only one bit of information. They now contain 10^3 bytes, and therefore have 1000 times more memory.

While that is a big jump, it is expected that the next jump (decrease in unit-bit size) will occur much further in the future when bit-memories might be decreased in magnitude to 10^{-10}. This memory is considered to be a 'hybrid', since its unit size is between atomic-size and Planck-length sizes. The first application of this memory might be used in global weather computer forecasting systems.

- Ten thousand (tiny) parallel processors will perform the calculations. The data captured by its satellite sensor will consist of ocean water flows, near-Earth gas flows, and Earth-to-Sun gaseous and unseen space energy flows.
- Information on the unseen space energy flows will be gathered by a system of forty-eight satellite systems in space.
- It is expected that forecasts be accurate to 2% variance in time and all basic measurements—barometric, temperature and the time each forecast is expected to occur.

Designing Systems that 'Copy' the "Ultra-Sensitive" Systems in the Human Body: There are extremely sensitive systems in the human body. An example of such a body-system is the sensitivity we have in our mouth —where a millimeter bulge on a tooth's surface can 'feel like a mountain'. And there are the areas in the body where a slight distortion or disruption in the surface can induce tremendous pain. Another area is the sensitivity of the eyesight system.

The algorithms that would be used in this class of quantum computers are part of the detailed simulation of the *complete* human body system. Human Body systems engineers will define and copy the algorithms in human systems, to make inexpensive sensing systems for practical use—outside of the body's system. A 'first sensor' is a 'smoothness' detector, which can have multiple applications in factories, construction industries and the home in general.

If you have an interest in this subject—the interface between human physical activities and the future, the author suggests reading source 24, *The Fabric of Reality, The Science of Parallel Universes*, a book by David Deutsch, an Israeli-British visiting professor at Oxford University. England. And there is also source 68, *Physics of the Future: How Science Will Shape Human Destiny and Our Daily Lives by the Year* 2100 by Professor Michio Kaku.

LIST OF APPENDICES

1. The Controversial Histories of Copernicus, Galileo, and Friar Bruno—and
The Church of Rome
2. Inventors, Developers and 'Commercializers' of the Steam Engine.
3. Why Engines Are Inefficient
4. Isaac Newton's *Anni Mirabilis*
5. Differences Between the Unit Measuring System in the U.S. and Europe
6. The Big Bang, According to the European Space Agency,
7. Does Dark Matter Really Exist in Space?
8. Planck Length and other Planck Constants
9. Events Leading Up to Explosion of the First Atomic Bomb, August 1945
10. Chronology of Einstein's papers of 1905, The *Anni Mirabil8is*
11. Frames of Reference in Space
12. The Ether and the Michelson-Morley Experiments
13. The Lorentz (or Einstein) Transformations
14. The 29 Parameters Required for Each Running of the Standard Model
15. The Nobel Prize, and the Fields and Abel Medals for Mathematics.
16. Seventeen Numbers That are Helpful to Understanding Physics
17. Laws of Physics Invoked by Physicists and Cosmologist to Prove, Disprove

Appendix 1. The Controversial Histories of Copernicus, Galileo and Friar Bruno—and The Church of Rome.

Introduction and Summary: Starting from about the age of ten, American and European history was my major interest. I became aware of the various forces and ideas and religions that have interplayed in history as 'one big soup'. -

I may have been interested in Copernicus (1473-1543), a Renaissance mathematician and astronomer born in what is now known as Poland, and Galileo (1564-1642) and his generally poor relationship with the Church of Rome, But I do not remember reading any specific books about his relationship with the Church. Also, this would not be a subject that would be discussed in the public schools I attended in New York City and in Washington D.C. Thus, my so-called knowledge was based on hearsay, discussions and 'indirect references'.

What was made clear was that Copernicus advocated that Earth and the planets go around the Sun, whereas the Church advocated the Earth is the center of it all. I also remember believing that the Church of Rome had treated Galileo very poorly for his advocating that the Earth is not the center of the Universe—the theory put forth by Copernicus. How poorly? Possibly prison, excommunication, ...? That was *my* folklore.

And, now, in the 21st Century, when I ask friends and family how the two, Copernicus and Galileo, were treated by the Church, I usually get answers that are similar to my early ideas on the subject, which I had kept from my early years to the present. Possibly prison, excommunication,...? That is their folklore.

I learned in preparing this appendix, however, that Galileo were not treated as poorly as commonly believed. I also learned that Galileo was punished for writing a play that deliberately insulted a Pope on the matter of the

Cosmos; a play he had been told *not to write*. For that insult, Galileo was brought before the Roman Inquisition, and his sentence was that he be confined to his home for the rest of his life. No bodily harm, no excommunication, to the scientist Galileo.

In recent years I have learned of a second (true) victim, Giordano Bruno (1548-1600), who was burnt at the stake in 1600 due to being found guilty by the Roman Inquisition. (Bruno lived in the years between the other two forenamed.) Reading some of the web sites, I got the (wrong) impression that it was for Bruno's advocating the Copernican system. No. It was for his heretic writings, talks, and teaching and dealing with people in Protestant nations such as England, Germany and Switzerland, which he was found guilty by the Roman Inquisition.

But his life, its varied experiences, his personal drive, etc., convinced me that he should also be included in this discussion of Copernicus and Galileo, but that the author does not discuss his being guilty or innocent, but rather to show the 'potential intellectual activities' a person of those times could be involved in.

For example, Bruno did have an idea that was novel at that time: the Universe is infinite. As will be seen, the issue of the infinite Universe was but one, minor issue—among six or seven, strong heretical categories—for which he was convicted of heresy and burnt at the stake.

So, the question is now: Where or when did the Church of Rome maltreat Galileo, primarily for his advocating the Copernican system? It did not. But it forms part of the legend, or folklore, about these two men and the Church of Rome. The author considers the Church of Rome has gotten a bad rap for this episode of science-meeting-religion.

Introduction to Contemporary Church Doctrine During This Time-Period

It is helpful to describe first what Church doctrine was, who Copernicus was, and what his theory consisted of. Much of the following about Copernicus has been gathered from Sources 6, 7, 43, 60, 62 and the *Encyclopedia Britannica*, 15th edition.

The reason for having that many sources concerning Copernicus is the following: The author sought to track down the *definite* or *specific* reaction of Pope Clement when, in 1533, he was given an explanation of Copernicus's heliocentric (sun-centered) concept of the solar system.

Church doctrine, at that time, considered the Earth to be center of the Universe. This view is called the geocentric system, with 'geo' representing Earth. It was firmly entrenched in Christian doctrine as of the 16th Century. This system is based on the concepts of Ptolemy, a Greek, who live in *circa* 140 AD.

Ptolemy published his overall concepts in his book, *Almagest*, or, *Mathematical Composition*. The book contained the ideas of three theoretical Greek astronomers, Apollonius, Hipparchus and his own. Within that book, Ptolemy is said to have chosen the system that was mathematically the simplest, *and was not necessarily the true system design*. That is, in his system, the Earth was the center of the Cosmos—a geocentric system.

At the same time, Ptolemy had developed geometric tools, or small *sub*systems, that are called *epicycles*, *equants* and *deferents*, account for the motions of the planets. These helped Ptolemy explain the motions of the planets where the observed motions did not conform to his desired geocentric concept of the Cosmos.

Also, remember that the telescope, which might have shown that there was a very different system of movement in space, was first introduced in the time of Galileo, *circa* 1610. It is assumed that the presence of telescope 'forced' the Church to change to, or admit to, the heliocentric system, as it did in 1822.

The Histories of Copernicus, Galileo and Friar Bruno

Copernicus (1473-1543) was born in Poland and initially studied at the University of Cracow, Poland, where he was interested in astronomy. Through the assistance of an uncle who was a bishop in the Church, Copernicus became a *canon* in the Church at the age of 24 and was first located in the Cathedral of Frauenberg, Poland (now Frombork). This provided Copernicus with sufficient income to enable him to study.

Shortly thereafter, in 1496, Copernicus went to the then-center of modern civilization, Northern Italy, where he studied at the Universities of Bologna, Padua and Ferrara. He returned to Poland in about 1500, where he lived, until his death.

In Poland he continued to study the heavens and developed the famous, or infamous, (depending on when and where you lived), sun-based system, called the Heliocentric System. He gave lectures on this system, and, in about 1512, he also wrote a short manuscript describing his ideas of the heliocentric system in a document titled *Commentariolus*, or *Little Commentaries*. He gave written copies of this short work to a few trusted friends. (This work was first published the end of the 19th Century, and was based on two manuscript copies that survived until then.)

In 1543, during the last year of Copernicus's life, he agreed to publish his masterwork, *On the Revolutions of the Celestial Spheres*. This, different document, was written over the preceding years. It is a long and detailed document. One of Copernicus's much-later biographers, Angus Armitage, considers it ranks in stature with Newton' *Principia* and Darwin's *Origin of the Species (Source 6)*.

And to add to the above plaudits of the biographer, Angus Armitage, Professor Rosen, the author of *Three Copernican Treatises*, source book 7, writes that the reading the Little Commentaries and the book about to be described, *Narratio Primo*, is sufficient for a *complete* understanding of Copernicus's work, *On the Revolutions of the Celestial Spheres*.

The book, *Narratio Primo*, was written by the scholarly Italian professor, George Joachim Rheticus (1514-1574). Rheticus had gone to study with Copernicus in 1539 and also recognized that Copernicus was "a genius of the first rank." He also observed that Copernicus was reluctant to publish his work, so Rheticus wrote a detailed description of that work by Copernicus, whose title is *Narratio Primo*. Professor Rosen's translation of *Narratio Primo* is also presented in his book, source 7, where it is just under one hundred pages.

It appears that written copies of this short work, *Commentariolus* got into the hands of Church in Rome—specifically into the hands of John Widmanstad, the papal secretary. In 1533 Widmanstad made a lecture (or presentation), presumably based on *Commentariolus* (according to the *Encyclopedia Britannica*) "before Pope Clement VII—who approved of it."

The approval by the Pope Clement VII did not mean—in any way, however, that the Church adopted the Copernicus's theory—especially, since it was in conflict with Church doctrine. Even only discussing it, was banned until much later, in 1611. It could have been *just interesting* to Pope Clement.

Or, it should be understood that astronomer at that time, *sometimes used false models of the Universe to facilitate their calculations*. While these models might have been in conflict with church doctrine, they were allowed beyond their purpose of facilitation. (Perhaps Clement thought of it that way.)

Clement VII was Pope from 1523 to 1534. He may be considered to be 'different' from most other Popes of that period. This may be part of the reason for his 'possibly accepting' Copernicus's system theory of how the Universe is organized. He may have been more independently minded—or, possibly, he did not understand the full 'religious' significance, or implication, of Copernicus's theory. The *Encyclopedia Britannica's* comment about him is that he, Clement VII, was "financially extravagant and unsystematic".

The following is a quote from Wikipedia about Pope Clement VII and Copernicus's system: "in 1533, Johann Widmanstetter, secretary to Pope Clement VII, explained Copernicus' heliocentric system to the Pope and two cardinals. The Pope was so pleased that he gave Widmanstetter a valuable gift." Even though Clement VII liked and probably understood Copernicus's sun-centered system, he did not change Church policy, which favored the earth-centered system. That change occurred later.

An aside about the interesting time period of the reign of Clement VII: He was the bastard son of a member of the famous Medici family and was raised by his uncle Lorenzo, the Magnificent. While the *Encyclopedia Britannica* says he was weak and vacillating in some struggles between kings who were fighting for domination of Europe, he was involved in two other, *major* issues: His not-granting Henry VIII a divorce, and a growing reformation that was flamed/started by his predecessor, Leo X.

Allan Karson

A note about Italy at the time of these three men: Italy was not a unified country. It consisted of what might be called 'small nations', such as the Republic of Florence, the Duchy of Milan, The Papal States and so on. Each nation had its diplomatic ambassadors representing the nation to the other nations, similar to what nations do now. In making any trip, individuals had to check in with the authorities of their 'nation' and with the nation he was visiting. End of note.

The attitude of the Church of Rome toward the heliocentric system is well known. The Church was against it, at least implicitly against it. It made the ban explicit in the early 17th Century when it banned Galileo's work—which advocated the heliocentric system,

You may think that the new Protestant religion, founded by Martin Luther, might favor that concept of the heliocentric system. This was not the case. The new religion rejected the Copernican system even more decisively that the Catholic Church did. Copernicus had been severely criticized by Luther, … " (Angus Armitage in source 6.)

Note: In the case you wish to read one or more of the following source book on Copernicus, I note that those books was written by authors with different backgrounds.

- Source 6's author, Angus Armitage, is acclaimed as being one of the world's foremost Copernican scholars. He taught history of astronomy at University College, London.
- Source 7's author, Edward Rosen, was Professor of history, City College of New York. In addition to writing the source book, he also translated two of Copernicus's papers from the Italian to English.
- Source 60's author, Jack Recheck, writes about leading scientists and economists.

Galileo (1564-1642). The story or history concerning Galileo and the Church has long been a much-discussed issue in both classical and scientific history. It tells of one of the first instances where science tried to break out (unsuccessfully) from the prescribed bonds of church doctrine.

When Galileo wanted to make a trip to Rome, he first had to get permission from his patron, Cosmo, to make the trip, and while in Rome, he had to deal with Tuscany's (the Italian state in which Galileo lived) ambassador to

371

Rome, who was charged with finding a place for Galileo to reside in during his visit.

And the organization Galileo had to deal with, the Church in Rome, was probably more complicated than it is today. Byzantine, complex and intricate, would be appropriate. There was:

- Paul V, Pope from 1605-1621 and Pope Urban VIII, Pope from 1623-1644. Pope Urban was a friend and sympathetic to Galileo.

- The many cardinals, some sympathetic and friendly, some not,

- The Jesuit organization and the Roman College,

- The Curia (The administration for the Vatican),

- The Holy Office (A modernized version of the Tribunal of the Inquisition) and the Congregation of the Index (to censor books)

In addition, in order for Galileo to publish *anything*, he was required to ask for approval of the manuscript from a designated person in the Vatican.

Galileo's personality: While these factors are difficult to evaluate after four centuries, Galileo appears to have been pleasant, articulate and resourceful. He also thought highly of himself, and was sure of himself. He could be arrogant and impatient with people who did not listen or agree with him—especially over the Copernican idea. He also appears not to be 'a good listener' when told not to do something'. This latter point probably led to his downfall, in the eyes of the Church.

The following is a simplified description of what transpired during those six trips, which consisted of an involved set of discussions, meetings and communications—within a Byzantine world.

Allan Karson

Galileo's Six Trips to Rome

First trip, 1587: At this time he was young and unknown, and the trip was uneventful. He may have learned about Copernicus's ideas, but there is no record that the issue came up.

In 1610 Galileo, using his telescope, observed that the planet, Venus, followed Copernicus's predicted path—and not the path predicted by the Church's geocentric theory. Note: In various source books, Galileo is credited with being the first user of the telescope. Since there were probably only a very few telescopes at that time, that statement about being the 'first' user of the telescope is probably true. In any case, he was, at least, 'one of the first users.' End of note

An aside:, prior to Galileo's second trip, he wrote to Johannes Kepler declaring that he 'was a Copernican'. It is doubtful that the Vatican knew of this, since Kepler was a Protestant.

Second trip, 1611: Since his first trip, he had become a celebrity, known for his astronomical discoveries—we assume, thanks to the telescope. These were not related to Copernicus's theory but rather to the *satellites around Jupiter*. (Amazing!!! —the sighting of those objects in 1611) But during this trip he spoke positively about the Copernicus's heliocentric system.

During this trip he was wined and dined by Vatican society, met with many cardinals and had an audience with Pope Clement VIII at the Vatican. Also, the Jesuits gave him an honorary doctorate at the Roman College. He returned to Florence after his triumphal visit.

But celebrities, by their nature, build up animosities. This, and his heliocentric ideas, led to the development an anti-Galileo sentiment in some quarters in the Vatican. After he returned to his home in Florence, he learned that there had been a great deal of activity concerning Copernicus's ideas— and Galileo's. For example, he learned that the Holy Office had denounced Copernicus.

Third trip, 1615-1616: This time, his reception was far from favorable. He was told, for example, that the Pope was not interested in 'intellectual fireworks', clearly meaning Copernicus and his heliocentric ideas. Also, Copernicus's book, *On the Revolutions of the Celestial Spheres had* been taken out of circulation and placed on the Church's Index (a list of banned books)—adding a new, strong point *against* Galileo. And finally, while in Rome toward the end of his third visit, Galileo was given a certificate by Cardinal Francesco Bellarmine (that was agreed to by the Pope) that said that Copernicus's theory is "contrary to Holy scriptures and cannot be defended or held".

We shall now see how Galileo (unwisely) persisted in supporting the theory. He thought he could present the Copernican system as a tool, to represent what a system might look like, or to use for some other reason. That ploy, or approach, did not work.

Fourth trip, 1624: A new Pope had been recently selected, Urban VIII. Galileo previously had excellent relations with Pope Urban VIII, since he had previously knew him, and got along well with him, when the new Pope was previously known as Cardinal Barbarini. As a matter of fact, the new Pope, Pope Urban VIII, appointed some of Galileo's friends to the Curia, and he also granted Galileo six interviews during this trip to Rome.

Insofar as our interest in Galileo is concerned, however, during this trip nothing positive transpired for Galileo. The Copernican system was still not accepted as truth. And Bellarmine's certificate to Galileo was still in force. On Galileo's return to Florence, he undertook the writing of *Dialogue on the Two Chief World Systems*. It (stubbornly) presented the arguments for the heliocentric/Copernicus system.

Galileo's book, "The Dialogue…" consists of a discussion between three men, one of whom is a simpleton—and it is that person whom Galileo chose to articulate some of the Pope Urban's views! What a gaff by Galileo! Thus, it appears, (at least to the author) that Galileo deserved the mild sentence, or reprimand, he later received—and the Church gets a bad rap—to this day, for Galileo's sentence.

Fifth trip, 1624: The purpose of this trip was to get Vatican approval for printing and publishing Galileo's newly written book. The book was a 'hot potato' as it was presented to the clergy in Rome. Amazingly, Galileo did get the approval for his book. It was approved by a cardinal, who later admitted to the Pope that he had not read all of it. This was the beginning of a direct action by the Pope Urban to control the 'Galileo situation': A Trial.

Sixth trip, 1633: Galileo was called to the Vatican the beginning of October 1632, but he arrived there in February 1633. The delay was due to the plague, which had entered Tuscany in 1630, and to Galileo's health and age, sixty-six. (Galileo frequently delayed his trips due to ill health.) Galileo had to wait around (in comfortable quarters) until his 'trial' began in April 1633. The trial was by the Tribunal of the Inquisition, where the system, as defined by the Inquisition, *did not allow a defense*. The only option for Galileo was to 'acknowledge errors and recant'.

Galileo was called a few times before the court, and his testimony was not that admirable. He even said that in his book, *Dialogue on the Two Chief World Systems*, 'He had argued against the Copernicus system'—which was clearly not true. We may ask, "What caused this confusing behavior? Age? The imposing nature of the Tribunal?"

In June the court met without the presence of Galileo, with Pope Urban presiding. Galileo's book was condemned and Galileo was given a prison sentence plus penance. But later, in June 1633, Galileo signed a declaration saying he did not support the Copernicus system. The next day, his prison sentence was commuted to house arrest (in his home in Florence). During all this time in Rome he was treated respectfully and provided with comfortable living quarters.

He returned to Florence where he died in 1642, (the same year as the year of the birth of Isaac Newton). He lies entombed in the Basilica di Santa Croce in Florence, along with the tombs of Michelangelo, Dante and Rossini. (The Basilica di Santa Croce is the largest Franciscan church in the world.)

I end this discussion of Galileo with a two sequential remarks made by Nobel Laureate, Leon Lederman, in his co-authored book, *The God Particle*, source 19—which is not about God:

"And it took until 1822 that a reigning Pope officially declared that the sun could be at the center of the solar system. And it took until 1985 for the Vatican to acknowledge that Galileo was a great scientist and that he had been wronged by the Church."

And in the next section of his book, Dr. Lederman cites other 'masterpiece findings-in-space' by Galileo, derived with the aid of his homemade telescope, plus Galileo's other findings and physics-developments here, on earth, such as Galileo's investigations and findings about the *nature* of motion, (mass, gravity, velocity and acceleration). And in that section, Nobel Prize winner, Lederman, provides an appropriate, wise (humorous) remark—that is a hallmark of Lederman, who was Director Emeritus of Fermi National Accelerator Laboratory:

…. "Galileo wanted to leave something for Newton to do."

Giordano Bruno (1564-1642): Giordano Bruno was a devout Catholic most of his life. He had a remarkable memory and a remarkable mind. He was an independent thinker. Articulate, brilliant, worldly, and observant... He may have been naïve, but, as you will now learn of his many activities, it is interesting to speculate where and how he might have participated in today's world.

Almost all of the following is from source book 62, *Giordano Bruno, Philosopher/Heretic,* by Ingrid D. Rowland.

Bruno was born in the town of Nola, which is close to Naples, in the then-Kingdom of Naples, in the overall territory of Italy. Spain ruled the Kingdom of Naples, and at that time only Constantinople, Cairo, Tabriz and Paris had more inhabitants than the city of Naples.

When he first spoke, he always spoke in full sentences. He stayed in Nola till the age of fourteen, when he went to live in Naples in 1562. Bruno developed his own ideas about politics and philosophy—calling it the *Nolan philosophy* and took to the art form of writing plays. His plays were frank and strong, reflecting on life in Naples. They were not tragedies; they were mostly comedies.

He sought the priesthood; he lived in the convent where 'rich kids' came to be taught by about sixty Dominican friars—who could be unruly, outspoken, politically active, non-celibate and at times, thieves and forgers. It was the Dominicans who staffed the Spanish Inquisition in Naples.

Bruno had done and said a number of (minor?) things that were observed by the friars, and they developed a cautionary attitude to what were called, his 'Protestant' ideas. One action was that he removed pictures of the Madonna and Saint Catherine from his room, leaving only a crucifix. And added to that was that he had spoken with many different people, (apparently

376

'uncommon' at the time) even though it was mainly to friars in other convents. He also read Plato and adopted many of the Greek's ideas. One might say that, "This was the beginning of Bruno's rap sheet."

Learning the Hebrew language was part of his theological education—and Bruno learned it quickly. He did so well that his superiors at the convent sent him to perform that language before Pope Pius V. In Rome, in 1569, he recited a psalm, and then recited it backward—a feat due to his extraordinary memory. (Not known to his superiors, he also studied the Jewish religious book, *The Kabbalah*.)

At this point in his life, according to source 62, "Bruno had already decided in his private thought that Jesus of Nazareth could not have been the Son (original has 'son') of God incarnate… (During his trials and imprisonment in both Rome and Venice, which extended from 1593 to 1600, he expressed similar thoughts to his Venetian inquisitors. Throughout Bruno's life he made similar and many other heretical statements; these will not be quoted or described in this text.)

In 1571, at age 23, he became a deacon. In 1579, Bruno went to the Protestant city of Geneva. From 1579 to 1581 he was living in the French (Catholic) cities of Lyon and Toulouse.

It was in 1580 that he adopted the concept of the sun-centered cosmos of Copernicus, "and then, more radically still, *for an infinite universe without center or limits of any kind*." This latter concept, his own, became a *Nolan* hallmark. He also knew of the work of astronomer, Tycho Brahe. He was what we call today, "A Worldly man"

While in Toulouse "Bruno learned that he had been excommunicated *in absentia* and defrocked by the Dominican provincial (the head or chief of a province or of a religious order in a province) of Naples. Also, while in Toulouse, he attended philosophy lectures at the University of Toulouse and, in 1582, Bruno became the Professor of Philosophy at the University of Toulouse.

He went to Paris and was there from 1582 to 1583. He may have grown concerned about his excommunication. Even so, "Whipped and skittish he may have felt, he continued to move with the aristocratic confidence of a friar…" It is to be noted that in the coming years Bruno made repeated requests to have his excommunication annulled, and returning back into the Catholic faith. In all cases, he was refused.

In Paris, Bruno gave a series of lectures on philosophy—and through the lectures, he came to the attention of the French King, Henry III, whose Italian-born mother was Catherine de' Medici. Henri was impressed by Bruno and appointed him *Lecteur Royale.* In Bruno's courses on logic on metaphysics, he also introduced his unique course, "The Art of Memory", for which he had developed sub-courses and course material.

But in Paris, France, Henri II was losing his tolerance for Protestantism and becoming more Catholic in his feelings. This may have meant, since he had been excommunicated and 'had a record', that France would not be the best place to be teaching and living. Bruno then set his eyes on 'tolerant' England, "where Queen Elizabeth had earned a reputation for her culture and her tolerance."

Bruno then went to London in 1583 and set his 'new' sights for Oxford, where his lectures would be in Latin, as was the norm there. His first lecture, however, turned out to be, for him, a disastrous lecture. One reason out of many: his *Latin-accented* English. Also, at that time the English did not take to foreigners. (One of the aspects of their 'island mentality'. The English can be brutal in such areas.) He, thus, forgot his aim for Oxford, Cambridge or any academic career. At the same time, however, he obtained a reading position at a well-placed English lord's home, who was a regular attendee at Queen Elizabeth's court and who had also assembled his own court.

To demonstrate the modernity of Giordano Bruno, and to show just a small part of his wide-ranging thoughts, the words of one of the characters in a Nolan-London play are paraphrased in source 62, and they are worth reading for their evaluations of our recent and present world:

> He (Teofilo) goes on... to make stinging condemnations of European colonialism, before contrasting the effects of the Nolan philosophy with capitalist greed. At that time, England had only begun its colonial explorations; Bruno was writing primarily about Spain and Portugal. He must have read the denunciations of Spanish cruelty in Mexico made by the Dominican...

378

As we see, he was far from a 'true heretic. He was a 'positive', *Universal Man*. And there are other writings by Bruno that add to that evaluation. But the Church of Rome was considered Bruno to be a heretic at the time. For various reasons, Bruno returned to Paris in 1585.

In Paris, as usual, he met many 'new' people, saw old acquaintances, and asked for reentry into Catholicism—which was, once again, unsuccessful. Thus thwarted (again), in 1586, "he inscribed himself in the register of the University of Wittenberg, was ready to see what the Lutheran version of the Reform might offer." According to source 62, "Wittenberg, in Saxony, was considered the "Athens of Germany.""

In Wittenberg, his life was happier than before. "He responded to his German colleagues and their hospitality with unusual warmth. His admiration for German philosophers... but he was no less impressed by the openness of his colleagues at Wittenberg."

But due to what he was told by a friend, that there was higher teaching salaries in "Golden Prague", he left Wittenberg for Prague in 1588, but he was disappointed to learn, by experience, that his friend was wrong. He then returned to Saxony in 1889, to the Lutheran city of Helmstad, where he could count on the financial support of Duke Julius, the ruler of Helmstad.

Duke Julius died shortly thereafter, and at his funeral Bruno offered an oration in which he made a frontal attack on the Catholic Church. Shortly after that, Bruno, lost favor with the Lutheran authorities and was excommunicated. (Two down; the third, you're out.)

In 1589 he went on to Frankfurt, a city that had a strong attraction for Bruno, owing to its book-publishing industry. From Frankfort he went with friends to Zurich, and returned after five months. And then, in 1591, he made a very stupid move: he returned to Italy—first to the Padua/Venice area. What could he have been thinking?

As his contemporaries may have guessed would happen, in May 1592, he was placed under arrest by the Venetian Inquisition, and in 1593, at the request of the Roman Inquisition, he was brought before that Inquisition. He suffered *eight* years of confinement in prison, but, thankfully, in not the worst of conditions.

In 1598 the Roman Inquisition finished its deliberation with the sentence of death by burning. He was given the opportunity to recant some of the statements he had made, but he would not. In January 1600, Rome's *civil* authorities carried out the sentence. The Church could then say, "They had no blood on its hands."

The charges: Source 62 presents a list of eight charges that are a combination of two sets of accusations. This combination is not the actual consolidated set made by Cardinal Bellarmine, the Roman Inquisition leader. Note: Cardinal Bellarmine was also leader of the Roman Church's Inquisition's prosecution of Galileo.

Rather than my trying to understand Giordano Bruno's complicated sentence, I quote directly from the writing's of it author, Ms. Ingrid D. Rowland, at Chapter twenty-nine, The Sentence, Rome, 1600:

> Giordano Bruno's conviction of heresy hinged on two points: his refusal to believe that the bread of Communion was literally transformed into the body of Christ; and his refusal to renounce as heretical the eight propositions distilled from his writing by Robert Bellarmine. But the eight propositions in themselves did not motivate his sentence. In his last defenses, Bruno declared that the inquisitors had no right to dictate what was heresy and what was not. It was this denial of their authority that sealed his fate.

Finally, an historical note: "It is most likely that the original records for Bruno's trial in Rome no longer exist. Napoleon Bonaparte, who had ordered the razing of an (unknown) named-site in Venice), also attacked the memory of the Inquisition in Rome."

(Author's note: Any time that I finish reading about Giordano Bruno, I feel that I had just run a mile.)

Appendix 2. Inventors, Developers and 'Commercializes' of *The* Steam Engine

Thomas Savery (1650-1715)

Savery was an English military engineer and inventor who, in 1698, patented the first crude steam engine, based on Denis Papin's Digester (a 'pressure cooker' of the time). Savery had been working on solving the problem of pumping water out of coalmines. His machine consisted of a closed vessel filled with water into which steam under pressure was introduced.

James Watt (1736-1819)

Before James Watt, the design of the steam engine was difficult to apply to different applications. James Watt patented an improved version of the steam engine that enabled the steam engine to be used for many different applications, such as textile machines to make cotton and wool clothing. Watt's version of the steam engine became one of the major steam engine-driven machines that started the Industrial Revolution in England.

Peter Cooper (1791-1883)

Cooper was an inventor who designed and built in 1830, what was called, *The Tom Thumb* railroad locomotive,. This model locomotive was the first American-built steam locomotive to be operated on a common-carrier (used by the general public) railroad.

John Fitch (1743-1798)

Fitch made the first successful trial of a steamboat on the Delaware River in 1787. Fitch later built a larger vessel that carried passengers and freight between Philadelphia and Burlington, New Jersey. Fitch, however, failed to pay attention to construction and operating costs, and was unable to justify the economic benefits of steam navigation.

Robert Fulton (1765-1815)

Fulton built his first boat after Fitch's death, and became known as the "Father of Steam Navigation." He was accredited with turning the steamboat into a commercial success. In 1807, Robert Fulton's boat, the Clermont, went from New York City to Albany, a 150-mile trip, in 32 hours, for an average speed of 5 miles-per-hour.

Appendix 3. Why Engines Are Inefficient.

Let us first look at the heat transfers in the steam engine. Its operation contains multiple examples of the heat losses (a gain in entropy) throughout the process:

The Steam Engine, in the burning or conversion or transaction from coal to heat:

- Coal starts out with a certain amount of "built-in" energy, which is called BTUs (British Thermal Units).
- Part of the coal's innate heating energy (BTUs) is lost due to the inefficient process of just *burning* coal. That is, not all of the coal burns and there is wasted coal-power.
- As the steam in the boiler get hot, the walls of the boiler get hot and some heat goes to the outside air. The heat that went outside does not contribute to the steam engine's job. This can be a lot of heat. More losses.
- The engine moves, but it has to move against friction in the moving parts, such as the gears and wheels. Moving parts rub against each other (friction) and these parts heat up. Again, lost energy.
- The engine makes noise as it moves. The noise is caused by air being pushed. (Sound waves.) Noise is lost energy.
- The moving engine also has to 'push' against the air. Again, lost energy.

We can also make a similar examination of the automobile engine in order to get an idea of its efficiency. This engine also involves moving a piston, but instead of injecting steam, gasoline is injected into the cylinder holding the piston. A spark ignites the gasoline, the gasoline explodes, and the explosion's hot air-pressure pushes the piston. In an automobile engine:

The conversion from inert gasoline to an explosive gas to drive the pistons in the motor involves a considerable amount of lost energy. That is, the original (inert) energy in the gasoline just cannot *all* be converted into heat without some loss of the basic energy that the gasoline contains—the work that the gasoline is *intrinsically* capable of doing. In addition:

- The noise generated by the engine is lost energy.
- As the wheel axel turns, in what is called 'bearings', metal parts rub against one another—causing friction, which causes the wheels to heat up.
- The metal outside structure of engine itself heats up, as well as the hood of the auto. (Once again, entropy increases.)
- And many more places or actions in an auto lose energy.

Finally, an important general statement: In *ALL* heat transfers or transaction there is *always* a loss of energy—or, conversely, a gain in entropy.

Efficiency of Engines and Efficiency of the Complete Moving System

A note describing the efficiency of an engine: The energy (or work) the engine actually performs work *is divided* by the energy making the engine work. A simple relationship. Restated: Work Performed divided by Energy of Fuel (coal, oil, gas…) is the efficiency of the system.

Efficiency is conceptually simple. It will also be shown to be illuminating when the efficiency in various types of engines is discussed below. End of note.

The Engine Alone: If you expect that an engine of any type to be near 100% in efficiency, you will be disappointed. For example, an ordinary steam engine's efficiency is less than 10%. A well-designed steam engine with special reheating features is in the 30% range, and a modern steam-driven turbine, such as are used in some coal-generator power plants, is in the 50% range.

(A repeat about the turbine: A turbine is different from the ordinary steam engine. In a turbine, steam, or water, or wind push the many fan-like blades, all connected to a common axel, that turn the turbine's axel at very high speeds.)

Automobile engines are in the 20 to 30% efficiency range and diesel engines are in the 40% range.

Allan Karson

The Complete System Must Be Considered.

There are many losses that are *in addition* to the engine/piston itself, at such places as the gears and wheels There is the frictional rubbing of metal on metal generating heat, energy going to make wasteful noise, heating the outside of the vehicle to a higher temperature—all increasing wasted energy, or entropy.

Thus, most *overall* system engines (the engine itself, the gears, all the moving parts, all the heated parts that give off heat to the atmosphere)—all operate in the ranges from 5% to 20%,. What high *inefficiency* for our supposedly *modern* systems! What a waste of the available energy of our fuels—wood, coal, gasoline and diesel oil.

Appendix 4. Isaac Newton's Description of His Accomplishments During the Period 1665-1666, Known as Newton's *Anni Mirabilis*

The years of Newton's *Anni Mirabiles* were also the years the plague devastated England. At that time Newton was a student at Cambridge University, and the school closed down due to the plague. Newton went home to study—and create. The plague subsided in 1667, and Newton returned to the reopened Cambridge.

The following, written by Isaac Newton, is from source 18, page 38, *The Life of Isaac Newton*, Richard Westfall, Cambridge University Press, 1993.

In the beginning of the year 1665 I found the Method of approximating series & the Rule for reducing any dignity of any Binomial into such a series. The same year in May I found the method of Tangents of Gregory & Slusius, & in November had the direct method of fluxions and the next year in January had the Theory of Colors & in May following I had entrance into ye inverse method of fluxions. And the same ear I began to think of gravity extending to the ye orb of the moon & (having found out how to estimate the force with which (a) globe revolving within a sphere presses the surface of the sphere) from Kepler's rule of the periodical time of the Planets being in sesquialterate proportion of their distances from the center of their Orbs, I deduced that the forces which keep the Moon in her Orb with the force of gravity at the surface of the earth & found them answer pretty nearly. All this was in the two plague years of 1665-1666. For in those days I was in the prime of my age for inventions & minded Mathematics and Philosophy more than any time since.

Allan Karson

Clarifications:

Fluxions = the rate of change of a variable quantity; a derivative.; the derivative relative to the time

Philosophy = astronomy or physics

Sesquialterate – Numbers in a ratio of one and a half to one; e.g., 9 and 6.

Appendix 5. The Difference Between the Unit Measuring System Used in the U.S. and the Measuring System Used By the Rest of the World

Measurement systems are basic and ubiquitous to just about all-human activities in the world: science, commerce, manufacturing, transportation systems—and households... Measurements systems measure distance, mass, time, temperature, force, volume....

All countries in our so-called 'One world' use the International System of Units system (ISU)

—with exception of the United States and two other nations, Liberia and Myanmar, which use their 'own' measuring system,

The ISU system's basic measuring units are meters, kilograms and seconds. To increase or decrease those basic units, all that is necessary is to divide or multiply by 10s, 100s, or by thousands. What could be simpler?

This U.S. system uses feet/yards/miles; pounds for weight, and pints-quarts and gallons. Nowhere is the U.S.'s measuring system more annoying than in this 'homey' subject of weights and mass. Everyone in the U.S. knows the problems of going from inches to feet to yards and to miles, and dealing with temperature measurements, that go from 32 degrees for freezing water, to 212 degrees for boiling water. Gothic!

In 1964 the National Bureau of Standards (NBS) started to promote the use of the metric system, known throughout the world as the International System of Units (ISU). ISU uses the MKS system, of meters, kilograms, seconds system—that usually involve conversion factors of 10, 100.... But the NBS's promotion effort failed miserably.

Allan Karson

It appears that when the U.S. continues to speak of the international system of trade, and how to improve our participation within that worldwide system of commerce, one of the fundamental features of having a world-compatible, simple measuring system, is rarely mentioned.

Appendix 6. The Big Bang

According to the European Space Agency, 2007 (This chart describes the Big Bang according to the Inflationary theory of Professor Guth)

http://www.esa.int/esaSC/SEMC6TS1VED_index_0.html April 2007

A millionth of a second after the Big Bang

The Universe is roughly the size of our Solar System today, and the temperature drops to 1 GeV (equivalent to 10 thousand million degrees). It is cool enough for quarks to combine and make the particles in the atomic nucleus, protons and neutrons. These particles are called hadrons, so this period is often called the 'quark-hadron transition'.

From one second to three minutes after the Big Bang

At one second, the Universe grows to about a thousand times the size of our Solar System today and the temperature drops to 1 MeV, equivalent to 10 thousand million degrees. Neutrons and protons combine to form the first nuclei: first deuterium, then helium and other elements. This is called 'primordial nucleosynthesis' and it lasts several minutes.

Three minutes after the Big Bang the temperature is a thousand million degrees. It is still too hot, however, for the atomic nuclei to capture electrons and form real elements.

From three minutes to 300,000 years after the Big Bang

The Universe keeps expanding. It is too hot yet for electrons to be captured by the atomic nuclei. Electrons wander freely and are therefore able to interact with the photons (light 'particles'). As a result, light is trapped and cannot propagate more than a very short distance before encountering an electron. Therefore the Universe is opaque.

But about 300,000 years later, the Universe has cooled enough (to some 3000°C) to allow protons to capture electrons, and form neutral hydrogen atoms (in a phenomenon called 'recombination' or 'decoupling').

The Universe is about 1000 times smaller than its present size. Light can propagate without hindrance: the Universe suddenly becomes transparent. The Cosmic Microwave Background radiation that we detect today is that 'first light'.

Soon after recombination (The Dark Age) The matter is now cool and luminous. Initial 'clots' of matter start to grow by gravitational attraction. This process is still unknown, but involves both the matter that we can see ('baryons') and so-called 'Dark Matter'.

The first stars form when the lumps of matter grow to about 10 million times the mass of our Sun, when the Universe is about one thirtieth of its current size. The lumps of matter (containing stars) coalesce to form galaxies and clusters of galaxies.

The first stars produce lots of ultraviolet radiation, which ionizes most of the neutral hydrogen (that is, liberating the electrons from the hydrogen's protons. The protons are now free to 'do their job', which is to light up the universe—thereby ending the so-called 'Dark age' of the Universe.

1,000 million years after the Big Bang

The Universe is a fifth of its present size. Observations indicate that there are already fully formed galaxies. Therefore galaxy formation must have started much earlier. When the Universe is half its present size, the nuclear reactions inside the stars have already produced most of the chemical elements that are needed to make Earth-like planets.

10,000 million years after the Big Bang

About 5000 million years ago, our Sun was formed from the collapse of a cloud of dust and gas, producing a very average- looking star.

The *remnants* from the formation of the Sun, swirling in a disk around *our* infant star, gradually coalesce into the planets that form part of our Solar System. 4500 million years ago, the Earth and the inner planets form with rocky mantles and molten interiors, while more distant planets become gaseous giants.

Around 700 million years later, life begins on Earth. The oldest fossils of living organisms (bacteria) found on Earth are 3800 million years old.

Studying space science at its largest scale therefore provides some of the deepest insights into physics on the smallest scale. ESA Science is now providing access to the largest science laboratory we have ever known: our Universe!

Appendix 7. Does Dark Matter Really Exist in Space?

This appendix is about the writings of the physicist, John W. Moffat, Professor Emeritus in Physics at the University of Toronto. Professor Moffat provides a different meaning to some of the findings of the NASA's space probes, such as the Wilkinson Microwave Anisotropy Probe (WMAP). Much of what follows is from Professor Moffat's relatively recent book of 2008, *Reinventing Gravity, A Physicist Goes Beyond Einstein*, source 61.

Professor Moffat's book provides a special insight that is very different from books written by many other physicists: Many books, not all, discuss accepted theories; Professor Moffat presents his arguments *against* two generally accepted theories—those theories describing what Dark Matter and Dark Energy is. In this appendix we will only consider Dark Matter. For completeness, we will mention Dark Energy, but only in the beginning of the presentation.

Let us first review the truly 'revolutionary' findings of the WMAP. As stated in Chapter 4A, the findings say:

Ninety-three percent of the Universe is made up of unknown forms of matter and energy, called Dark Matter and Dark Energy—both of which are *totally invisible*.

Dark Matter, therefore, *outweighs all the stars by a factor of 10*! (When the gravity of dark matter is considered, it acts as normal matter does; it attracts.)

Seventy-three percent of the Universe is made up of an unknown energy called *Dark Energy*! Dark Energy acts with a repulsive force as opposed to gravity, which brings matter together.

It is beyond the scope of this book for us to analyze how the cosmologists have arrived at these *controversial-but fascinating* findings. We note, however, that they measured many parameters in space such as the velocities and accelerations of various large galaxies in far off outer space, then performed calculations on their measurements, and then derived these varied, important parameters.

And now, a key issue: In their performing their calculations, cosmologists employ "Universal Constants' in their calculations, such as the numbers that represent the force of gravity, the speed of light, and others. One of these, a very important one in this case, is the Universal Constant, 'G', used in calculating the force between two bodies.

It was Isaac Newton, in the seventeenth century, who identified that there is a 'gravitational force', and realized that a number, a proportionality number, such as 'G', would eventually be needed to 'make numerical sense' of the other numbers used to equate his force of gravity. The value of G was established at the now-Cavendish Laboratory at Cambridge in 1778.

G is used to calculate the force between two bodies. The bodies can be of modest size such as a human body, large size such as earth and planets, and (presumably) far-off galaxies. That is, it is supposed to be the correct value of G *everywhere* in the Universe. Its value, in what is called the *mks* system, is:

$(6.67428 * 10^{-11} * (meters)^3 *) / (kilograms * (seconds)^2)$

You probably have seen this number, usually represented by 'G', or 'g' applied in the simple, but powerful, equation: Force $= (m_1 * m_2 * G) / R^2$

where 'm' represents the mass of each body and 'R' is their distance from one another. Once again, the bodies can be of modest size such as a human body, large size such as earth and planets, and far-off galaxies. Now to Professor Moffat.

Professor Moffat postulates the 'standard' value *G is not applicable to such huge bodies, or in all parts of the Universe*. Thus, he considers that the supposition that there is Dark Matter in the Universe should be investigated much further. And within this investigation, Professor Moffat also considers that considerable consideration should be given to investigating (or questioning) *the universality* of 'G' with the numeric value of $6.67428 * 10^{-11}$.

Professor Moffat refers to his work on gravity as the *Modified Gravity Theory*, or MOG. This is all rather straightforward, but expectedly complicated in calculations. Most of his book is concerned with redefining Dark Matter, or perhaps, better said, *doing away with the concept of Dark Matter.*

Advanced readers (and, possibly, Intermediates) should be interested in this book by Professor Moffat since it provides considerable explanatory background in the presentations of his ideas.

Note: Dr. Lee Smolin, author of *Life of the Cosmos*, source 23, also speaks of there being a kind of Darwinian progression in space, which bring about *changes* in what are called Universal Constants. Dr. Lee Smolin is a founding member of the Perimeter Institute for Theoretical Physics in Waterloo, Canada. Professor Moffat is also associated with the Perimeter Institute.) End of note.

Appendix 8. Planck Length and other Planck Constants

Max Planck provided insights into the quantum world and predicted that if further studies would be made of small particles, the smallest distance that they would find would be 1.6×10^{-33} centimeters. It is known as *the* Planck Length, or Lp.

Planck calculated the Planck Length by using a simple equation consisting of three *fundamental* parameters: 1. The speed of light 'c', 2. Newton's gravitational constant (G), and 3. Planck's constant (h). The formula he used is: $Lp=((G*h)/c^3)^{1/2}$.

It is called a *natural* number since it is based on what appears to be universal constants (Repeating the three: Newton's gravitational constant, Planck's constant, and speed of light). It is the smallest possible length that can be (presently) considered in quantum mechanics. (As we saw in the description of String Theory in chapter 12, Lp is considered to be the general size of strings.)

Planck went on to define the other units that are listed below, and they, too, were based on numbers that have no human construct, i.e., they are fundamental or *natural*. The five are presented here to show their size, with comments-in-parentheses comparing the size of the Planck Constants with the size of every-day measurements.

Planck Length: 1.6×10^{-33} centimeters (*Very* short)

Planck Mass: $2 * 10^{-8}$ kilograms (kgs) (Very light-in-weight in the every-day world, but very heavy in the atomic world, where the mass of an electron if approximately 9×10^{-31} kgs.)

Planck Time: $5 * 10^{-44}$ seconds (Very short period of time)

Planck (electric) Charge: $1.87 * 10^{-18}$ Faradays (Very little)

Allan Karson

Planck Temperature: $1.4 * 10^{32}$ degrees Kelvin (Very hot)

A significance of these numbers is that we mortals cannot presently measure — with any accuracy, *any* of those five items (or numbers) that are listed above.

To compare the Planck Length with other objects, we can compare it to the size of a simple atom, such as the hydrogen atom. An example: Niels Bohr proposed that the hydrogen atom has a radius of approximately $5.3*10^{-9}$ centimeters. That small atom is larger than the Planck length by a factor of about 10^{+24}.

The physical parameters that were used by Planck to derive these numbers can be seen at http://en.wikipedia.org/wiki/Planck_units.

Appendix 9. People and Events Leading Up to the First Atomic Bomb, August 1945.

Part 1. A List of the First Persons Who Were Involved in the First Nuclear Fission That Led to the Making of the First Atomic Bomb

Much of the 'inside' information for Part 1 is attributed to the biography, written by Ruth Lewin Sime, of *Lise Meitner, A Life in Physics*. Meitner was one of the first physicists to design an experiment that produced nuclear fission within a laboratory environment, source 21.

James Chadwick (1891-1974) was an English physicist who discovered the neutron and was awarded the Nobel Prize for Physics in 1935. Later, during World War II, he was a member of the Manhattan Project in the United States.

Enrico Fermi (1901 -1954) Italian physicist. One of his outstanding abilities was the rare ability to be an expert in both theoretical and experimental physics.

Otto Hahn, German chemist (1879 -1968) The Kaiser Wilhelm Chemistry Institute, in Dahlem-Berlin, an area in the southwestern part of Berlin. Hahn worked in collaboration with Lise Meitner for thirty years.

Lise Meitner (1878-1968) Austrian (Later Swedish) physicist, Originally Jewish and converted to Protestantism in 1908. Meitner was Director of physics section at the Kaiser Wilhelm Institute for Chemistry, Berlin. She was the first woman physics professor in Germany and worked for, and in collaboration with, Otto Hahn for thirty years. As will be seen below, she was the person who designed a physics experiment that produced the first fission process—and she was also the sole person, at the time, who understood implications of her physics experiment.

In 1938 Meitner was forced to leave Germany due to the Nazi laws against Jews—even though she had converted to Protestantism. She first went through Holland to visit with Niels Bohr in Copenhagen, Denmark, but later located to Stockholm. She eventually became a Swedish citizen and member of the Nobel Committee. In 1960 she retired in England to be near the family of Otto Frisch, her nephew.

In 1946, Meitner was feted by President Harry Truman and was honored as "Woman of the Year" by the National Woman's Press Club. Note: Lisa Meitner is probably the only woman to have an element named for her, the Meitnerium.

Otto Frisch (1904-1979) Austrian physicist. University of Vienna. Frisch was a nephew of Lise Meitner, and, for a brief time, he was a member of Lise Meitner' team working in Copenhagen and Stockholm. He later relocated to the U.S. to work on the Manhattan project and moved later to England.

Paul Rosbaud (1896-1963) Austrian by birth, trained in physics in Berlin, was and Scientific Editor for German's leading scientific journal, *Naturwissenschaflen*. In this role, he was in close and frequent contact with, and confidence of, many leading German scientists who opposed the decline of science under the Nazis. During the coming war he communicated with Allied scientists via the Norwegian underground and neutral channels.

Fritz Strassmann German chemist (1902 – 1980) University of Hanover, University of Mainz, and Director of the Chemistry Department at the Max Planck Institute for Chemistry. Anti-Nazi, who freely expressed his opinions, making it difficult for him to make a living wage, even after the war, as a Director of the Max Planck Institute.

Part 2 Key Events Leading to the First Fission Process

In 1932 James Chadwick made a fundamental discovery in the domain of nuclear science: he discovered the particle in the nucleus of an atom that became known as the neutron. Later, in 1939, it was shown that the neutron proved capable of penetrating and splitting the nuclei. In this way,

Chadwick—in 1932, prepared the way towards the fission (breaking of the nucleus) of uranium 235—among the heaviest of elements.

(Note: The neutron is part of the nucleus of an atom. The neutron and the proton comprise the nucleus. . The neutron has nearly the same weight as the proton, but is slightly more massive in size. The neutron has no electric charge, while the proton does have positive electric charge.

The neutron (and the proton) is 2,000 times heavier than the electron. Thus, most of the mass of an atom is in the nucleus. Remember that in Einstein's energy equation, $E=mc^2$, the energy in matter depends on the mass of an object. End of note.)

With that discovery of the neutron in 1932, speculation arose in the scientific community that it might be possible to create elements heavier than uranium (atomic number 92) in the laboratory. (Making a bomb was not considered. Probably no one suspected that this research would eventually culminate in an atomic weapon.)

A scientific race began between Ernest Rutherford in Great Britain, Irene Joliot-Curie in France, Enrico Fermi in Italy and the Meitner-Hahn team in Berlin. All team members believed that this was abstract research project for the probable honor of a Nobel Prize.

In 1934 Enrico Fermi won the race. Fermi bombarded the heavy element, uranium, with neutrons. He hoped that it would cause the uranium to emit a beta particle and become a new, artificial element—above uranium in the periodic table (meaning, heavier than uranium). This was the first use of neutrons as active 'bullets', used to bring about changes in an atom. It appeared to Fermi—and to the world's physicists, that Fermi had created a new element. Fermi won the Nobel Prize for this work in 1938.

Three items concerning Fermi: 1. In the process Fermi showed that slow-moving neutrons were more effective than high-energy neutrons for this task—changing the characteristics of an element to that of another element.

2. Fermi left Italy after receiving the Nobel Prize since Mussolini, a fascist, who had taken over Italy in the 1920s, as Il Duce. Fermi had a Jewish wife, which could have put him in jeopardy if Mussolini followed the anti-

Semitic policies of his fellow fascist, Hitler. Fermi came to the United States and became a professor at Columbia University.

3. When Fermi submitted his now-famous paper describing the results of his work, on what was called *beta decay*, to the prestigious English journal, *Nature*, the journal's editor turned it down. The editor considered "It contained speculations which were too remote from reality". *Nature* eventually did publish Fermi's report on January 16, 1939.

Going back to 1934, when Fermi announced his results, the teams that had been working on the problem could not understand Fermi's results—that a new element was created. The teams were the Curie team in France and the physics/chemistry team consisting of Lise Meitner, Otto Hahn and Fritz Strassmann in Germany. For both teams, Fermi's work had produced more questions than answers.

Both teams replicated the work of Fermi; they both performed hundreds of experiments. They, too, as had done the Fermi team, used neutrons to bombard heavy elements such as uranium—to see what was the outcome of the bombardments. Both teams detected that they produced 'new elements', but they could not determine what these new elements were!

The teams worked as tough competitors from 1934 to 1939. From time to time, they would communicate with one another —but more as competitors than collaborators. During this period, in 1938, all hell broke loose in Europe. Germany annexed Austria. Meitner and Otto Frisch (Meitner's nephew, also a physicist) who had converted from Judaism, had to flee Germany. With the help of Otto Hahn and Paul Rosbaud, they illegally crossed the German-Dutch border. where the Dutch physicist, Dick Coster, helped them. At that time, Holland was still a free country; later, in 1940, Germany invaded Holland.

Niels Bohr, the Danish physicist in nearby still-not-invaded Denmark, helped obtain positions for Meitner and Frisch in Stockholm, Sweden, which was neutral during the entire war. The split-up team of Hahn and Meitner continued to communicate and meet secretly. Those meeting stopped in September 1939, when Germany invaded Poland, which was followed by Germany invading France and attacking England by air, thus starting World War II.

During one such meeting before the outbreak of the war, Meitner had designed an experiment to perform which involved a form of uranium, and she gave the experiment's procedural description to her former German associates, Hahn and Strassmann who were still in Germany at their laboratory. They did perform the experiment according to her instructions, but since they were chemists, they did not *fully* understand the experiment's outcome—which was about the *physics* of the experiment.

Nor could they fully (or correctly) interpret the results. But they did see that another, known, but lighter element, barium, was created in the process. This was puzzling to them since barium was not expected. Rosbaud, the science editor of the German magazine, *Naturwissenschaflen*, knew of the experiment and pressed Hahn to publish the results of the (German) team's analysis. Hahn agreed. The results were published in *Naturwissenschaflen* in January 1939.

Meitner, in the meantime, was informed of the results of the experiment and had a much more different and powerful interpretation—on both the outcome of Fermi's experiment and her German-team's experiment. *Her analysis of the results of her team's test showed, implicitly, that the test performed by Fermi had split the uranium atom—resulting in nuclear fission*, a term created by Otto Frisch.

Meitner was led to this (far-out?) conclusion by an unexpected occurrence within the experiment: *a tremendous amount of heat was generated in her team's (Hahn and Strassman) experiment*. She attributed this to Einstein's as-yet-unproven equation, $E=mc^2$, which strongly suggested that there was a *nuclear explosion*, albeit, a *small* explosion.

But first, a digression about Paul Rosbaud and the post-war Hahn-Meitner relationship. Throughout the war, Paul Rosbaud was actually a highly effective British spy/agent. His effectiveness was partially due to his using his strong contacts with top German scientists. As we just learned, for example, he pressed Hahn to publish the results about the successful nuclear transaction in his (Rosbaud's) journal, *Naturwissenschaflen*, making sure the world-at-large would learn of it.

This news—which was later published in the English journal, *Nature* by Meitner, electrified the (relatively small-in-size) scientific world, including

402

and was one of the sparks that ignited the chain of Szilárd-Einstein-Roosevelt-Oppenheimer and the U.S.'s Manhattan Project, whose "product' was the atomic bomb that helped the United State win and end WW II. (Part 2, below, describes the role of Szilárd in this chain, leading to the Manhattan Project.)

Back to Meitner and Hahn: Hahn continued his research in nuclear physics in Germany during World War II and is believed to *not* have worked on Germany's unsuccessful atomic bomb project. Even so, Hahn was captured by the Allied Forces and sent to England in 1945. He returned to Germany in 1946 and was awarded the Nobel Prize for Chemistry (!!!), becoming, soon after, president of the Max Planck Institute in Göttingen, Germany.

Hahn was also called the "founder of the atomic age" by his contemporaries and, officially, by the Max Planck Society. Even Glenn T. Seaborg (A former chairman of the U.S. Atomic Energy Commission) deemed Hahn "the father of nuclear chemistry".

Many historians of science of this period believe Meitner should have been awarded the Nobel Prize, to be shared with Hahn. For example, a 1997 *Physics Today* study concluded that Meitner's omission was "a rare instance in which personal negative opinions apparently led to the "exclusion of a deserving scientist from the Nobel Prize."

Since I find this period to be especially interesting, I continue to quote about the issue of the Nobel Prize being awarded to Hahn and not to Meitner. I provide a footnote from source 27, page 41, note 5, which shows the-good-and-the-bad of Hahn in a very difficult situation. (Source 27, titled, *Heisenberg and the Nazi Atomic Bomb Project, A Study in German Culture*, deals with the unsuccessful German atomic bomb war effort, which will be discussed at the end of this appendix.) And remember, Meitner and Hahn had a collegial relationship for thirty years! In the following, the author's inserts are in italics.

" ...even his (*Hahn's*) behavior toward Lise Meitner (*his coworker*) in 1938 is open to reevaluation. Hahn had originally caved in to Nazi pressure to abandon Meitner, and it was only because Paul Rosbaud shamed him into changing his stance that he helped his old colleague and friend to escape from Germany."

Some would regard his (sole) acceptance of the Nobel Prize in 1946 as being indecent. That he knew his action (to accept the prize) to be unjust, personally and scientifically to Meitner—Meitner who had given to the mystified-Hahn the physical explanation of what precisely had occurred in his crucial experiment, may be reinforced by his donating a portion of the prize money to his former colleague, Meitner.

Finally, years later, in 1966, Lise Meitner was awarded a share of the Fermi Award for her contribution to physics (*along with Hahn and Strassmann*). She was eighty-eight years old and declined to make the difficult trip to the U.S. to receive the reward directly, since she was then living in London. It was Glenn Seaborg (former chairman of the U.S. Atomic Energy Commission) who brought the prize to her in London.

If you wish to learn more about this work by Meitner, and her professional and personal relationships, I suggest the biography, *Lise Meitner, A Life in Physics*, by Dr. Ruth Lewin Sime, source 21. Also, to just help you understand more about this personal situation between Meitner and Hahn, the web site, http://en.wikipedia.org/wiki/Otto_Hahn, presents a much more favorable presentation about Otto Hahn than is presented here.

Part 2 The Commencement of the U.S. A-Bomb Manhattan Project.

Part 2 is a brief description of the role of five physicists, Leo Szilárd, Enrico Fermi, Edward Teller, Eugene Wigner and Albert Einstein in the commencement process of producing and A-Bomb.

Leó Szilárd (1898-1964) was a Hungarian-Jewish physicist who played an important role in initiating America's atomic bomb project. During the 1920s and early 1930s, Szilárd had been associated with Humboldt University of Berlin and Berlin's Institute for Theoretical Physics. Also, at one time, Szilárd taught a seminar on nuclear physics and chemistry with Lisa Meitner.

Due to the fearful anti-Semitic environment in Germany, Szilárd moved to London in 1933. It was during this time in London that he conceived of the key action in an atomic bomb that later became an important part of the American bomb design, the chain reaction.

Allan Karson

Szilárd remained in London until 1938, when he came to the U.S. to work at Columbia University. After coming to the U.S., Szilárd probably learned about the Hahn/Meitner results that were published in the technical journals *Naturwissenschaflen* and *Nature*. He sought to warn America of the potential of a German bomb.

He realized that a letter from him to any high U.S. official or scientist, a relatively unknown physicist in the U.S., would not carry much weight. He, therefore, led the composition of a letter—performed as a joint project with his colleague at Columbia University, Enrico Fermi, and two former Hungarian physicists, Edward Teller, a professor of physics at George Washington University in Washington, D.C. and Eugene Wigner, a professor of physics at Princeton University.

The letter they composed advocated that the U.S. should build the atomic bomb. Szilárd personally brought this letter to the internationally respected physicist, Albert Einstein, in Princeton, New Jersey. Szilárd convinced Einstein to send their warning letter to President Roosevelt. (There were two additional letters sent by Einstein to Roosevelt, but we will not go into them.)

As we know, President Roosevelt acted positively on the letter and from there on, the rest is history. The atomic bomb project, named *The Manhattan Project*, was the established, and Dr. J. Robert Oppenheimer (1904-1967) was chosen to be its overall Technical Director.

We should also be thankful that the U.S. Army had an officer who had the qualifications to manage the *entire* large project, Lieutenant General Leslie Groves 1896 – 1970. He had recently overseen the construction of the Pentagon building for the Department of Defense. The Pentagon building was completed and dedicated in January 1943, and was, at that time, the world's largest office building.

Groves' *Wikipedia* site says he was "an officer of high intelligence, tremendous drive and energy, and great organizational and administrative ability, as well as considerable ruthlessness, arrogance, and self-confidence." It was Groves who commissioned the Smythe report, described below, and requested that it be ready when the first bomb was used.

405

The Smyth Report 1945

The *Smyth Report* was the name given to an administrative history written by physicist Henry DeWolf Smyth about the Allied World War II effort to develop the atomic bomb, the Manhattan Project.

The full title of the report was the unwieldy *Atomic Energy for Military Purposes; The Official Report on the Development of the Atomic Bomb under the Auspices of the United States Government*, 1940-1945. It was released to the public on August 12, 1945, immediately after the atomic bombings of Hiroshima and Nagasaki, on August 6 and 9, 1945.

One additional note: I have mentioned a number of times that physics-time is slow as a glacier. But here we have an instance where a tremendous project and 'product' were accomplished in what, in the world of physics, would be comparable to the speed of light. The theory of nuclear fission was proven in 1939 and a physics product was produced in 1945. Amazing.

As you can learn from reading about the Manhattan Project, it was a *huge* project, involving thousands of people, research and manufacturing operations in about a half-dozen states and a mix of production, manufacturing, quality control, security, logistics, physics....

Finally, a brief review of the atomic bomb effort in Germany before and during the war.

As for the Germany nuclear program, it never went anywhere. The position of head physicist was given to Werner Heisenberg. I will tell you, only briefly, that they made a major error in the calculation of how much material they would need to build a bomb. To learn the details of that inept program and leader, I suggest source 27, *Heisenberg and the Nazi Atomic Bomb Program*.

A final note about this time period: When Germany was defeated, its leading atomic physicists were rounded up by the British and brought to Farm Hall, Cambridge, England to be interrogated. It appears that no one of the group felt he was planning a bomb for an evil regime. This is all recounted in the source 27's first seventy-five pages.

The author of that book, Paul Lawrence Rose, also shows that the German physicists—including, and primarily their leader, Werner Heisenberg, refused to admit to the British that he "did not do his physics well".

While the book describes what they told the British, it turns out that the British had also set up a recording system to catch what the German physicists were saying in private, amongst themselves. General Leslie Groves first revealed this a few years later after the 1945 period. It appears that the recordings frequently contradicted the Germans' testimonies to the British.

Appendix 10. A Chronology of Einstein's Articles Appearing in Volume 13 of The *Annalen der Physik* of 1905, Einstein's *Annui Mirabilis*, The Miraculous Year

March 17, 1905 *On a Heuristic Point of View Concerning the Production and Transformation of Light*: Einstein completes the paper on the light-quantum hypothesis. This is an introduction to the light-quanta hypothesis with the assistance of an argument based on the Boltzmann's work—which, naturally, was on thermodynamics.

Note: It was later, in 1909, that Einstein first made a statement about the possibility of particle-wave duality in light, which is a follow-on to his March 17, 1905 paper. The light-quantum hypothesis was not generally accepted within the world physics community until 1923, when the theory was proven by what is known as the *Compton effect*.

May 11, 1905 *On the Motion of Small Particles Suspended in Liquids at Rest Required by the Molecular-Kinetic Theory of Heat*:

The paper on Brownian motion. This was the first paper that *Annalen der Physik* received on this subject—and it is the most important among them. *Annalen der Physik* received a second paper from Einstein on December 19, 1905, that is also on Brownian motion.

A sequel followed it in 1906 that included a discussion of rotary Browning motion, a brief comment on the interpretation of the mean velocity, published in 1907, and a semi-popular account of the whole subject in 1908.

June 30, 1905 *On the Electrodynamics of moving Bodies*:

Einstein's first paper on Special Relativity—and on light consisting of quanta *OR* waves.

Allan Karson

September 27 1905 *Does the Inertia of a Body Depend on Its Energy Content?*

The second paper on Special Relativity, containing the derivation of E = mc^2.

A P.S. Einstein's PhD thesis, dated April 30, 1905, was not sent to the *Annalen.* Its title was

On a new determination of Molecular Dimensions. The thesis was prepared in Bern, Switzerland and submitted to the University of Zurich, It is dedicated to his friend, Dr. Marcel Grossman. A correction to the paper was published in 1911, and a minor comment was added in 1920.

Appendix 11. Frames of Reference in Space

Standing Still, Moving (Velocity), and Changing Your Velocity (Acceleration), or Newton, Mach and Einstein.

Just about all the topics in this book are about 'Big' topics, such as thermodynamics, electromagnetism, and nuclear reactions both on Earth and in the stars. This one is different. It is about what you, the reader, and I do every day. It is about 'our moving in space.'

You may also find that it is counter-intuitive, even though it is basic.

Frame of Reference and Constant Speed

Refer below to figure 37. First look at the top three graphs, in which, compared to the speed shown in each chart, you and I move much slower. These first three show the speed you might move in a car, a boat and an airplane. Each of those coordinates of x, y, and z represent a moving coordinate system for the persons in each of those conveyances. We will call the set of coordinates *Frames of Reference*.

410

Figure 37 Frames of Reference in Space

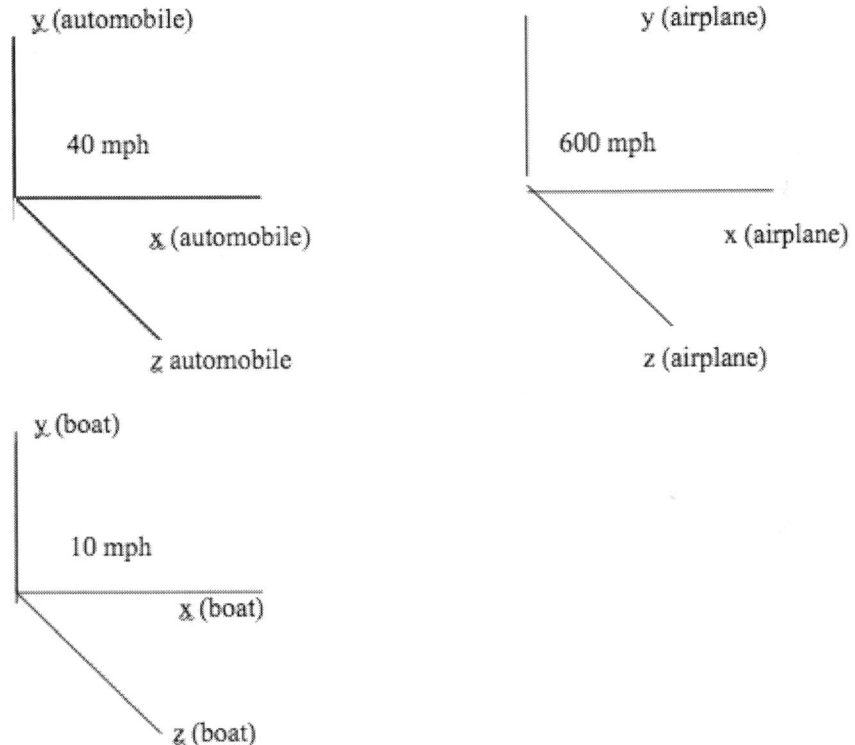

Now look at the paragraph that follows this one. It describes other coordinate system, or different *Frames of Reference*. It describes how fast you are moving by just standing still on your spot on Earth—a Frame of Reference. The Earth revolves once around its North-South axis every twenty-four hours.

Thus, you can be moving as fast as 1000 miles per hour depending on where you are located on Earth: 1000 miles per hour at the equator, 700 to 800 mph for most of the world's population, and zero at both Poles. (You are actually *accelerating* inward, toward Earth at 2 miles per hour, but it is so insignificant that you do not feel it.)

Now you can add to that the speed of the Earth traveling around the Sun, another Frame of Reference. Here you go 67,000 mph. And with this unit, just Earth going around the Sun, it does not matter where you are located on Earth. 67,000 mph.

Finally, for us Earthlings, who cannot see far beyond our galaxy, we now see our speed in our galaxy, turning around the center of our galaxy at the approximate speed of 558,000 miles per hour!!

We could go on and on investigating how fast our galaxy is moving with respect to another neighboring or far-off galaxy, but I believe the point is made. We are traveling fast. But how come we do not feel this motion!

The non-feeling effect is due to one action that is not shown in any of the *Frames of Reference*—the car, boat, airplane, Earth rotating and moving, the solar system moving, and so on. That has to do with the *lack of significant acceleration in any of the frames. If there were any significant acceleration or deceleration,* you would definitely feel it. And this applies to both linear acceleration and circular acceleration.

And this holds for *any* Frame of Reference, no matter how fast it is going. The first person who realized this and was the first person to write about this phenomena, was the Italian scientist, Galileo (1564-1642). He wrote about constant, non-accelerating motion. He was the first person to realize this physical occurrence—which we refer to as an *inertial* system.

This 'load of facts' about inertial systems is from Roger Penrose's encyclopedic math book, *The Road to Reality*, source 44. You will see that Galileo certainly makes his point that things (objects such as you and me, and other objects), *are at rest* in their *non-accelerating* Frame of Reference. That is, *they only feel motion when accelerating*. It was Galileo who wrote the following:

> Shut yourself up with some friend in the main cabin below decks on some large ship, and have with you some flies, butterflies, and other small flying animals.... have the ship proceed with any speed you like, so long as the motion is uniform and not fluctuating this way and that... The droplets will fall... into the vessel beneath without dropping toward the stern, although while the drops are in the air the ship runs many spans... the butterflies and flies will continue their flights indifferently toward every side, nor will it ever happen that they are concentrated toward the stern, as if tired out from keeping up with the course of the ship...

Steady motion is one thing; we do not feel it. Acceleration is a wholly different matter and experience.

Acceleration: A change in velocity—either speed or direction.

We feel acceleration. We know it is there. Acceleration is a *change* in speed and/or direction.

Examples of when we feel acceleration:

- Moving in a car (bus, train, bicycle, airplane that momentarily moves quicker or slower, or makes a turn.

- Being on a merry-go-round, where we are moving at constant speed, but changing our direction parallel to the perimeter of the big wheel we are on.

- Orbiting Earth in a space ship, when we have constant speed and the space ship is 'falling to Earth', but the space ship is given a constant 'boost' (acceleration) to keep it in a circular orbit. Thus, the space ship is changing its direction continuously. Under this situation, we may feel weightless, so we know something is definitely different from standing still.

• Fighter pilots who dive their plane down toward Earth, and then 'recover' and turn their direction away from Earth, can experience multiples of the pull of gravity, expressed as "g's". For example, 3g's, 5g's, 20g's and higher. This means that if they weighed 190 pounds, in the first two cases they would feel that they weighed 3*190 pounds, or 570 pounds, and for 5 g's, 950 pounds.

We all know that Newton introduced the idea that a body accelerates when a force acts on it. His equation says that, F =Mass x Acceleration. He applied this law to the notion of gravity, watching falling bodies accelerate as they fell. (Of course, he may have applied it to bodies that were lying on Earth, and not falling—such as a body sliding on an icy lake.)

Similarly, when Einstein applied this law to gravity to derive his General Theory of Relativity, he, too, described falling bodies—and acceleration.

Newton lived in the mid-1600s. I ask you: How many examples of acceleration did Newton see, or experience around him in the normal course of a day? Not many. There were no accelerating automobiles, buses, railroads, airplanes or elevators. I consider this amazing, since he had so few personal examples to feel. And yet he did introduce a way to understand acceleration in the field of physics.

Absolute Space and Relative Space

Newton introduced the idea of Absolute Space. This would seem to imply there is one true, set of space coordinates. This essentially contradicts the concept of there being more than one Frame of Reference. Much has been written about this 'no mention' by Newton of relative coordinate system and Newton's disagreement with Leibniz over this issue. (Leibniz (1646-1716), a co-founder of the mathematics, calculus, but independently from Newton), but was a contemporary of Newton (1643 – 1727).

Galileo had previously described 'relative space coordinates (or frames of reference) as presented above. Thus, apparently Newton did not take this into account, and it has been held as a demerit against Newton. Much has

414

been written (an understatement) about this issue. It turns out, however, that Newton *did believe* in relative coordinate systems.

It is assumed that, for Newton to get his fundamental ideas across, about motion, mass, acceleration, force, he did not want to 'carry along the idea of 'more than one coordinate system'. Certain writers have ascribed it to the fact that Newton had many issues to present to his broad audience of physicists and mathematicians of the time. Others consider that Newton did not consider it essential at this time.

Thus, to try to convince his (Newton's) audience about coordinate systems in general (at that time in the mid-1600s), he decided to forego *any* discussion about relative coordinate systems. Some physicist and mathematicians consider, therefore, that Newton's not speaking of the fact that a 'full set of relative coordinates' should be be considered that he, Newton, personally, did not believe in relative systems.

Since Newton did not say specifically what we on Earth should consider for 'our 'largest frame of reference', we had to wait for Ernst Mach (1838-1916) to make that selection. That is, Mach was the one who said that it is the stars that form our largest, relative (near-to-us, on Earth) Frame of Reference.

A word about Newton. Newton is unquestionably considered one of the greatest scientists who ever lived. There is only one close competitor—or equivalent, Albert Einstein. While later physicists may have replaced many of Newton's theories, Newton's range of subject–investigations was immense and included celestial mechanics, centrifugal principles, color theory, Earth theory, force theory, sound theory, the calculus—and many more.

The following is a summary of some of the 'speed effects' that we encounter daily:

1. You are rotating around the planet Earth at speeds between 1000 mph (at the equator) '0" at the North and South Poles. Most of the world's population moves in the 700 to 800 mph range.

2. Earth goes around the Sun at about 67,000 mph. If a person on Earth is going at 700 mph in the same direction as the Earth is moving around the

Sun, the person is moving with the sum of two speeds, 67,700 mph. If the directions are different, the directions must be taken into account. (As we shall see, the additional speed that is gained in each prior step is inconsequential.)

3. Our Solar System (Sun, Earth, Mars, Jupiter…) goes around the center of our Galaxy at about 483,000 miles per hour. Thus, when we, on Earth, are moving in the same direction as our Galaxy and Sun adding the two previous speeds is inconsequential when we consider our total, possible-maximum speed. This speed, in miles per hour, is approximately 567,700 miles per hour. Note: Since there is no linear acceleration in any of these speeds, we do not sense this high speed. (Appendix 11 describes the 'different feeling' a person has between acceleration and a constant velocity.)

Appendix 12. The Ether, the Michelson-Morley Experiments and the Lorentz Transformations.

Chapter 8 describes how, in the 1890s, the Dutch physicists Hendrick Lorentz (1853-1928) and Peter Zeeman developed a system of equations that are usually referred to as the Lorentz Transformations. (They are referred to below as Group 1.)

In their examination of the phenomena of measuring how light goes through the then-believed ether, they were focused on their belief that the lengths of moving bodies along the ether path were transformed into smaller distances. They believed if such a transformation occurred, the ether theory would be justified.

Some people, however, hold the position that Einstein gained a great deal of insight and information from these earlier physicists. We will see that the two groups were attacking entirely different problems—Group 1. Seeking to prove the ether exists, and Group 2. Einstein alone—Seeking to prove the actual transformation of moving bodies—without any involvement of the ether.

We will, therefore, be gracious and kind and exact in relieving them of this false belief. To do this, we shall consider the overall history of the study of the hypothesized, but never proven, ether. Note: We will not discuss *why* they believed that the ether was necessary. Also, many scientists spelled it 'aether'. We will use 'ether' throughout the discussion, except where 'aether' is quoted. End of note.

First, we must go back to the early 19th Century, when people first seriously started investigating the speed of light (which is now known to be 186,000 miles per second). Such investigations were now, in the 19th Century, possible since instruments had become sufficiently accurate to measure such speeds accurately.

The story starts earlier, however, in 1818, when two French physicists, Augustin-Jean Fresnel and Dominique François Arago, postulated that there is a *medium* that light travels in. They proposed that the medium is invisible to humans. It would be everywhere—in far off space, on the Earth in any sort of weather, in a closed room, in a basement... Wherever light goes, this medium goes. They called this medium, 'aether'.

Most people in science believed what Fresnel and Arago postulated: Even the great analytic scientist, Clerk Maxwell, of electromagnetic fame, believed in the ether.

In 1887 two Americans, **Albert A. Michelson** (1852-1931) and **Edward Morley** (1838-1923) performed experiments to learn more about the ether. (The details of Michelson-Morley experiment are well-described on the web.)

In 1887 Michelson was not a neophyte in the measurement of light, by anyone's evaluation. He was a graduate of Annapolis, the U.S. Navel Academy, and after graduation, was a science instructor at the Navel Academy where he conducted his first experiments measuring the speed of light. He left the Academy to study in Europe and worked with the great German physicist, Hermann von Helmholz, with whom he learned more about the measurement process associated with the measuring of the speed of light

Michelson returned to the U.S. in 1882, now to teach at the then-recently founded Case Western Reserve University in Cleveland, Ohio—and to continue his light-measuring experiments. Since his return to the school, he had designed a formidable set of measuring equipment that he located in a small, well-controlled space in the University. It was all set up in this basement at what was the new university, where Michelson's experiments were performed in 1887.

I quote what Michelson said after he had analyzed the experiment's data: "The interpretation of these results is that there is no displacement of the interference bands. ... The result of the hypothesis of a stationary ether is, therefore, shown to be incorrect."

As opposed to what many people would say, even today, they did not actually disprove the theory or the presence of the ether. What they did prove, or find, was that the Earth, at any part of its orbit around the sun, was not moving with respect to the ether. And that the velocity of light was the same at any part of the earth's orbit.

The results of the experiment were not, however, universally accepted. (Even Michelson was surprised by the results.) Michelson repeated the experiment many times—and always obtained the same *null* result. But there remained many non-believers, as we shall see.

The Michelson-Morley experiment stirred the pot for physicists trying to understand what goes on in the theory of light—'and what was the role of the ether'. The next person's work that we know of is the Irish physicist, George Francis FitzGerald (1851-1901).

FitzGerald was an ether-believer. As seen from his paper, quoted below, Fitzgerald accounted for the ether being there (even after the Michelson-Morley experiment) by a *shrinking*, or *shortening*, in the physical length of an object (in its direction of travel) as it move at high speeds going through the ether. FitzGerald presented his thoughts in the *American Journal of Science* of 1889. The following is from his article *The Ether and The Earth's Atmosphere*. (He is referring to the Michelson and Morley experiment.)

> "Their result seems opposed to other experiments ... I would suggest that almost the only hypothesis that can reconcile this opposition is that the length of the material bodies changes.... by an amount depending on the square of the ratio of their velocities to that of light.... the motion through the aether affects the dimensions of the solid molecular aggregations."

We will, therefore, see that Fitzgerald (and others, as described below) came up with the now-established physical phenomena—*contraction*—but for an *incorrect* reason) Their hypothesis was 'that the ether exists and here is what happens to bodies that are passing through the ether'—as opposed to Einstein's later-developed reason, 1905, for the contraction—the *Speed of Light is Constant* (i.e., 186,000 miles per second).

Fitzgerald provided that novel, now familiar ratio, v^2/c^2 that would *adjust the length of a body* as it moves quickly through the ether. The letter 'v' stands for the speed of a body going through the ether. The speed of light is always a small 'c', represented as: c.

We see that Lorentz and Larmor, and Einstein, all developed equations using that important ratio, v^2/c^2. The first two, Lorentz and Larmor, did it to prove that the ether exists. Einstein, on the other hand, never believed in the ether—and independently developed those same equations and established a *completely different* theory, later called by Einstein, the *Theory of Special Relativity* (of light).

Later, toward the end of the 19th Century, the physicists Joseph Larmor and Hendrick Lorentz and the French physicist-mathematician, Henri Poincaré, picked up and continued FitzGerald's idea that when a body goes through a measured space, the body contracts in length along the axis of movement.

The equation below shows the contraction; it was developed by the Irish physicist, James Larmor (1857-1942) in 1897 and revised and published by Lorentz in 1905. (Larmor was at Cambridge University where he held the Lucasian Chair of Mathematics from 1903 to 1932. Lorentz held the Chair of Theoretical Physics at Leyden, Holland from 1877 until 1912.)

The equations show the decrease in length in direction of v_f to be: $[1 - (v_f)^2/(c)^2)]^{1/2}$

A repeat: The purpose of this equation was to show the *actual decrease* in the length that would happen to a fast moving body—in the direction of travel. The reason that 'motivated' Larmor and Lorentz to develop these equations was to justify the ether *and*, thereby, to enable the world to continue to believe in the stationary ether. (Yes, Larmor and Lorentz, too, were ether-believers, as was Fitzgerald.)

If you find this discussion difficult to understand, do not feel lost. After all, FitzGerald, Larmor and Lorentz were trying to prove something that did not exist in reality. Ether existed only in their minds'—and in the minds, presumably of many other scientists. So the conjectures of the three of them —and many of the world's scientist's at the time, would be difficult for anyone, today, to understand.

Also, at this time in scientific history, the beginning of the 20th Century, it would have been very difficult to attempt to prove directly such a theory of contraction. The reason for this is that few 'bodies' could be made to move at a speed that is a significant fraction of the speed of light.

In 1892 Michelson became the head of the Physics Department at the new University of Chicago and president of the American Physical Society. And in 1907 Michelson won the Nobel Prize for Physics, *thus making him the first American to win the Nobel Prize for any science award.* He was second American prizewinner, following President Theodore Roosevelt's Nobel Prize for arranging the armistice of the Russian-Japanese war in 1906.

A final note on the Michelson-Morley experiment and its aftermath: As we have already seen above, in Chapter 11, Part 2, how Albert Einstein 'solved' all the dilemmas and problems that Fitzgerald and Michelson and Morley were attempting to resolve.

Now, a continuation dealing, again, with Light

Appendix 13. The Lorentz (or Einstein) Transformations

The Equations to Make the Calculations of: A <u>Stationary Observer</u> of a <u>Speeding Body</u> (approaching the speed of light) and the <u>Observed Changes</u> in Length, Time and Mass of that Speeding Body.

Since you have gotten this far with Einstein, you probably realize that he must have developed a set of equations that can tell him how a body becomes smaller in length (as observed by a stationary or slower-moving observer) when it is approaches the speed of light.

Einstein also applied these equations to show how the time and mass of fast moving bodies change. The table of numbers above is the results of those equations, for the 100-foot space ship. (The author made the calculations, based on the various speeds (a % of C) of the space ship.)

Some rules:

- The speed of light is usually referred to as c (c in italics)

- The speed of body (the spaceship, in this case) is usually referred to as i_f.

- The initial condition, in Length, Mass and Time is shown with an '0' as $Length_o$, $Mass_o$ and $Time_o$.

The basic equations are:

Length (reduction in direction of v_f) = $(Length_o)[(1 - (v_f)^2/(c)^2]^{sqrt}$

And time clocks on the moving vehicle would slow down according to the same transformation equations—in the same way that distance in the direction of travel is shortened. (The clock could be any number of atomic clocks, wristwatches, or large grandfather clocks.)

Allan Karson

$$\text{Time} = (\text{Time}_o)[(1 - (v_f)^2/(c)^2]^{sqrt}$$

But for Mass, which gets heavier, the inverse of the basic equation is used:

$$\text{Mass} = (\text{Mass}_o)/[(1 - (v_f)^2/(c)^2]^{sqrt}$$

Later in this chapter, it will be shown that these equations actually are simple equations, so do not be overwhelmed by their apparent complexity; they just look complex, but really are not.

People moving on the fast vehicle, however, would see no change in their dimensions along the direction of travel, nor any 'slowness' in their clocks! As far as they are concerned, their world has not changed. Thus, the faster moving person's clock would not show the same time as a standing-still person's clock. Two clocks were both correct within each independent 'system of reference'. This was the *death knell* to the concept of *Absolute time*, which had stood a fundamental concept since the era of Isaac Newton.

A digression: Let us examine the significance of the increase in mass of the moving frame of reference. As v_f approaches c, the fraction $((v_f)^2/(c)^2$ approaches '1'. And the denominator, which is: $[(1 - (v_f)^2/(c)^2)]^{sqrt}$ approaches '0'.

The numerator stays at '1' times the mass. Any number divided by a very small number (that nears zero), becomes very, very large. It might be said to be approaching infinity.

The 'new' mass of the vehicle just becomes too, too large. To make the vehicle move at a speed close to the speed of light, would require a *tremendous* amount of energy. This simple result strongly says that nothing (with any weight) can go near the speed of light. It is a universal-wide limit to velocity.

A note of trepidation: The author is aware of so many other conclusions that have been drawn by physicists in the past have eventually been proven wrong. Thus, I am hesitant to say that nothing (with any mass) can go near the speed of light.

It can be said, however, that based on today's limitation in energy for such experiments or travel, such a prediction may be true. But we should not predict that situation will last for all time. End of note.

The 'Simple' Equations

Let us now see that these equations are actually quite simple to calculate for different speeds. As we saw earlier in this chapter, the length in the direction of travel of light decreases by the factor:

Decrease in Length$_o$ (in direction of v_f) = $[(1 - (v_f)^2/(c)^2]^{sqrt}$

(Please excuse the fact that format $[1 - (v_f)^2/(c)^2)]^{sqrt}$ is used.

You may have used a format in high school where there was the line/bracket to denote a square root operation. The two are equivalent.)

That square root equation looks formidable, but is actually quite simple; the reason is as follows:

V_f represents the speed of the something (a space ship, Superman…); it can be represented as a decimal number times c. For example, if v_f equals 20% of c, (37,200 miles per second) V_f is replaced with .2c.

The most important fraction in the Lorentz equation is $(v_f)^2/(c)^2$. For the case where V_f is .2c, replace $(v_f)^2$ with $(.2c)^2/(c)^2$. The c is factored out in the fraction, leaving the number .04. Simple? Yes.

The equation above, $[1 - (v_f)^2/(c)^2)]^{sqrt}$, becomes $(1 - .04)^{sqrt}$, or $(.96)^{sqrt}$ or . 97979, or approximately, .98 In this case, the time clocks go 2% slower and the spaceship's length is 2% shorter.

The mass of the space ship increases as (Mass$_o$)/.9797, or 2% heavier than when not moving.

Let us repeat the process for a vehicle traveling at 60% of the speed of light. The faction is $(.60c)^2/(c)^2$ or .36. The equation becomes $[1 - .36]^{sqrt}$ or $[.64]^{sqrt}$ and finally, 0.8

In this case, the time clocks go *80% slower* and the spaceship's length is *80% shorter.*

The mass of the spaceship increases as (Mass$_o$)/.8, or *125% heavier* than when not moving.

424

Appendix 14. The Twenty-nine Parameters That Are Required For Each Running of the Standard Model.

These 29 parameters represent certain details that exist in nature. They are not man-made parameters. These are such things as the weight of an electron, its charge, the weight of a nucleus, its charge, the strength of the electromagnetic force and the other forces. I wish to make two points concerning these parameters.

Point 1. As noted above in describing the preparation for running the Standard Model, it is noted that:

Whenever a physicist seeks a solution of the equations of the Standard Model, she must first calculate a subset of these 29 parameters. And this is usually *not* a straightforward calculation. For example, within the text of the Standard Model where it describes the data necessary for looking for aspects of the Higgs particle, about twenty calculations are required.

We would think that these numbers have been used many times in the past and could be 'reused'. The problem is, however, that these numbers are not derived from First Principles. (First Principles means that they are numbers that come directly from the established laws of physics. These parameters are not so derived.) Thus, I repeat once again, these parameters must be calculated for the specific model being tested. (This recalculation process does not apply, or is not necessary for certain items within many experiments for particles and parameters that are that are recognized to be always the same, such as the electron's mass, charge, etc.)

At the same time, we may ask, "Do these 29 'fundamental' parameters ever change over time. Why do we have to measure them if they do not change." The answer to that is 'we have not observed them change—*but they might.*'

While the physicist may know the specific numbers that represent these various particles, they do not (or may not) understand the underlying reason for the value (or number) for each of these parameters. And, as importantly, the interrelationships (ratios, difference between these number for two particles, etc.) are also not well understood. Some examples are:

- Why is the electron's charge $1.6021917 * 10^{-19}$ coulombs?
- Why is the ratio of the Strong Nuclear Force to the Weak Nuclear Force 10^{16}?
- Why is the ratio of the weight of a proton 1836 times heavier than the weight of an electron?

A list of examples of this type can go on and on and on. Thus, an important goal of physicists working on the evolving development of the Standard Model is the achievement of understanding why or how these fundamental parameters are "chosen" in nature.

Digression for the reader:

1. In source 23, *Life of the Cosmos*, Dr. Smolin raises the question whether the cosmos *was ever* or *is* a self-regulating system that adjusts itself, as we know living system do in our living world. That is, to follow the principles similar to those in Darwin's Theory of Evolution. Dr. Smolin first develops the argument that if we had around us 'a more vibrant and dynamic cosmos', more black holes would have been created than are presently considered to be in the Universe. He then explores this topic by examining how some (a very small number) of the 29 parameters that might have *adjusted themselves* to increase the rate of creation of Black Holes.

2. The article written Dr. Robert N. Cahn of the Lawrence Berkeley National Laboratory. Its title is *Eighteen Arbitrary Parameters of the Standard Model in your Everyday Life*.

http://www.hep.yorku.ca/menary/misc/eighteen_parameters_of_sm.pdf

Within that article, Dr. Cahn explores what our daily world would be like *if* some of the parameters had been different at the time of the Big Bang from what they are now. As he says in the introduction to that article, "…the purpose of physics is to understand the everyday world"—and showing the direct impact of such differences in the Standard Model to our daily life, would surely—after a period of analysis of the change, show the *cause* and *relationship* between the previous characteristic and the *newly observed* characteristic, i.e., the observed difference. End of digression.

Appendix 15. The Nobel Prize, and the Fields and Abel Medals for Mathematicians

Alfred Nobel (1833 – 1898), a very industrious Swedish chemist, had transformed the handling of liquid nitroglycerine—a high used, but highly dangerous substance, into a stable explosive. He called hid stable explosive, dynamite. During his career he obtained more than 350 patents. And he spoke six languages: English, French, German, Italian, Russian and Swedish.

Nobel became very wealthy from this invention, was became one of the richest men in Europe. It was in his will that he left his fortune for the highly esteemed prize named after him, the Nobel Prize. The prize has been given since 1901. It is given every year to leaders in peace, literature, physics, chemistry, physiology or medicine and economics. You can find the names of the winners of the prize in numerous source books and at http://nobelprize.org/.

Some interesting facts:

- In 1901, each award was worth about $20,000. In 2009, the price-value rose to $1.4 million.

- No posthumous prizes are allowed. Before 1974, someone who had been nominated but later died could receive a prize. The rules were changed in 1974 so a prize can only go to a deceased person who had actually won the prize, but died before receiving it. This was the case of William Vickrey, who won the economics prize in 1996.

- The youngest winner is Lawrence Bragg, who was 25 years old when he received the physics prize with his father in 1915.

- The oldest winner is Raymond Davis Jr., who was nearly 88 years old when he received the award for physics in 2002.

- Linus Pauling is the only person to have been awarded two unshared Nobel Prizes (chemistry in 1954 and peace prize in 1962).

- George Bernard Shaw is the only person to have won both a Nobel Prize and an Oscar (Nobel for literature in 1925 and the Academy Award for best screenplay for the film adaptation of Pygmalion in 1938).

- The University of California at Berkeley has graduated more future Nobel Prize winners than any university in the world, and the Bronx High School of Science, located in Bronx, a borough of New York City, has graduated more future Nobel Prize winners than any high school in the world.

Alfred Nobel was not that interested in theories, but rather in 'discoveries'. Mathematics, which is mostly a theoretical world, was not set up (or allowed) by Nobel to receive a prize.

Professor J. C. Fields, a Canadian mathematician, established an award and medal for achievement in mathematics. The award is called the Fields medal and is the highest scientific award for mathematicians. It is presented every four years at the International Congress of Mathematicians, together with a prize of 15,000 Canadian dollars. The first Fields Medal was awarded in 1936 at the World Congress of Mathematicians in Oslo.

A more recent award in mathematics, called the Abel Prize, was established by the Norwegian government in 2002 with a fund of $22 million. It was first awarded in 2003 and First prize was $816,000.

The award is named after one of history's greatest mathematicians, Niels Hendrick Abel (1802-1829). He was Norwegian and worked and studied in Copenhagen (called Christiania at the time), Berlin and Paris. His life was short, dying while only in his 20s, and life is told by the astrophysicist and author, Mario Livio, in "The Equation That Couldn't Be Solved", source 48. (Chapter Four of Mario Livio's book consist of eleven pages, is only about Abel and is titled "The Poverty Stricken Mathematician".)

When looking at the Nobel Prize, Fields Medal and Abel Prize, I, personally contend that that there are two types of awards—according to whether the awards will lead to many future benefits coming directly from the award, or, where the findings leading to the award will not affect us in any way..

- First, concerning prizes awarded for achievements in mathematics, physics, chemistry, physiology or medicine and economics. I believe that it is expected that the achievement will lead to greater understanding of the subject. Also, it is expected that it will lead to new and better products, procedures or theories in the subject.
- Secondly, concerning prizes awarded for achievements in peace and literature. This work has been accomplished and has its present effects, and I believe that there is no expectation for greater work in that specific area of peace and literature. It may come to pass, but it is far from certain.

Appendix 16. Seventeen Numbers That Are Helpful to Understanding the Cosmos and Physics

1. Cosmic Numbers

- The speed of light is 186,000 miles (300,000 km) per second
- Our Sun has enough fuel for another 5×10^9 years.
- Some massive stars will use their fuel up in 10^8 years.
- The suspected age of the Universe 14.7 billion years
- Neutron stars have a radius of 10 miles and a density on the order of 10^8 tons per cubic inch.
- Large Black Holes can have masses of 100 times—or even more than 100, the mass of our Sun.
- The Inflationary Theory, according to MIT's Professor Alan Guth, says that the Universe expanded 10^{50} times within about 10^{-30} of a second.

2. Classical Physics: Electrons, neutrons, some subatomic particles, work and energy and electron-volts…

- Mass of a Neutron is approximately equal to the Mass of a Proton
- Mass of a Neutron = 1836 x Mass of an Electron
- Charge of an Electron, e, is $4.80325 \times 10\text{-}10$ electrostatic units (esu), 1.602176×10^{-19} coulombs,
- An Electron Volt, eV, is 1 volt, multiplied by the electron charge (1.602176×10^{-19} coulombs). (One volt is one joule divided by one coulomb.)
- One electron volt (eV) is equal to $1.60217653(14) \times 10^{-19}$ joules.

One Joule is a measurement of:

- Work energy (In SI units) equal to the work required to move a 1 kg mass against an opposing force of 1 Newton

431

- Electrical energy equal to the work done when a current of 1 ampere passes through a resistance of 1 ohm for one second, or 1 watt-second.

Electron volts, eVs, is calculated by using Einstein's equation for equating mass to energy, $E = mc^2$, where 'm' is the particle's rest mass.

- Electron energy = .5110041...MeV
- Neutron energy = 939.6... MeV
- Proton energy = 938.3... MeV

Appendix 17. Laws of Physics that Physicists and Cosmologists Invoke to *Prove, Disprove,* or *Rationalize* their Theories

In the many books that are referred in this book and in my general readings of works written by physicists, I frequently find interesting phenomena:

The physicist or scientist is discussing a subject and she or he has something to prove the theory she is discussing. Frequently she introduces another, very different theory in order to prove a point—that is, we might say, 'brought in from left field'. Once that point is proven by the new, *outside* rule, the author goes on with the discussion.

That is, for a moment or two, we may be playing according to a new set of rules. I find nothing objectionable about this. It just amazes me each time it is done! It shows ingenuity and complete control and understanding of the subject being examined.

I would also believe that the actual analysis performed on the subject by the physicist, long before his or her writing about it, was done in a more trial-and-error or, *indeterminate* manner. But in reading about it, it just seems so fresh—especially if the theory's development- is proven correct.

As future theories are developed, we should expect this occurrence. And we should enjoy the "bringing in from left field" such a "new to the issue" theories. The following is a partial list of some of the theories that can be 'clinchers' or 'zingers'—(OR, conversely, *destroyers)* of the new, proposed idea.

- Entropy: It always holds, as seen with Black Holes as shown by Stephen Hawking 'invoking' entropy to show that Black Holes actually do transmit information (in the form of heat or temperature) to the outside of the region of the Black Hole.

- Einstein's Theories of Relativity and Special Relativity, used to predict presence of Black Holes and variations in the paths of the our Universe's planets.
- Wolfgang Pauli's Exclusion Principle, which states that no two electrons could exist in the same quantum state. This covers energy levels in an electron's orbit and its angular momentum and spin.
- Heisenberg's Uncertainty Principle.
- De Broglie's Equivalency of Particles and Waves in Matter.
- And many others, that are left to the reader to add to this list.

Allan Karson

Sources: Physics, Mathematics, and Cosmology—and Histories

Books are listed in order of publication date. Given, however, the glacial pace of changes in both mathematics and physics, rarely are any of these books 'out of date'.

At the end of each book's listing, following the year published, I placed a 'rating' of the book, according to the minimum scientific background of the reader. It is a subjective rating and it is only based on my own reading of the book.

By "Beginner", I am thinking of the person who may not have had a high school or college course in physics, but has at least, as a minimum, read a book such as this one.

Certain books, that may have a significant amount of mathematics, may also be interesting for the 'beginner'—such as source 52, *Einstein's Miraculous Year: Five Papers That Changed the Face of Physics*. In that case I rated it 'Beginner/Intermediate'

By "Advanced", usually means the book also contains a significant amount of mathematics and discussion of theories. In any case, I advise you to 'look before you buy'. The criterion is whether you will feel comfortable reading the material.

1. *Scientific Autobiography and other papers*, by Max Planck, Philosophical Library, 1949, Beginner
2. *Theory of Relativity,* Wolfgang Pauli, Dover Press, 1958, written originally in 1921. Advanced
3. *Einstein's Theory of Relativity,* Max Born, Dover Publications, 1962, Advanced.
4. *Einstein, the Life and Times*, Ronald W. Clark. Avon Books, 1971, Beginner, Bio

435

5. *The Born-Einstein Letters 1916-1955*, Max Born, Macmillan Press, 1971, Beginner/Intermediate
6. *Copernicus: The Founder of Modern Astronomy*, Angus Armitage, Dorset Press, 1971, Beginner
7. *Three Copernican Treatises*, by Edward Rosen, 1971, Octagon Books, Intermediate
8. *Subtle is the Lord... The Science and Life of Albert Einstein*, Abraham Pais, Oxford University Press, 1982, Intermediate/Advanced.
9. *Mathematics, The Loss of Certainty*, Morris Kline, Oxford University Press, 1980, Intermediate.
10. *The Cosmos*, Carl Sagan, Random House, 1980, Beginner
11. *Inward Bound, Of Matter and Forces in the Physical World*, Abraham Pais, Oxford University Press, 1982, Advanced.
12. *The Cosmic Code, Quantum Physics as the Language of Nature*, Heinz R. Pagels, Simon & Schuster, 1983, Beginner.
13. *The Dilemma of an Upright Man, Max Planck as a Spokesman for German Science*, u. L. Heilbron, University of California Press, 1986, Beginner
14. *Quantum Reality*, Nick Herbert, Anchor Books, Div. of Random House, 1987, Beginner/Intermediate.
15. *The Emperor's New Mind, Concerning Computers, Minds and the Laws of Physics*, Roger Penrose, Oxford University Press, 1989, Advanced
16. *Schrödinger, Life and Thought*, Walter Moore, Cambridge University Press, 1989, Beginner/Intermediate
17. *From Quarks to the Cosmos*, Leon Lederman and David Schramm, Scientific American Library, 1989, 1999, Beginner/Intermediate (Author's comment: Encyclopedic and great bang for the buck.)

18. *The Life of Isaac Newton*, Richard Westfall, Cambridge University Press, 1993, Beginner
19. *The God Particle*, Leon Lederman with Dick Teresi, a Delta Book, 1993, Beginner (maybe), Intermediate/Advanced
20. *Hyperspace*, Michio Kaku, Anchor Books, Div. of Random House, 1994, Beginner
21. *Lise Meitner, A Life in Physics*, Ruth Lewin Sime, University of California Press, 1996, Beginner/Intermediate

22. *Einstein, Bohr and the Quantum Dilemma*, Andrew Whitaker, 1996, Cambridge University Press, Intermediate/Advanced
23. *Life of the Cosmos*, Lee Smolin, Oxford University Press, 1997, Intermediate
24. *The Fabric of Reality, The Science of Parallel Universes*, David Deutsch, Penguin Group, 1997, Beginner/Intermediate
25. *A Tale of Two Continents, A Physicist's Life in a Turbulent World*, Abraham Pais, Princeton University Press, 1997, Beginner
26. *Fermat's Enigma, The World's Greatest Mathematical Problem*, Simon Singh, Anchor Books, Div. of Random House, 1998, Beginner
27. *Heisenberg and the Nazi Atomic Bomb Project, A Study in German Culture*. Paul Lawrence Rose, University of Chicago Press, 1998, Intermediate
28. *A Brief History of Time*, Stephen Hawking, Bantam Books, Second Edition, 1998, Beginner/Intermediate
29. *Paul Dirac, the Man and his Work*, A. Pais, M. Jacob, D. Olive and M. Atiyah, Cambridge University Press, 1998, Advanced
30. *Ludwig Boltzman, The Man Who Trusted Atoms*, Carlo Cercignani, Oxford University Press, 1998, Intermediate/Advanced
31. *Einstein's Miraculous Year*, Princeton University Press, John Stachel, 1998
32. *Quantum Philosophy, Understanding and Interpreting Contemporary Science*, Roland Omnès, Princeton University Press 1999, Intermediate/Advanced
33. *Quantum Generations, A History of Physics in the Twentieth Century*, Helge Kragh, Princeton University Press, 1999, Intermediate/Advanced
34. *Just Six Numbers*, Martin Rees, Basic Books, 1999, Beginner
35. *Niels Bohr's Times, In Physics, Philosophy, and Polity*, Abraham Pais, Oxford University Press, 1999, Intermediate
36. *The Accelerating Universe*, Mario Livio, John Wiley & Sons, 2000, Beginner
37. *Boltzmann's Atom*, David Lindley, 2001. The Free Press, a division of Simon & Schuster, Beginner
38. *$E=MC^2$, A biography of the World's most Famous Equation*, David Bodanis, Berkley Books, 2000, Beginner, Biography
39. *Three Roads to Quantum Gravity*, Lee Smolin, Basic Books, 2001, Intermediate/Advanced

40. *The Fabric of the Cosmos, Space, Time and the Texture of Reality,* Brian Greene, Alfred A. Knopf, 2003, Beginner/Intermediate
41. *The Elegant Universe,* Brian Greene, Vintage Press, Div. of Random House, 2003, Beginner
42. *Faster Than the Speed of Light,* Joao Magueijo, Penguin Books, 2003, Intermediate
43. *Galileo in Rome,* William R. Shea and Mariano Artigas, Oxford University Press, 2003, Beginner
44. *The Road to Reality,* Roger Penrose, Vintage Books, 2004, Advanced
45. *Heavenly Intrigue, Johannes Kepler, Tycho Brahe, and the Murder Behind One of the History's Greatest Scientific Discoveries,* Joshua Guilder and Anne-Lee Gilder, Doubleday, 2004, Beginner
46. *Symmetry and the Beautiful Universe,* Leon M. Lederman and Christopher T. Hill, Prometheus Books, 2004, Intermediate
47. *The Big Bang,* Simon Singh, Fourth Estate, an imprint of Harper Collins, 2004, Intermediate, but also Beginner for the second half of the book.
48. *The Equation That Couldn't Be Solved,* Mario Livio, Simon & Schuster, 2005, Beginner
49. *Parallel Worlds,* Michio Kaku, Anchor Books, Div. of Random House 2005, Beginner
50. *The End of the Certain World, the Life and Science of Max Bor*n, Nancy Thorndike Greenspan, Basic Books, 2005, Beginner
51. *Incompleteness, the Proof and Paradox of Kurt Godel,* W. W. Norton, in the Great Discovery Series, 2005, Beginner
52. *Einstein's Miraculous Year: Five Papers That Changed the Face of Physics, Albert Einstein,* John Stachel, Editor, foreword by Roger Penrose, Princeton University Press, 2005 Beginner/Intermediate
53. *The Cosmic Landscape: String Theory and the Illusion of Intelligent Design,* Leonard Susskind, Back Bay Books, Division of Little Brown, 2006, Intermediate
54. *The Theory of Almost Everything: The Standard Model, the Unsung Triumph of Modern Physics,* Robert Oerter, 2005, Penguin Group, Beginner
55. *The Trouble With Physics, The Rise of String Theory, the Fall of Science, and What Comes Next,* Lee Smolin, Houghton Mifflin Company, 2006, Intermediate

56. *Programming the Universe, A Quantum Computer Scientist Takes on the Cosmos*
Seth Lloyd, Vintage Books, 2006, Beginner

57. *Endless Universe, Beyond the Big Bang*, Paul J. Steinhardt and Neil Turk, Doubleday, 2007, Beginner/Intermediate/Advanced

58. *Uncertainty: Einstein, Heisenberg, Bohr, and the Struggle for the Soul of Science*, David Lindley, Doubleday, 2007, Beginner

59. *One World or None*, many scientist-authors, The New Press, (Originally 1946), 2007, Beginner

60. *Copernicus' Secret, How the Scientific Revolution Began*, Jack Repcheck, Simon & Schuster, 2007, Beginner/Intermediate

61. *Reinventing Gravity, A Physicist Goes Beyond Einstein*, John W. Moffat, HarperCollins, 2008, Advanced (and possibly Intermediate).

62. *Giordano Bruno, Philosopher/Heretic*, Ingrid D. Rowland, University of Chicago Press, 2008, Beginner

63. *Extrasolar Planets and Astrobiology*, Caleb A. Scharf, University Science Books, 2009, Intermediate

64. *The Strangest Man, The Hidden Life of Paul Dirac, Mystic of the Atom*, Graham Farmelo, Basic Books. 2009, Beginner/Intermediate

65. *Encyclopedia Britannica*, Fifteenth Edition, 1982, for varied historical and scientific facts.

66. *Einstein in Love*, Dennis Overbye, Viking, 2000, Beginner

67. *The Black Hole War*, Leonard Susskind, Back Bay Books, 2008, Beginner/Intermediate

68. *Physics of the Future: How Science Will Shape Human Destiny and Our Daily Lives by the Year 2100*, Michio Kaku, Random House, 2011, Beginner and beyond

69. *Gravity's Engines, How Bubble-Blowing Black Holes Rule Galaxies, Stars, and Life in the Cosmos*, Caleb Scharf, Scientific American/Farrar, Straus and Giroux, 2012 Beginner/Intermediate/Advanced.

70. *Quantum: Einstein, Bohr, And The Great Debate About The Nature of Reality*, Manjit Kumar, Norton Paperback, 2008. Intermediate/Advanced. (Presents many comprehensive discussions of important theories.)

I also wish to thank, acknowledge, and recommend the use of the Internet. I appreciate its standard format and the reliability of the varied sets of information (names, title, dates, locations, university positions, equations, etc,. When I seek different types of information, such as lists, comparison of populations or countries, or symbols to use—I go to the web as my first, easy accessible, general source of information.

Three time-charts, located via the web:

No. 2 Physics Nine of Physics' Disciplines:

http://bit.ly/akchart1

No. 3 Why Science Slept Sixteen Centuries:

http://bit.ly/akchart2

No. 23 Quantum Fathers> *Circa* 1895 to *circa* 1945+:

http://bit.ly/akchart3

35756831R00250

Made in the USA
San Bernardino, CA
03 July 2016